Advances in Optics: Reviews

Book Series, Volume 4

Sergey Y. Yurish

Editor

Advances in Optics: Reviews

Book Series, Volume 4

International Frequency Sensor Association Publishing

Sergey Y. Yurish
Editor

Advances in Optics: Reviews
Book Series, Vol. 4

Published by International Frequency Sensor Association (IFSA) Publishing, S. L., 2019
E-mail (for print book orders and customer service enquires): ifsa.books@sensorsportal.com

Visit our Home Page on http://www.sensorsportal.com

ISBN: Print: 978-84-09-09014-3
e-Book: 978-84-09-09013-6
BN-20190715-XX
BIC: TTB

Acknowledgments

As Editor I would like to express my undying gratitude to all authors, editorial staff, reviewers and others who actively participated in this book. We want also to express our gratitude to all their families, friends and colleagues for their help and understanding.

Contents

Contributors

Flávia Beltrão Pires
Associate Program of Masters and PhD in Rehabilitation Sciences for Universidade Estadual de Londrina (UEL) and Universidade do Norte do Paraná (UNOPAR)/ Londrina – Paraná, Brazil

Damien Bigourd
Institut FEMTO-ST, Département d'Optique, UMR 6174 CNRS-Université Bourgogne Franche-Comté, 25030 Besançon, France

Do Tan Si
Association of Physicists, Ho Chi Minh City, Vietnam

Larissa Dragonetti Bertin
Associate Program of Masters and PhD in Rehabilitation Sciences for Universidade Estadual de Londrina (UEL) and Universidade do Norte do Paraná (UNOPAR)/ Londrina – Paraná, Brazil

Peter Eiswirt
AMETEK GmbH BU Taylor Hobson/Luphos, Weiterstadt, Germany

Zhigang Fan
Harbin Institute of Technology, Harbin, Hei Long Jiang, China

Stheace Kelly Fernandes Szezerbaty
Associate Program of Masters and PhD in Rehabilitation Sciences for Universidade Estadual de Londrina (UEL) and Universidade do Norte do Paraná (UNOPAR)/ Londrina – Paraná, Brazil

Sergey A. Fomchenkov
Samara National Research University, Samara, Russia

Coralie Fourcade-Dutin
Institut FEMTO-ST, Département d'Optique, UMR 6174 CNRS-Université Bourgogne Franche-Comté, 25030 Besançon, France

Er'el Granot
Department of Electrical and Electronics Engineering, Ariel University, Ariel, Israel
E-mail: erel@ariel.ac.il

Svetlana N. Khonina
Image Processing Systems Institute – Branch of the Federal Scientific Research Centre 'Crystallography and Photonics' of Russian Academy of Sciences, Samara, Russia, Samara National Research University, Samara, Russia

Cuifang Kuang
Zhe Jiang University, Hang Zhou, Zhe Jiang, China

E. N. Lallas

Technical Educational Institute of Sterea Ellada, Lamia, Greece

Vannhu Le

Le Quy Don Technical University, Cau Giay, Ha Noi, Viet Nam
Zhe Jiang University, Hang Zhou, Zhe Jiang, China

Xu Liu

Zhe Jiang University, Hang Zhou, Zhe Jiang, China

Jéssica Lúcio da Silva

Associate Program of Masters and PhD in Rehabilitation Sciences for Universidade
Estadual de Londrina (UEL) and Universidade do Norte do Paraná (UNOPAR)/
Londrina – Paraná, Brazil

Gennady Medvedkin

General Molded Glass, Inc., Torrance, CA 90502, USA

M. Nazrul Islam

State University of New York at Farmingdale, USA

Rodrigo Franco de Oliveira

Centro Universitário de Anápolis (UNIEVANGÉLICA), Anápolis - Goiás, Brazil

Priscila Daniele de Oliveira Perrucini

University Anhanguera-UNIDERP (Universidade para o Desenvolvimento do
Estado e da Região do Pantanal), Campo Grande-Mato Grosso do Sul, Brazil

Davor Peruško

Institute of Nuclear Science Vinča, University of Belgrade, Mike Petrovića Alasa
12-14, Belgrade, Serbia

Suzana Petrović

Institute of Nuclear Science Vinča, University of Belgrade, Mike Petrovića Alasa
12-14, Belgrade, Serbia
E-mail: spetro@vin.bg.ac.rs

Jürgen Petter

AMETEK GmbH BU Taylor Hobson/Luphos, Weiterstadt, Germany

Deise Aparecida Almeida Pires-Oliveira

Centro Universitário de Anápolis (UNIEVANGÉLICA), Anápolis - Goiás, Brazil

Regina Célia Poli-Frederico

Associate Program of Masters and PhD in Rehabilitation Sciences for UEL and
UNOPAR/ Londrina – Paraná, Brazil

Alexey P. Porfirev

Image Processing Systems Institute – Branch of the Federal Scientific Research
Centre 'Crystallography and Photonics' of Russian Academy of Sciences, Samara,
Russia, Samara National Research University, Samara, Russia

Sucheta Sharma

AMETEK GmbH BU Taylor Hobson/Luphos, Weiterstadt, Germany

Andrey V. Ustinov

Image Processing Systems Institute – Branch of the Federal Scientific Research Centre 'Crystallography and Photonics' of Russian Academy of Sciences, Samara, Russia

Xinping Zhang

College of Applied sciences, Beijing University of Technology, Beijing 100124, P. R. China

Preface

It is my great pleasure to introduce the fourth volume from our popular open access Book Series *'Advances in Optics: Reviews'* started by the IFSA Publishing in 2018. The Vol. 4 of this Book Series is also published as an Open Access Book in order to significantly increase the reach and impact of this volume, which also published in two formats: electronic (pdf) with full-color illustrations and print (paperback).

Like the first three volumes of this Book Series, the fourth volume also has been organized by topics of high interest to offer a fast and easy reading of each topic, every chapter in this book is independent and self-contained. All chapters have the same structure: first an introduction to specific topic under study; second particular field description including sensing or/and measuring applications. Each of chapter is ending by well selected list of references with books, journals, conference proceedings and web sites.

The fourth volume is devoted to optics, lasers and optical sensors, and written by 29 authors from academia and industry from 10 countries: Brazil, China, France, Germany, Greece, Israel, Russia, Serbia, USA and Vietnam.

This book ensures that our readers will stay at the cutting edge of the field and get the right and effective start point and road map for the further researches and developments. By this way, they will be able to save more time for productive research activity and eliminate routine work.

I hope that readers enjoy this book and that can be a valuable tool for those who involved in research and development of various physical sensors, chemical sensors, biosensors and measuring systems.

I shall gratefully receive any advices, comments, suggestions and notes from readers to make the next volumes of *'Advances in Optics: Reviews'* Book Series very interesting and useful.

Dr. Sergey Y. Yurish

Editor
IFSA Publishing   *Barcelona, Spain*

Chapter 1
Fiber Optical Parametric Amplifier for Ultra-short Pulses

Coralie Fourcade-Dutin and Damien Bigourd

1.1. Introduction

Fiber optical parametric amplifier (FOPA) has been widely investigated in the context of telecommunication [1] for its interesting features as large gain bandwidth [2, 3], low noise [4], and very high gain [5]. The principle is based on a degenerated phase-matched four-wave mixing process involving one strong pump pulse with a relatively narrow bandwidth, a weak signal seeding the amplifier, and a generated idler wave during the process. Since recently, strong efforts are currently being achieved to develop ultra-fast sources by taking benefit from the FOPA performances. In this case, the signal is a stretched pulse to decrease the peak power to avoid any optical damage or spurious nonlinearities [6]. Then, the amplified signal is compressed to the Fourier transform limited pulse duration, as no spurious phase should be generated during the amplification. In order to reach ultra-short pulse duration, the FOPA requires a broad spectral gain bandwidth without significant gain ripples to limit pedestal after recompression. The spectral bandwidth can be enhanced by adapting the nonlinear process by using two pumps [7, 8], the dispersion profile of the optical fiber [9-11] or the pump pulse parameters [12, 13]. In this latter case, we demonstrated an optical method for ultra-broadband amplification up to 36 THz at high gain (~70 dB) by using a single pump stretched pulse with a relatively broad spectrum [12, 14]. These characteristics are of prime interest to overcome gain narrowing and pulse distortion occurring in ion-doped fiber amplifiers. By adapting the chirp rate of the signal owing a very large bandwidth, we have demonstrated the possibility to amplify pulses with a duration shorter than 30 fs [15]. Therefore, the FOPA is a promising system to pave the way toward the ultrashort pulse amplification.

In this chapter, a mathematical description of the FOPA is introduced to provide a significant understanding of the parametric process. Then, several FOPA configurations

Coralie Fourace-Dutin

Institut FEMTO-ST, Département d'Optique, UMR 6174 CNRS-Université Bourgogne Franche-Comté, 25030 Besançon, France are analytically detailed for specific cases when the pump shape is modified. In the last part, a new concept of FOPA for ultra-short pulses [16] is presented and numerically investigated highlighting the possibility to amplify ultra-short pulse with a duration of 300 fs by combining a continuous wave-CW and a chirped pulse.

1.2. FOPA Theory

1.2.1. General Theory

The origin of the FOPA process is an efficient four-wave mixing involving two strong pumps P, a signal S and an idler I at the angular frequencies ω_P, ω_S and ω_I, respectively. In the following description, the degenerate FOPA is only considered in which the two pump pulses are identical. The photon energy conservation is written as:

$$2\omega_P = \omega_S + \omega_I \tag{1.1}$$

The propagation of light in an optical fiber is well described by the nonlinear Schrödinger equation (NLSE) for a slow varying complex amplitude of the electric field $E(z,\tau)$ [17, 18]:

$$\frac{\partial E}{\partial z} = \gamma |E|^2 E + \sum_{k \geq 2} i^{k+1} \beta_k \frac{\partial^k E}{\partial \tau^k} \tag{1.2}$$

To focus our attention on the four wave mixing process (the FOPA), the Raman contribution and the self-steepening are not taken into account. The first term in the right hand side represents the nonlinearity while the second one is a linear term; i.e. the dispersion of the fiber. In the following, the dispersion coefficients (β_{20}, β_{30}, β_{40}) at the pump central frequency ω_{p0} are limited up to the fourth order, but can be generalized to higher order terms. In order to obtain the analytical form the signal and idler amplitudes as a function of the parameters, the modulation instability (MI) is introduced in which a small temporal perturbation $e(z,\tau)$ is added to a stationary $E_{stat}(z)$ amplitude:

$$E(z,\tau) = E_{stat}(z) + e(z,\tau), \tag{1.3}$$

with $|e(z,\tau)| \ll |E_{stat}|$.

The time independent-stationary $E_{stat}(z)$ is obtained by resolving Eq. (1.2) and the solution is

$$E_{stat}(z) = |E_0| e^{i\phi_{P0}} e^{i\gamma |E_0|^2 z}, \tag{1.4}$$

$|E_0|$ and ϕ_{p0} are the module and the phase of the field at the fiber input. In our case, $P_0 = |E_0|^2$ corresponds to the pump power. By inserting Eq. (1.3) in Eq. (1.2) and by keeping only the linear terms of the perturbation, the propagation of $e(z,\tau)$ is described by [19, 20]

$$
\frac{\partial e}{\partial z} + \frac{i}{2}\beta_2 \frac{\partial^2 e}{\partial \tau^2} - \frac{1}{6}\beta_3 \frac{\partial^3 e}{\partial \tau^3} + \frac{i}{24}\beta_4 \frac{\partial^4 e}{\partial \tau^4} =
$$
$$
= i2\gamma \left| E_{stat} \right| e + i\gamma \left| E_{stat} \right|^2 e^{i\left(2\gamma \left| E_{stat} \right|^2 z + 2\phi_p \right)} e^*
$$

(1.5)

The next step is to analyze the perturbation in the frequency domain Ω by performing the Fourier transform of Eq. (1.4). A new variable $c(z,\Omega)$ is then introduced:

$$
\tilde{e}(z,\Omega) = \tilde{c}(z,\Omega) e^{i\left(\phi_{p0} + \gamma \left| E_0 \right|^2 z\right)},
$$

(1.6)

to get a matrix form (with its conjugate) as

$$
i\frac{\partial}{\partial z}\begin{bmatrix} c(z,\Omega) \\ c^*(z,-\Omega) \end{bmatrix} =
$$

$$
= \begin{bmatrix} -\gamma P - \frac{1}{2}\beta_2\Omega^2 - \frac{1}{6}\beta_3\Omega^3 - \frac{1}{24}\beta_4\Omega^4 & -\gamma P \\ +\gamma P & \gamma P + \frac{1}{2}\beta_2\Omega^2 - \frac{1}{6}\beta_3\Omega^3 + \frac{1}{24}\beta_4\Omega^4 \end{bmatrix}\begin{bmatrix} c(z,\Omega) \\ c^*(z,-\Omega) \end{bmatrix}
$$

(1.7)

The eigenvalues and the solutions are then calculated from this matrix and the final solution can be written with the following form:

$$
\tilde{e}(z,\Omega) = U(z,\Omega)\cdot \tilde{e}(z=0,\Omega) + V(z,\Omega)\cdot \tilde{e}^*(z=0,-\Omega),
$$

(1.8)

where the complex function $U(z,\Omega)$ and $V(z,\Omega)$ are written as [14, 19-20]

$$
U(z,\Omega) = e^{i\left(\gamma P_0 z + \frac{1}{6}\beta_{30}\Omega^3 z\right)} \times \left\{ \cosh\left(g(\Omega)z\right) + i\frac{\gamma P_0 + \beta_s(\Omega)}{g(\Omega)}\sinh\left(g(\Omega)z\right) \right\},
$$

$$
V(z,\Omega) = ie^{i\left(\gamma P_0 z + \frac{1}{6}\beta_{30}\Omega^3 z + 2\phi_{p0}\right)} \times \frac{\gamma P_0}{g(\Omega)}\sinh\left(g(\Omega)z\right),
$$

(1.9)

with

$$
g(\Omega,t) = \sqrt{\beta_s^2(\Omega) + 2\gamma P_0 \beta_s(\Omega)} \quad \text{and} \quad \beta_s(\Omega) = \frac{\beta_{20}}{2}\Omega^2 + \frac{\beta_{40}}{24}\Omega^4,
$$

(1.10)

g is the parametric gain coefficient and β_s is the low power (linear) propagation mismatch. We can notice that only the even order terms play a role in this mismatch since the frequencies of the three waves are symmetrically positioned from the pump.

In the case in which only the strong pump and the weak signal are injected in the fiber $\tilde{e}(-\Omega, z = 0) = 0$, the evolution of the signal S(Ω,z) and the idler I(-Ω,z) with the propagation is given by:

$$S(z,\Omega) = \tilde{e}(z,\Omega) = U(z,\Omega) \cdot S(z = 0,\Omega),$$

$$I(z,-\Omega) = \tilde{e}(z,-\Omega) = V(z,-\Omega) \cdot S^*(z = 0,\Omega) \qquad (1.11)$$

The unsaturated gain is obtained from Eq. (1.8) and (1.10):

$$G(\Omega,t,L) = \frac{|S(\Omega,t,z = L)|^2}{|S(\Omega,t,z = 0)|^2} = 1 + \left[\frac{\gamma P(t)}{g(\Omega,t)} \sinh\left(g(\Omega,t)L \right) \right]^2, \qquad (1.12)$$

g can also be written as

$$g(\Omega,t) = \sqrt{ \left[\gamma P(t) \right]^2 - \left[\frac{\kappa(\Omega,t)}{2} \right]^2 }, \qquad (1.13)$$

and we can show that the gain is maximum when all the waves, i.e. the pump, signal and idler, are phase matched with $\kappa = 0$.

$$\kappa(\Omega,t) = \beta_{20} \cdot \Omega^2 + \frac{\beta_{40}}{12} \cdot \Omega^4 + 2\gamma P(t) = 0 \qquad (1.14)$$

1.2.2. FOPA Pumped by a CW, a Structured Quasi-CW or a Chirped Pulse

The properties of the FOPA are imposed by the characteristics of the fiber (the dispersion and the nonlinear coefficient) and the one of the pump pulses [21]. In the following, we aim to discuss how the gain value and the spectral gain bandwidth is modified through the pump parameters. Table 1.1 displays some examples of gain curves with several conditions but we do not aim to give complete explanations already detailed in the cited references. In all cases, the optical fiber owns a frequency independent nonlinear coefficient γ equals to 35 $W^{-1}.km^{-1}$ and $\beta_{20} = -1 \times 10^{-28}$ s^2/m, $\beta_3 = +1 \times 10^{-40}$ s^3/m, $\beta_{40} = -1 \times 10^{-55}$ s^4/m at a wavelength of 1030 nm corresponding to $\Omega = 0$ THz. The fiber length is 2 m. The pump profile has a top-hat profile with a maximum power of 100 W and a duration of 60 ps (Full Width at Half Maximum-FWHM).

When a CW or a quasi-CW pumps the optical fiber, the FOPA is analytically well described by the previous equations (Table 1.1-column 1). Two distinct side lobes are generated symmetrically around the pump frequency; at +/-24 THz (or ~951 and 1122 nm). These maximum gain values of 55 dB are obtained when all the waves are phase-matched ($\kappa = 0$; Eq. (1.14)).

Table 1.1. Parameters to include in the equations derived in Section 1.2.1 describing the parametric gain shapes for three types of pump profile.

Pump shape	CW or quasi-CW with a constant peak power.	Quasi-CW with a time dependent power	Chirped pulses with a large spectral bandwidth
Pump frequency	$\omega_p = \omega_{p0}$, $W_p = \omega_p - \omega_{p0} = 0$	$\omega_p = \omega_{p0}$, $W_p = \omega_p - \omega_{p0} = 0$	$\omega_p = \omega_{p0} + \alpha_p \cdot \tau$, $W_p = \omega_p - \omega_{p0} = \alpha_p \cdot \tau$
Signal Frequency	Ω	Ω	$\Omega - W_p$
Dispersion	$\beta_2 = \beta_{20}$, $\beta_3 = \beta_{30}$, $\beta_4 = \beta_{40}$	$\beta_2 = \beta_{20}$, $\beta_3 = \beta_{30}$, $\beta_4 = \beta_{40}$	$\beta_2(W_p) = \beta_{20} + \beta_{30} \cdot W_p + \frac{\beta_{40}}{2} \cdot W_p^2$, $\beta_3(W_p) = \beta_{30} + \beta_{40} \cdot W_p$, $\beta_4 = \beta_{40}$
Power P or $P(\tau)$			
Gain (dB)			

When a quasi-CW is used to pump the FOPA, the pulse shape can also be structured by modifying its amplitude (or its phase) by using several systems as an electrooptic modulator driven by an arbitrary wave generator, for example [13]. In this case, the power changes instantaneously in the temporal envelope (Table 1.1, column 2). Therefore, the characteristics of the FOPA evolves with time impacting the instantaneous gain value, its time-wavelength distribution and its spectral bandwidth [22, 23]. In our example, the power decreases in the center of the profile to 56 W. Therefore, the instantaneous gain

23

value decreases in the middle of the pulse while it is higher at the leading and trailing edges (Table 1.1, column 2, last raw). The vertical dashed lines at ± 25 ps correspond to the power of 100 W and therefore the gain shape is the same as for a constant power (Table 1.1, second column).

To add another degree of freedom, the spectral bandwidth of the pump pulse can be of great benefit to obtain a broader phase matching condition [12, 14-16, 24]. In this case, the pump is created from an ultra-short pulse that gives the spectral bandwidth. However, the pulse is often stretched to decrease the peak power avoiding spurious nonlinearity and optical damages in the optical fibers. In our example (Table 1.1, column 3), the pump pulse has a bandwidth of 2.8 THz (9.9 nm at 1030 nm) and is stretched to 60 ps (FWHM) with a chirp rate α_p ~21.4 ps/THz. At 0 ps (vertical black line), the instantaneous frequency is the same as for a CW pump and thus the gain is the same as in the column 2. However, the instantaneous pump frequency evolves with time and it instantaneously modifies the phase matching condition (Eq. (1.14)). Accordingly, the spectral gain has a temporal distribution (Table 1.1, column 4, last raw) and each pump frequency (at a given time) provides a spectrally shifted gain curve enabling very large bandwidth (~35 THz in the example). Therefore, this method is of great interest since it provides both a very large bandwidth and a high gain and thus, it is very promising to amplify ultra-short pulses [15].

1.2.3. Some Other Contributions

In the previous section, the derivation was performed for the parametric process in the unsaturated regime in which the pump is not depleted. The equations are interesting to understand the bases of the parametric amplification. However, for high conversion efficiency, the signal and idler extract both an important energy part from the pump. The depletion significantly impacts the dynamic of the amplification through the phase matching condition and therefore needs to be taken into account [25]. For example, the gain decreases at saturation [32] and the maximum shifts toward the pump frequency. In addition, the Raman response contributes to the amplification and it has usually an important effect on the gain for a frequency offset higher than 10 THz from the pump [26]. In this case, the nonlinear response $R(\tau)$ includes a delayed contribution (Raman effect) in addition to the instantaneous one already in Eq. (1.2) (Kerr effect) [26]. $R(\tau)$ is written as $R(\tau) = (1-f_r)\delta(\tau)+f_r\,h_r(\tau)$ with $f_r = 0.18$ the Raman contribution and $h_r(\tau)$ the delay Raman response [17]. The longitudinal evolution of the dispersion has also an important impacts on the parametric amplification. For example, the spectral gain bandwidth can be enlarged when the dispersion of the fiber is periodically evolving along the length [9-11].

Although all these contributions can be included in some semi-analytical equation, we usually prefer to perform numerical simulations [18] as in Section 1.3 to get the complete understanding of the dynamic.

1.3. FOPA for Ultra-short Pulses

In this section, we present the possibility to perform the amplification of ultra-short pulses with a FOPA. In addition to the advantages given by the fibers (as robustness, compact and reliability), FOPA can provide outstanding amplification performances (very high gain and large gain bandwidth) enabling the creation of the next generation of ultra-fast laser systems. In fact, state of the art ultra-short bulk or fiber laser systems usually suffer from temporal broadening during the amplification process caused by gain-narrowing. Although low energy pulses are available from commercial lasers with pulse duration of 10-50 fs, their amplification by conventional lasers are always done at the expense of spectral narrowing and thus pulse duration broadening. In order to increase the average power of a laser, one efficient method is to use an ytterbium doped fiber amplifier, giving access to pulses with several 100 µJ energy at high repetition rate [27] since the thermal dissipation is distributed along the full fiber length. This repetition rate is well above "standard" techniques, up to a few 100 kHz compared to the ~1 kHz rates found in leading research laboratories today. However, the ytterbium doped fiber amplifier, although efficient and robust, is based on "traditional" laser amplification and therefore suffers from broadening of the pulse duration together with a restriction of the available spectral region. The tunability is usually confined to a window around 1 µm with a bandwidth of about 40 nm [28] which limits the range of interesting potential applications sensitive to the wavelength. After amplification to the 100 µJ-1 mJ level, the pulse duration is usually broadened to more than 500 fs while at low energy, lasers can deliver pulses with duration of 50-100 fs. By combining the compact, robust high repetition rate fiber lasers (as ytterbium doped fiber amplifier) and ultra-broad band FOPA, ultra-fast photonic sources can deliver ultra-short pulses in a large range of wavelength and at high gain. Several numerical and experimental demonstrations have already been investigated to highlight this scheme of amplification [8, 12, 14-16, 29-35].

In the following, we will focus the section on the generation of broad pulses in a FOPA pumped by a chirped pulse and seeded by a continuous wave at the phase matching condition. In this configuration, a chirped idler is generated (Eq. (1.1)) together with the signal amplification. Its spectral bandwidth can be up to twice the one of the pump. The experimental implementation has recently been performed and W. Fu et al demonstrated the amplification of pulses at 1.03 µm with a duration of 210 fs at high gain [16]. In the following, we present some detailed simulations with similar conditions. In order to take into account the saturation process and the role of the propagation, the numerical simulations have been performed by integrating the nonlinear Schrödinger equation along the propagation (Eq. (1.2)) describing the evolution of the slowly varying total electric field in an optical fiber. The Raman contribution is not taken into account since it does not influence the phase matched parametric amplification occurring at a high frequency offset from the pump [14, 17] as it is in the following case (~50 THz). The equation is solved with the standard split-step Fourier method [17] including a pump pulse, a signal and a weak random initial condition that mimics the quantum fluctuation. The pump pulse has a spectrum centered at 1030 nm with a Gaussian shape and a bandwidth equals to 10 nm (FWHM). The pulse is stretched to 60 ps (FWHM) with a spectral phase introduced by a standard Öffner type stretcher. The pump energy is 550 nJ. The simulation also includes the dispersion of a commercially available fiber (NKT SC-5.0-1040) which has a zero dispersion wavelength closed to 1040 nm. The nonlinear coefficient is $\gamma = 35$ W^{-1} km^{-1} and the total fiber length is L = 3 cm.

Fig. 1.1(a) shows the parametric fluorescence spectrum when no signal is injected in the fiber. It has been obtained from an average over 50 shots. As the pump propagates in a weakly normal dispersion regime, the two parametric lobes are centered at 879 nm and 1243 nm. This spectrum is usually representative of the unsaturated gain shape when the fiber is pumped with a strong CW. However, as the pump pulse is chirped, the fluorescence has also a time-spectrum distribution according to Table 1.1-column 4 (Fig. 1.1(b)) [14]. When a CW signal is injected at a frequency shift of -50 THz from the pump (i.e. at ~1243 nm, white dashed line in Fig. 1.1(b)), an idler is generated at +50 THz (i.e. at 879 nm) in good agreement with Eq. (1.1) (Fig. 1.2(a)).

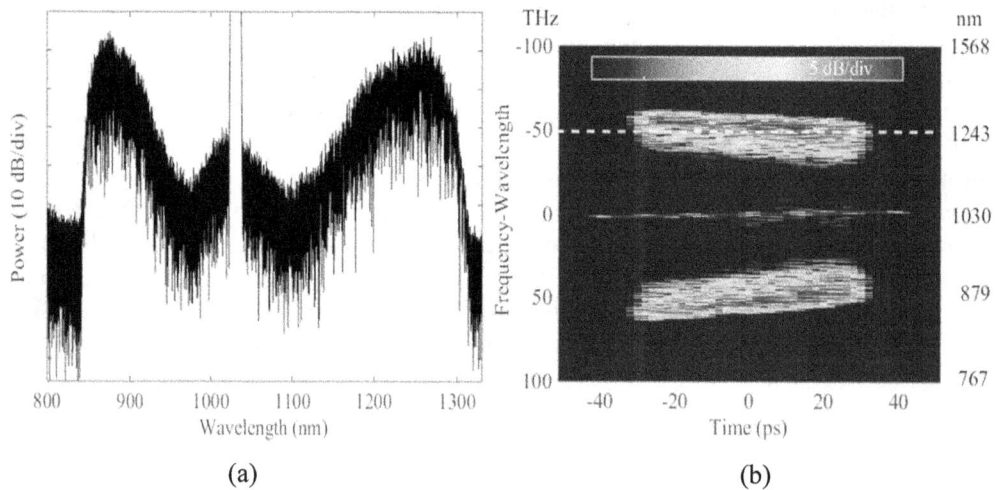

Fig. 1.1. (a) Average spectrum of the parametric fluorescence; (b) Spectrogram of the parametric fluorescence. The white dashed line corresponds to the CW signal injected for Fig. 1.2.

In this case, the signal power is 10 mW. At the PCF output, the pump spectrum is structured mainly due to the self-phase modulation and the saturation of the amplifier. The idler has a relatively broad spectral bandwidth of 4 nm (FWHM) resulting from the nonlinear interaction between the CW signal and the chirped pump. The large bandwidth is a key ingredient to generate an ultra-short pulse. However, the idler experiences also a chirp which needs to be compensated. Indeed, Fig. 1.2(b) shows the corresponding spectrogram of the total electric field with a CW signal, the chirped pump and idler owing a long pulse duration.

At the output of the PCF, the idler is selected by a spectral filter with a high order super-Gaussian profile and a width of 10 THz (FWHM). The idler energy is calculated as the function of the fiber length when the frequency offset of the signal is tuned from -40 THz to -50 THz (Fig. 1.3). For example, the idler energy is ~50 nJ for L = 3 cm and for a signal frequency offset of -50 THz leading to a conversion efficiency of ~10 %. From Fig. 1.3, the saturation of the amplifier is clearly observed even for the short fiber. The saturation is very important to extract the maximum of energy from the pump. However, the saturation depends on the frequency offset. Indeed, at -50 THz and -55 THz, the complete saturation occurs at around 3.5 cm while the amplifier weakly saturates for -40 THz. After propagation in 3 cm long PCF, the chirp of the filtered idler is compensated by a standard compressor with gratings (with 1200 line/mm). Fig. 1.4 displays the compressed idler pulse when the signal frequency offset is tuned from -55 THz to -40 THz. In all cases, the pulse duration is ~300 fs at the Fourier transform limit. The spectral phase has been fully compensated thanks to the optimization of the compressor that needs to match the stretcher parameters and the frequency offset [15].

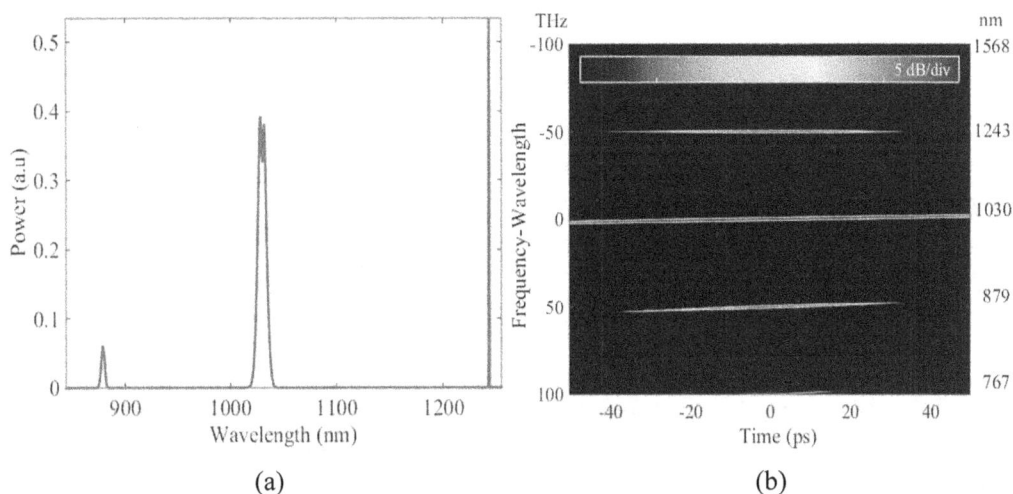

Fig. 1.2. (a) Spectrum when the chirped pump and a CW signal are injected in the PCF, (b) Corresponding spectrogram.

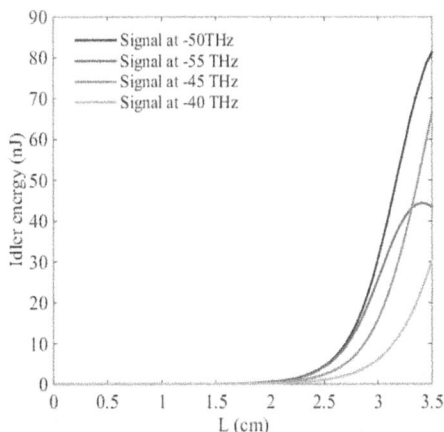

Fig. 1.3. Evolution of the idler energy with the fiber length. The frequency offset from the pump is tuned from -40 THz to -55 THz.

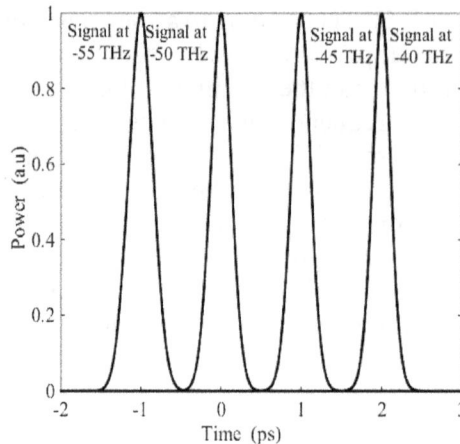

Fig. 1.4. Compression of the idler pulse. The frequency offset from the pump is tuned from -40 THz to -55 THz.

1.4. Conclusion

We have introduced different techniques for light amplification in optical fibers with a parametric process. The methods are very interesting to create the next generation of ultra-fast photonic source since they can amplify ultra-short pulses at high gain with very large bandwidth. For example, when a chirped pump is simultaneously injected in a PCF with a CW signal, an idler is generated at relatively high energy (~50 nJ) with large bandwidth. After the compensation of the spectral phase, the pulse duration reach a short duration (300 fs). In addition, this scheme is very interesting to develop a simple, robust and reliable fiber laser since no active synchronization is required between the signal and the pump.

Acknowledgements

This work has been supported by the Agence Nationale de la Recherche (FiberAmp projects ANR-16-CE24-0009), the EUR EIPHI program (ANR-17-EURE-0002) and the conseil Régional de Franche-Comté.

References

[1]. J. Hansryd, P. A. Andrekson, M. Westlund, J. Lie, P.-O. Hedekvist, Fiber-based optical parametric amplifiers and their applications, *IEEE. J. Select. Topics Quantum Electron.*, Vol. 8, 2002, pp. 506-520.
[2]. M. E. Marhic, N. Kagi, T.-K. Chiang, L. G. Kazovsky, Broadband fiber optical parametric amplifiers, *Opt. Lett.*, Vol. 21, 1996, pp. 573-575.
[3]. M.-C. Ho, K. Uesaka, M. Marhic, Y. Akasaka, L. G. Kasovky, 200-nm-bandwidth fiber optical amplifier combining parametric and Raman gain, *Journal Light. Technol.*, Vol. 19, 2011, pp. 977-981.

[4]. P. L. Voss, R. Tang, P. Kumar, Measurement of the photon statistics and the noise figure of a fiber-optic parametric amplifier, *Opt. Lett.*, Vol. 28, 2003, pp. 549-551.

[5]. T. Torounidis, P. A. Andrekson, B.-E. Olsson, Fiber-optical parametric amplifier with 70-dB gain, *IEEE Photon. Technol. Lett.*, Vol. 18, 2006, pp. 1194-1196.

[6]. D. Strickland, G. Mourou, Compression of amplified chirped optical pulses, *Opt. Comm.*, Vol. 56, 1985, pp. 219-221.

[7]. M. E. Marhic, Y. Park, F. S. Yang, L. G. Kasovsky, Broad band fiber-optical parametric amplifiers and wavelength converters with low-ripple Chebyshev gain spectra, *Opt. Lett.*, Vol. 21, 1996, pp. 1354-1356.

[8]. D. Bigourd, C. Fourcade Dutin, O. Vanvincq, E. Hugonnot, Numerical analysis of broadband fiber optical parametric amplifiers pumped by two chirped pulses, *J. Opt. Soc. Am. B*, Vol. 33, 2016, pp. 1800-1807.

[9]. K. Inoue, Arrangement of fiber pieces for a wide wavelength conversion range by fiber four-wave mixing, *Opt. Lett.*, Vol. 19, 1994, pp. 1189-1191.

[10]. M. E. Marhic, F. S. Yang, M.-C. Ho, L. Kazovsky, High nonlinearity fiber optical parametric amplifier with periodic dispersion compensation, *J. Lightwave Technol.*, Vol. 17, 1999, pp. 210-215.

[11]. C. Fourcade-Dutin, Q. Bassery, D. Bigourd, A. Bendahmane, A. Kudlinski, M. Douay, A. Mussot, 12 THz flat gain fiber optical parametric amplifiers with dispersion varying fibers, *Opt. Exp.*, Vol. 23, 2015, pp. 10103-10110.

[12]. D. Bigourd, P. Beaure d'Augères, J. Dubertrand, E. Hugonnot, A. Mussot, Ultra-broadband fiber optical parametric amplifier pumped by chirped pulses, *Opt. Lett.*, Vol. 39, 2014, pp. 3782-3785.

[13]. D. Bigourd, P. Morin, J. Dubertrand, C. Fourcade-Dutin, H. Maillotte, Y. Quiquempois, G. Bouwmans, E. Hugonnot Parametric gain shaping from a structured pump pulse, *IEEE Phot. Techno. Lett.*, Vol. 31, 2019, pp. 214-217.

[14]. O Vanvincq, C Fourcade-Dutin, A Mussot, E Hugonnot, D Bigourd, Ultrabroadband fiber optical parametric amplifiers pumped by chirped pulses. Part 1: Analytical model, *J. Opt. Soc. Am. B*, Vol. 32, 2015, pp. 1479-1487.

[15]. C Fourcade-Dutin, O Vanvincq, A Mussot, E Hugonnot, D Bigourd, Ultrabroadband fiber optical parametric amplifiers pumped by chirped pulses. Part 2: Sub-30 fs pulse amplification at high gain, *J. Opt. Soc. Am. B*, Vol. 32, 2015, pp. 1488-1493.

[16]. W. Fu, F. W. Wise, Normal-dispersion fiber optical parametric chirped-pulse amplification, *Opt. Lett.*, Vol. 43, 2018, pp. 5331-5334.

[17]. G. P. Agrawal, Nonlinear Fiber Optics, *Academic*, 2007.

[18]. J. M. Dudley, G. Genty, S. Coen, Supercontinuum generation in photonic crystal fiber, *Rev. Mod. Phys.*, Vol. 78, 2006, pp. 1135-1184.

[19]. M. I. Kolobov, A. Mussot, A. Kudlinski, E. Louvergneaux, M. Taki, Third order dispersion drastically changes parametric gain in optical fiber systems, *Phys. Rev. A*, Vol. 83, 2011, 035801.

[20]. M. J. Potasek, B. Yurke, Squeezed-light generation in a medium governed by the nonlinear Schrödinger equation, *Phys. Rev. A*, Vol. 35, 1987, pp. 3974-3977.

[21]. M. E. Marhic, Fiber Optical Parametric Amplifiers, Oscillators and Related Devices, *Cambridge University*, 2008.

[22]. C. Finot, S. Wabnitz, Influence of the pump shape on the modulation instability process induced in a dispersion-oscillating fiber, *J. Opt. Soc. Am. B*, Vol. 32, 2015, pp. 892-899.

[23]. A. Vedadi, A. M. Ariaei, M. M. Jadidi, J. A. Salehi, Theoretical study of high repetition rate short pulse generation with fiber optical parametric amplification, *J. Lightwave Technol.*, Vol. 30, 2012, pp. 1263-1268.

[24]. A. O. J. Wiberg, Z. Tong, E. Myslivets, N. Alic, S. Radic, Idler chirp optimization in a pulse-pumped parametric amplifier, *IEEE Photonics Society Summer Topical Meeting Series,* 2013, pp. 159-160.

[25]. K. Inoue, T. Mukai, Signal wavelength dependence of gain saturation in a fiber optical parametric amplifier, *Opt. Lett.*, Vol. 26, 2001, pp. 10-12.

[26]. A. S. Y. Hsieh, G. K. L. Wong, S. G. Murdoch, S. Coen, F. Vanholsbeeck, R. Leonhardt, J. D. Harvey, Combined effect of Raman and parametric gain on single-pump parametric amplifiers, *Opt. Express*, Vol. 15, 2007, pp. 8104-8114.

[27]. F. Röser, D. Schimpf, O. Schmidt, B. Ortaç, K. Rademaker, J. Limpert, A. Tünnermann, 90 W average power 100 µ J energy femtosecond fiber chirped-pulse amplification system, *Opt. Lett.*, Vol. 32, 2007, pp. 2230-2232.

[28]. R. Paschotta, J. Nilsson, A. C. Tropper, D. C. Hanna, Ytterbium doped fiber amplifiers, *IEEE J. Quant. Elect.*, Vol. 33, 1997, pp. 1049-1056.

[29]. C. Caucheteur, D. Bigourd, E. Hugonnot, P. Szriftgiser, A. Kudlinski, M. Gonzalez-Herraez, A. Mussot, Experimental demonstration of optical parametric chirped pulse amplification in optical fiber, *Opt. Lett.*, Vol. 35, 2010, pp. 1786-1788.

[30]. Y. Zhou, Q. Li, K. K. Y. Cheung, S. Yang, P. C. Chui, K. Wong, All-fiber based ultrashort pulse generation and chirped pulse amplification through parametric processes, *IEEE Photonics Technol. Lett.*, Vol. 22, 2010, pp. 1330-1332.

[31]. D. Bigourd, L. Lago, A. Mussot, A. Kudlinski, J.F. Gleyze, E. Hugonnot, High-gain, optical-parametric, chirped-pulse amplification of femtosecond pulses at 1 µm, *Opt. Lett.,* Vol. 20, 2010, pp. 3480-3482.

[32]. D. Bigourd, L. Lago, A. Kudlinski, E. Hugonnot, A. Mussot, Dynamics of fiber optical parametric chirped pulse amplifiers, *J. Opt. Soc. Am. B*, Vol. 28, 2011, pp. 2848-2854.

[33]. V. Cristofori, Z. Lali-Dastjerdi, L. S. Rishj, M. Galili, C. Peucheretand, K. Rottwitt, Dynamic characterization and amplification of sub-picosecond pulses in fiber optical parametric chirped pulse amplifiers, *Opt. Express*, Vol. 21, 2013, pp. 26044-26051.

[34]. R. Herda, Fiber optical parametric amplifier pumped by chirped-femtosecond pulses, in *Proceedings of the Laser Congress 2017 (ASSL, LAC), OSA Technical Digest,* 2017, p. JTu2A.49.

[35]. P. Morin, J. Dubertrand, P. Beaure d'Augères, Y. Quiquempois, G. Bouwmans, A. Mussot, E. Hugonnot, µJ-level Raman-assisted fiber optical fiber parametric chirped pulse amplification, *Opt. Lett.,* Vol. 43, 2018, pp. 4683-4686.

Chapter 2
Polarization Transformation Using Thin Optical Elements

Svetlana N. Khonina, Alexey P. Porfirev,
Andrey V. Ustinov, Sergey A. Fomchenkov

2.1. Introduction

Polarization is a well-known property of the electromagnetic waves which specifies the direction of the oscillations. Nowadays, along with full control of amplitude and phase distributions, control of the polarization state of the generated laser radiation is crucial for many applications in the field of optical manipulation [1-3] and laser material processing [4-8], for example, the use of heterogeneously polarized beams to shape matter with subwavelength features in a desired way [8]. The so-called cylindrical vector beams (CVBs), a class of axially symmetric laser beams with spatially variant polarization [9] which can be characterized by the so-called polarization singularities of vector fields [10, 11], also make it possible to realize a long-range tractor beam for airborne light-absorbing particles [12] and to increase the capacity of free-space and fibre optical communication systems [13, 14]. Many studies on tight focusing of singular laser beams [11, 15] with complex types of polarizations have showed the possibility to use it for overcoming the diffraction limit [16, 17] as well as the three-dimensional control of a light field formed in the focal region [18-20]. Shaping given the three-dimensional (3D) intensity distributions with a desired polarization state of focused laser radiation is also often required in optical microscopy [21] and high-capacity information storage [22].

Heterogeneously polarized beams can be generated using interferometric techniques [23], anisotropic crystals [24-26], subwavelength gratings [27], q-plates [28] and S-waveplates [29]. In two latter cases, both single plates and their cascades are widely used to generate higher-order cylindrical polarization [30-33]. In fact, q-plates are subwavelength periodic structures behaving as a uniaxial crystal with the optical axes parallel and perpendicular to the subwavelength grooves. Such q-plates tuned by the temperature control [34] or the external electric field [35, 36] can be used for efficient generation of higher-order CVBs at different wavelengths, however there are no tuned q-plates capable of performing the

Svetlana N. Khonina
Image Processing Systems Institute – Branch of the Federal Scientific Research Centre 'Crystallography and Photonics' of Russian Academy of Sciences, Samara, Russia

tunable generation of cylindrical polarization of different orders. For this propose, optical setups with spatial light modulators (SLMs) supporting implementation of multi-level phase profile [37-40] or a combination of such SLMs with a q-plate [41] are usually used.

When laser beams are sharply focused (when the numerical aperture (NA) is more than 0.7), polarization transformations occur simultaneously, often associated with the appearance of a strong longitudinal electric component of the laser field [42-46]. However, polarization transformations can occur in the paraxial regime [47] due to the spin-orbit interaction [11, 48–51]. The spin-orbital interrelation of photons was detected a long time ago [52–54], and it has been discussed in several recent reviews [9, 50, 51, 54-57]. The polarization conversion can take place when beams with phase singularity (for example, optical vortex beams) are tightly focused [20, 58, 59]. Conversion of the spin (related to the state of polarization) and orbital (relevant to phase distribution) angular momentum is also manifested in anisotropic media [26, 60-62]. The interaction of the polarization singularities and phase singularities is used to detect the polarization state of the laser beam [63, 64]. Note that the complete differentiation of polarization types is only possible with sharp focusing of light [64]. In this chapter, the 3D spatial intensity and polarization transformation, at the sharp and paraxial regimes, performed with the help of thin diffractive optical elements (DOEs), such as diffractive axicons, fork-shaped gratings and spiral phase plates (SPPs), were considered. We theoretically describe the principles of transformation of heterogeneously polarized laser beams using such diffractive optics. The numerical simulation and experimental results obtained demonstrate the efficient formation of hybrid high-order CVBs even under conditions of weak focusing.

2.2. Calculations Roadmap: Debye Approximation and Matrix Formulation

Under conditions of tight focusing, the cylindrical components of the electric field of a monochromatic electromagnetic wave can be calculated using the Debye approximation [65]:

$$\mathbf{E}(\rho,\varphi,z) = \begin{pmatrix} E_x(\rho,\varphi,z) \\ E_y(\rho,\varphi,z) \\ E_z(\rho,\varphi,z) \end{pmatrix} = \qquad (2.1)$$

$$= -\frac{if}{\lambda} \int_0^{\theta_{max}} \int_0^{2\pi} B(\theta,\phi)T(\theta)\mathbf{P}(\theta,\phi)\exp\left[ik(r\sin\theta\cos(\phi-\varphi)+z\cos\theta)\right]\sin\theta\,d\theta\,d\phi,$$

where $(\rho,\,\varphi,\,z)$ are the cylindrical coordinates in the focal region, $(\theta,\,\phi)$ are spherical angular coordinates of the focusing system's output pupil, θ_{max} is the maximum value of the azimuthal angle associated with the numerical aperture of the system, $B(\theta,\,\varphi)$ is the transmission function, $T(\theta)$ is the pupil's apodisation function ($T(\theta)=\sqrt{\cos\theta}$ for aplanatic systems), $k = 2\pi/\lambda$ is the wavenumber, λ is the wavelength of radiation and f is the focal length (see Fig. 2.1).

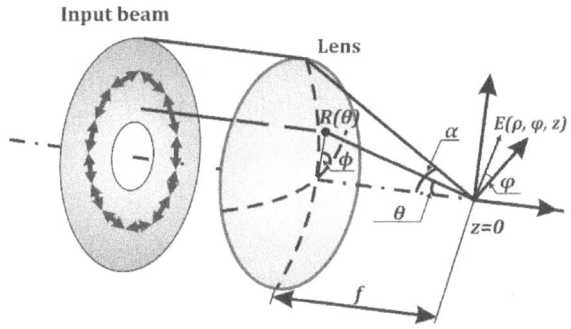

Fig. 2.1. Schematic representation of focusing of an azimuthally polarized laser beam through a lens of focal length f and maximum azimuthal angle α.

The vector of polarization transformation in the focal region for the Cartesian coordinates has the following form:

$$\mathbf{P}(\theta,\phi) = \begin{bmatrix} 1+\cos^2\phi(\cos\theta-1) & \sin\phi\cos\phi(\cos\theta-1) & \cos\phi\sin\theta \\ \sin\phi\cos\phi(\cos\theta-1) & 1+\sin^2\phi(\cos\theta-1) & \sin\phi\sin\theta \\ -\sin\theta\cos\phi & -\sin\theta\sin\phi & \cos\theta \end{bmatrix} \begin{pmatrix} c_x(\phi) \\ c_y(\phi) \\ c_z(\phi) \end{pmatrix}, \qquad (2.2)$$

where $\mathbf{c}(\phi) = (c_x(\phi), c_y(\phi), c_z(\phi))^T$ is the polarization vector of the input field.

In some cases, Eq. (2.1) can be simplified – for example, for the laser beams with a radial symmetry and m-th order vortex phase in the form of $B(\theta,\phi) = R(\theta)\exp(im\phi)$ Eq. (2.1) is represented as follows

$$\mathbf{E}_m(\rho,\varphi,z) = -ikf \int_0^{\theta_{max}} R(\theta)T(\theta)\mathbf{Q}_m(\rho,\varphi,\theta)\sin\theta \exp(ikz\cos\theta)\mathrm{d}\theta , \qquad (2.3)$$

where components of the vector $\mathbf{Q}_m(\rho, \varphi, \theta)$ depend on the polarization of incident laser beam $\mathbf{c}(\phi)$ and represent a superposition of Bessel functions of various orders [66]. So, for a vortex laser beam with circular polarization $\mathbf{c}(\phi) = \left(1/\sqrt{2}\right)\left(1,\pm i,0\right)^T$, the polarization vector has the following components [17]:

$$\mathbf{Q}_m^{circ\pm}(\rho,\varphi,\theta) = \frac{i^m e^{im\varphi}}{\sqrt{2}} \begin{bmatrix} J_m(t) + \frac{1}{2}\left[J_m(t) - e^{\pm i2\varphi} J_{m\pm 2}(t)\right](\cos\theta - 1) \\ \pm i\left\{J_m(t) + \frac{1}{2}\left(J_m(t) + e^{\pm i2\varphi} J_{m\pm 2}(t)\right)(\cos\theta - 1)\right\} \\ \mp i e^{\pm i\varphi} J_{m\pm 1}(t)\sin\theta \end{bmatrix} \qquad (2.4)$$

In the case of different types of CVBs with the polarization order p and the inner polarization rotation of the beam $\varphi_0\, \mathbf{c}(\phi) = \begin{pmatrix} \cos(p\phi + \phi_0) \\ \sin(p\phi + \phi_0) \end{pmatrix}$ [30]:

$$\mathbf{Q}_p(\rho, \varphi, \theta) = i^p \begin{bmatrix} J_p(t)\cos(p\varphi + \phi_0) - \\ \quad -\dfrac{1}{2}\left\{J_p(t)\cos(p\varphi + \phi_0) - J_{p-2}(t)\cos\left[(p-2)\varphi + \phi_0\right]\right\}(1 - \cos\theta) \\ J_p(t)\sin(p\varphi + \phi_0) - \\ \quad -\dfrac{1}{2}\left\{J_p(t)\sin(p\varphi + \phi_0) + J_{p-2}(t)\sin\left[(p-2)\varphi + \phi_0\right]\right\}(1 - \cos\theta) \\ iJ_{p-1}(t)\cos\left[(p-1)\varphi + \phi_0\right]\sin\theta \end{bmatrix}, \qquad (2.5)$$

with $t = k\rho\sin\theta$.

2.3. 3D Transformations of Light Fields Implemented by Different Types of Diffractive Axicons

2.3.1. Theoretical Analysis for Tight Focusing

Let us consider the focusing system containing a binary axicon with the following complex transmission function [67]

$$R(\theta) = \tau_{bin\_ax}(\theta) = \exp\left\{i\arg\left[\cos(k\alpha_0 f\sin\theta))\right]\right\}, \qquad (2.6)$$

where α_0 is the axicon parameter associated with the numerical aperture of the optical element [67-70]. Results of modelling of the focusing of a circularly polarized Gaussian beam with a binary axicon are shown in Fig. 2.2 ($f = 101\lambda$, the radius of illuminating Gaussian beam is 50λ, $\theta_{max} = 0.99$ (lens NA) and $\alpha_0 = 0.05$ (axicon NA)). Different colours indicate components of the electrical part of electromagnetic field: red for $|E_x|^2$, green for $|E_y|^2$ and blue for $|E_z|^2$.

As can be seen from Fig. 2.2, when using a binary axicon, a Bessel beam is formed before and after the focal plane, and a double ring is formed in the focal plane. This is explained in the following way: the transmission function of a binary axicon in the first order of diffraction can be described approximately as a superposition of two complex conjugated linear axicons (focusing and defocusing) [69, 70]:

$$\tau_{bin\_ax}(r) \approx \cos(k\alpha_0 r) = 0.5\left[\exp(-ik\alpha_0 r) + \exp(ik\alpha_0 r)\right] \qquad (2.7)$$

The field in the focal plane has a mixed elliptic-azimuthal polarization (Fig. 2.2d) and the intensity in the rings in the focal plane is different. The ratio of the ring width can be changed in the axicon period (fill factor) or the periodic structure shifted from the centre [69, 70] to redistribute the intensity in the focal rings. Note that the Bessel beam before and after the focus is formed with the circular polarization at the centre of the beams, with a mixed elliptic-azimuthal polarization on the periphery of beams (Fig. 2.2c and Fig. 2.2e). Thus, there is a 3D transformation, not only for intensity but also for polarization. Note that the 3D polarization structure [71] is of considerable importance in sharp focusing conditions. Fig. 2.3 shows the results of focusing by a binary spiral axicon [72] having the following complex transmission function:

$$R(\theta,\varphi) = \tau_{sp\_ax}(\theta,\varphi) = \text{sgn}\left[\cos\left(k\alpha_0 f \sin\theta + m\varphi\right)\right] \tag{2.8}$$

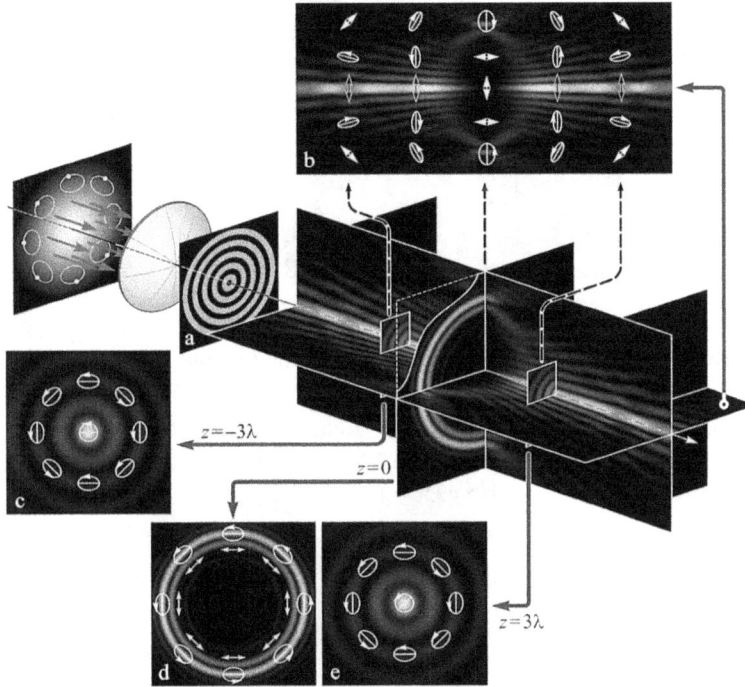

Fig. 2.2. Focusing of a circularly polarized Gaussian beam with a focusing system containing a binary axicon: (a) phase distribution of the axicon; (b) the longitudinal intensity distribution of the focused beam ($z \in [-10\lambda; 10\lambda]$, $y \in [-7\lambda; 7\lambda]$); (c) transverse intensity distribution before the focal plane ($z = -3\lambda$, picture size $4\lambda \times 4\lambda$); (d) transverse intensity distribution in the focal plane ($z = 0$, picture size $12\lambda \times 12\lambda$) and (e) transverse intensity distribution after the focal plane ($z = 3\lambda$, picture size $4\lambda \times 4\lambda$).

The binary spiral axicon defined by Eq. (2.8) actually contains two vortex complex conjugated axicons:

$$\tau_{sp\_ax}(r,\varphi) \approx \cos\left(k\alpha_0 r + m\varphi\right) = 0.5\left[\exp\left(-ik\alpha_0 r - m\varphi\right) + \exp\left(ik\alpha_0 r + m\varphi\right)\right] \tag{2.9}$$

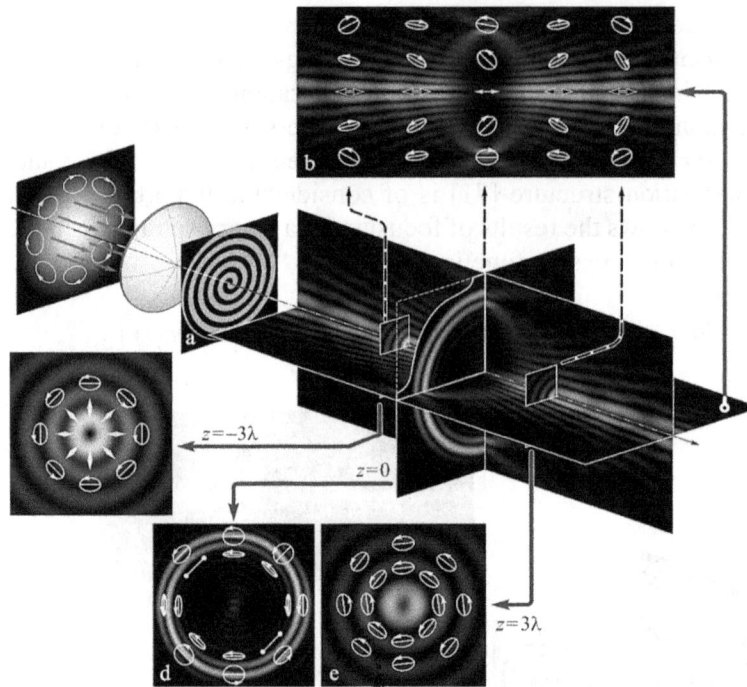

Fig. 2.3. Focusing of a circularly polarized Gaussian beam with a focusing system containing a binary spiral axicon ($m = 1$): (a) phase distribution of the axicon; (b) the longitudinal intensity distribution of the focused beam ($z \in [-10\lambda; 10\lambda]$, $y \in [-7\lambda; 7\lambda]$); (c) transverse intensity distribution before the focal plane ($z = -3\lambda$, picture size 4λ×4λ); (d) transverse intensity distribution in the focal plane ($z = 0$, picture size 14λ×14λ) and (e) transverse intensity distribution after the focal plane ($z = 3\lambda$, picture size 4λ×4λ).

For a vortex axicon of the first order ($|m| = 1$), which has the opposite direction with respect to the direction of polarization, a longitudinal component will be formed on the optical axis [46]. If the direction of the phase vortex is changed to the opposite (co-directional with polarization), then zero intensity will be on the optical axis. Thus, a binary spiral axicon makes it possible to obtain a hollow Bessel beam before the focus and a Bessel beam with a longitudinal component on the axis after the focus (Fig. 2.3). In the focal plane, instead of two rings (Fig. 2.2d), a double helix is observed (Fig. 2.3d). The polarization state before the focal plane is radial in the central part of a beam (Fig. 2.3c) and the polarization state changes to partially azimuthal after the focal plane (Fig. 2.3e).

2.3.2. Theoretical Analysis for the Paraxial Regime

In the paraxial case, taking into account in Eq. (2.4) that $\cos\theta \approx 1$ and the z-component is now insignificant, the light field in the focal region defined by Eq. (2.3) can be rewritten only with the transverse components [17, 73]:

$$\mathbf{E}_{\perp,m}^{circ\pm}(\rho,\varphi,z) = \begin{pmatrix} E_x(\rho,\varphi,z) \\ E_y(\rho,\varphi,z) \end{pmatrix} =$$

$$= -kf \frac{i^{m+1} e^{im\varphi}}{\sqrt{2}} \begin{pmatrix} 1 \\ \pm i \end{pmatrix} \int_0^{\theta_{max}} R(\theta) J_m(k\rho\sin\theta)\sin\theta \exp(ikz\cos\theta)d\theta \qquad (2.10)$$

Let us consider the focusing system with a linear phase axicon [74]:

$$R(\theta) = \tau_{ax}(\theta) = \exp(i\sigma k\alpha_0 f \sin\theta), \qquad (2.11)$$

where σ = "$\pm$" is the sign corresponding to the defocusing or focusing element.

It follows from the properties of Bessel functions [75] that there is a nonzero intensity on the optical axis (ρ = 0) only for m = 0 (in the absence of a vortex phase singularity):

$$\mathbf{E}_{\perp,m=0}^{circ\pm}(0,0,z) = -\frac{ikf}{\sqrt{2}} \begin{pmatrix} 1 \\ \pm i \end{pmatrix} \int_0^{\theta_{max}} \exp\left[ik\left(z\cos\theta + \sigma\alpha_0 f \sin\theta\right)\right]\sin\theta d\theta \qquad (2.12)$$

The integral in Eq. (2.12) can be calculated approximately by the stationary phase method [76]. Then, by following the change of variables, (x = $\sin\theta$, dx = $\cos\theta d\theta \approx d\theta$) can be written as follows:

$$G(z) = \int_0^{\sin(\theta_{max})} \exp\left[ik\left(z\sqrt{1-x^2} + \sigma\alpha_0 fx\right)\right]x\,dx \qquad (2.13)$$

The stationary point is determined by the vanishing of the derivative of the expression in brackets:

$$x_s = \frac{\sigma\alpha_0 f}{\sqrt{(\alpha_0 f)^2 + z^2}} \qquad (2.14)$$

It follows from Eq. (2.14) that when using the defocusing axicon (σ = "+"), the maximum value is formed to the right of the focus of the lens (the focus is located at point z = 0), and when using the focusing axicon (σ = "–"), the maximum value is formed to the left of the lens focus.

The total intensity of transverse components on the axis for the field given by Eq. (2.12) is defined as follows:

$$I(0,0,z) = kf \frac{2\pi z^2 f (\alpha_0 f)^2}{\left[(\alpha_0 f)^2 + z^2\right]^{5/2}} \qquad (2.15)$$

37

$I(0,0,z)$ is zero at the focal point, i.e., at $z = 0$, and also tends towards zero for larger values of z, when leaving far beyond the focal region. The maximum of $I(0,0,z)$ has the following relationship:

$$z_{max} = \pm\beta(\alpha_0 f) = \sigma\beta(\alpha_0 f), \tag{2.16}$$

where the sign is determined by the type of axicon, either defocusing or focusing, and β is the coefficient of proportionality. It can be seen from Eq. (2.16) that the maximum value point moves away from the focus position with the growth of the numerical aperture of the axicon (parameter α_0) and the increase in focal length. Thus, the variation of parameter α_0 makes it possible to control the longitudinal distribution in the focal region.

When using a binary axicon with the transmission function defined by Eq. (2.6), two maxima are generated simultaneously before and after the geometric focus, since the binary axicon contains both defocusing and focusing linear axicons. In this case, by controlling parameter α_0, it is possible to delete or bring together two maxima up to their merging into one line. This technique is usually used to increase the depth of focus.

In order to control the transverse distribution of the generated light field, one can use axicons with a vortex phase singularity of the m-th order. In this case, annular distributions with a zero central part are formed in the cross-section, the radius of which increases with the order of vortex singularity m, and the longitudinal distribution depends on parameterα_0.

For a light field with the azimuthal polarization $\mathbf{c}(\phi) = (\sin\phi, -\cos\phi, 0)^T$, the polarization vector defined by Eq. (2.5) has the following components [17]:

$$\mathbf{Q}_m^{az}(\rho,\varphi,\theta) = \frac{i^m e^{im\varphi}}{2}\begin{bmatrix} -\left[e^{i\varphi}J_{m+1}(k\rho\sin\theta) + e^{-i\varphi}J_{m-1}(k\rho\sin\theta)\right] \\ i\left[e^{i\varphi}J_{m+1}(k\rho\sin\theta) - e^{-i\varphi}J_{m-1}(k\rho\sin\theta)\right] \\ 0 \end{bmatrix}\sigma\beta(\alpha_0 f) \tag{2.17}$$

When m $= 0$, there is zero intensity on the optical axis, and the azimuthal polarization is conserved in the focal region:

$$\mathbf{E}_{\perp,m=0}^{az}(\rho,\varphi,z) = kf\begin{pmatrix} -\sin\varphi \\ \cos\varphi \end{pmatrix}\int_0^{\theta_{max}} R(\theta)J_1(k\rho\sin\theta)\sin\theta\exp(ikz\cos\theta)\sqrt{\cos\theta}\,d\theta \tag{2.18}$$

The nonzero intensity on the optical axis will develop in the presence of the first-order vortex phase singularity $|m| = 1$. In particular, for $m = 1$:

$$\mathbf{E}_{\perp,m=1}^{az}(\rho,\varphi,z) =$$
$$= -\frac{kf}{2}\int_0^{\theta_{max}} R(\theta)\begin{pmatrix} J_0(k\rho\sin\theta) + e^{i2\varphi}J_2(k\rho\sin\theta) \\ iJ_0(k\rho\sin\theta) - ie^{i2\varphi}J_2(k\rho\sin\theta) \end{pmatrix}\sin\theta\exp(ikz\cos\theta)\sqrt{\cos\theta}\,d\theta \tag{2.19}$$

Let us consider the apodisation of the focusing system by the linear phase axicon defined by Eq. (2.11) and analyse the light field distribution on the optical axis ($\rho = 0$) in the focal region, taking into account the paraxial approximation:

$$\mathbf{E}^{az}_{\perp,m=1}(0,0,z) = \frac{kf}{2}\begin{pmatrix}1\\i\end{pmatrix} \int\limits_0^{\theta_{\max}} \exp\left[ik\left(z\cos\theta + \sigma\alpha_0 f\sin\theta\right)\right]\sin\theta\,d\theta \qquad (2.20)$$

Equation (2.19) is up to a factor equal to Eq. (2.11), which corresponds to the field with circular polarization in the absence of a vortex phase singularity. Thus, the introduction or change of optical vortex order m makes it possible not only to change the transverse distribution but also to change the polarization state in the focal region.

2.3.3. Simulation Results

Numerical simulations allow one to illustrate the difference between tight focusing and paraxial case. The simulation parameters are follows: laser wavelength $\lambda = 532$ nm, focus length of the focusing lens $f = 40$ mm and parameter $\alpha_0 = 0.005$. The system consists of an axicon and the focusing lens is illuminated by a Gaussian beam with a waist radius of 1.5 mm. Fig. 2.4 shows the simulation results of the propagation of light fields generated by various axicons in the case of a circularly polarized illuminating laser beam. The white-filled circular labels reading "1", "2" and "3" indicate planes along the optical axis in which the transverse irradiance is analysed and displayed in the right-hand panel. It is clearly seen that when using a linear focusing axicon (see Fig. 2.4, first row), a zero-order Bessel beam is formed only in the region between an optical element and its focal plane. In the focal plane, a narrow light ring is formed, then the light is scattered, i.e., there is an area of shadow. In this region, although the amount of energy on the optical axis is much smaller (we enhanced the transverse picture) than in the region before the focal plane, the formed beam is similar to the Bessel beam, but its full width at half maximum (FWHM) of the central spot is 13 % smaller. When using a binary axicon that actually contains focusing and defocusing linear axicons, Bessel beams are formed on both sides of the focal plane (see Fig. 2.4, second row). The intensity distribution in the focal plane looks like a double ring with unequal intensity; that is, the inner ring is weak in intensity [69, 70]. Introduction of a vortex component into an axicon, for example, the use of a vortex axicon (see Fig. 2.4, third row), allows the generation of a first-order Bessel vortex beam. Moreover, a similar intensity distribution, but with a smaller spot size, is formed in the area of shadow (we also enhanced the transverse picture). A spiral axicon is a binary analog of the vortex axicon, allowing the generation of Bessel vortex beams on both sides of the focal plane (see Fig. 2.4, fourth row). In this case, the intensity distribution in the focal plane looks like a double spiral ring.

More complex 3D distributions can be obtained by varying the parameter α_0 of the binary spiral axicon (see Fig. 2.5). In this case, in contrast to the situation shown in Fig. 2.3, the beam remains hollow everywhere. This is due to the paraxial mode of the focusing system of Fig. 2.5. The non-zero value of the intensity on the optical axis after the focal plane (Fig. 2.3e) is related to the longitudinal component of the electric field, which has large values only at sharp focusing. Thus, in the paraxial case, a binary spiral axicon for a circularly polarized beam does not allow the above-mentioned effects. However, the variations of the axicon parameter α_0 make it possible to form "hourglass" type configurations (Fig. 2.5, second and third rows); such distributions can serve as selective optical traps for particles with a definite size.

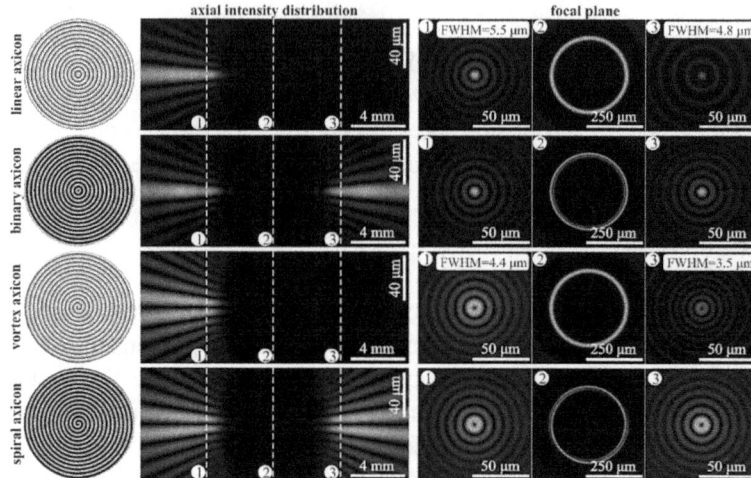

Fig. 2.4. Simulation results of 3D light field transformations by various types of axicons ($\alpha_0 = 0.004$) in the case of a circularly polarized illuminating laser beam.

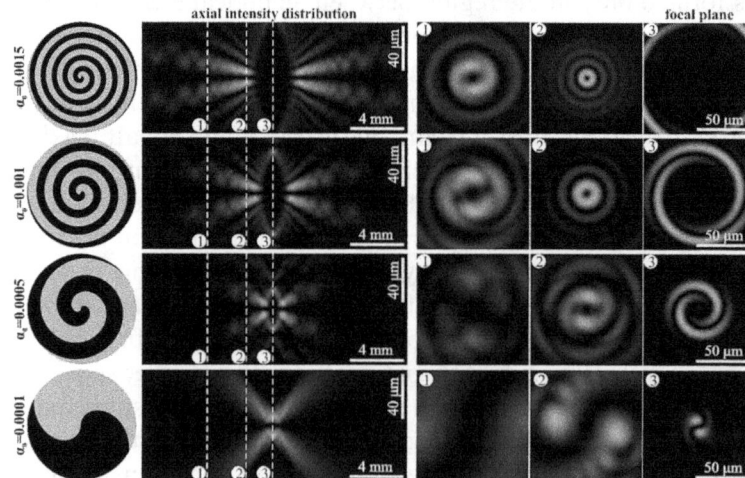

Fig. 2.5. Simulation results of 3D light field transformations by a binary spiral ($m = 1$) axicon in the case of a circularly polarized illuminating laser beam for different values of α_0.

Note that as the parameter α_0 decreases, the symmetry of the transverse distribution is lost (Fig. 2.5, two last rows). This effect can be explained by the following: for smaller α_0, additional diffraction orders (arising as a result of binarisation and containing higher vortices) propagate at smaller angles to the optical axis [69, 70] and begin to intersect, distorting the axial symmetry of the beam in zero diffraction order. Moreover, when α_0 approaches zero (Fig. 2.5, the last row), the characteristic annular structure of the axicon disappears/degenerates. In this case, the structure of the optical element is similar to a binary phase plate. Such a plate forms a Gaussian-Hermite mode (1,0) from the Gaussian beam, consisting of two maxima [77]. Such a picture of two maxima is shown in Fig. 2.5 (the focal plane in the last row).

As shown by the theoretical analysis, the action of various types of axicons is also significantly dependent on the polarization of illuminating radiation. Fig. 2.6 shows the simulation results of 3D light fields' transformations by various axicons in the case of an azimuthally polarized illuminating laser beam. In the case of using a linear focusing axicon (Fig. 2.6, first row), a Bessel beam with the azimuthal polarization is generated. The intensity distribution of the beam corresponds to the first-order Bessel function, but a phase does not have a vortex component. We enhanced the transverse picture in the shadow region (as for Fig. 2.3) to show that the central spot is smaller than in the light region. A binary axicon allows for the generation of such beams on both sides of the focal plane. In the case of the azimuthal polarization, a double ring in the focal plane changes its configuration in comparison with circular polarization; the outer ring becomes less bright (the second rows in Figs. 2.4 and 2.6). Introduction of a vortex component into an axicon (Fig. 2.6, third and fourth rows) allows for the generation of a zero-order Bessel beam, but its central spot is slightly larger. In accordance with Eq. (2.20), the polarization state of the beam changes: instead of being azimuthal, the polarization becomes circular. Thus, the simulation results are in full agreement with the conducted theoretical investigation.

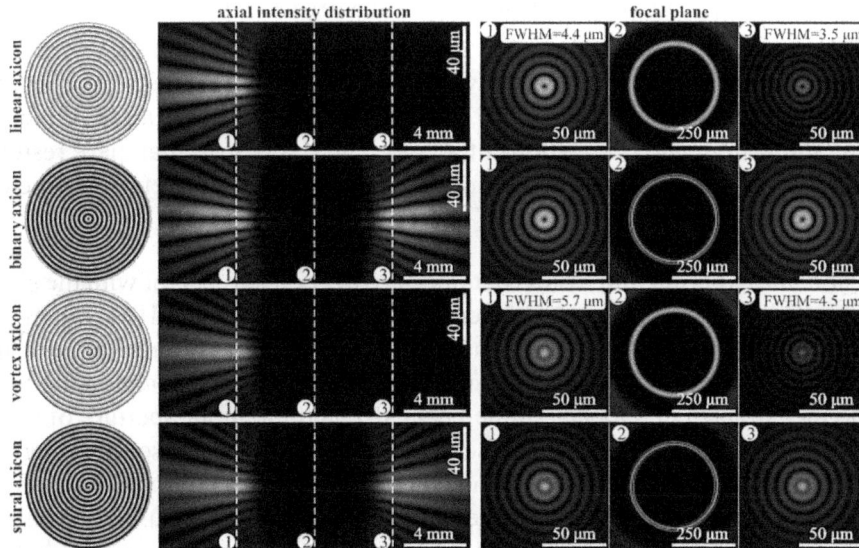

Fig. 2.6. Simulation results of 3D light field transformations by various types of axicons ($\alpha_0 = 0.004$) in the case of an azimuthally polarized illuminating laser beam.

41

2.3.4. Experimental Results

The experimental investigation was performed on a focusing system with a moderate numerical aperture, i.e., the focusing mode can be considered a paraxial one. Fig. 2.7 shows an optical setup that was utilized in order to experimentally investigate spatial polarization transformations of 3D light fields using diffractive axicons of various types. A phase-only PLUTO VIS SLM based on a reflective LCOSmicrodisplay with a spatial resolution of 1920×1080 pixels and a pixel size of 8 μm was used to implement phase profiles of the axicons. An initial laser beam (λ = 532 nm) was collimated and extended by a combination of a microobjective (MO$_1$)(20×, NA = 0.4), a pinhole(PH) and a lens (L$_1$). A combination of lenses (L$_2$ and L$_3$) and a diaphragm (D) was used to carry out spatial filtration of the laser beam reflected from the display of the SLM. A microobjective (MO$_2$) (4×, NA= 0.1) focused the generated laser beam. A microobjective (MO$_3$) (20×, NA = 0.4) and a CCD video camera, both mounted on the movable stage, were used to obtain intensity distributions of the generated laser beam near the focus of the microobjective MO$_2$. To transform the initially linearly polarized laser beam into a circularly or azimuthally polarized laser beam, a quarter-wave plate (P) or an S-waveplate (P) [29] was utilized.

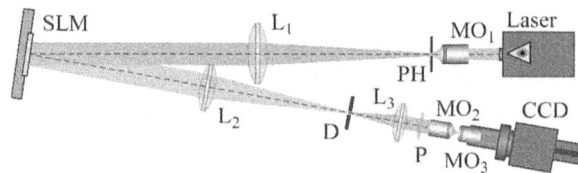

Fig. 2.7. Experimental setup to investigate spatial polarization transformations of 3D light fields using diffractive axicons.

Figs. 2.8 and 2.9 show the experimentally obtained intensity profiles of the laser beams generated by a linear, binary, vortex and spiral axicon, with α_0 = 0.005 in the cases of the circular and azimuthal polarizations of a laser beam. It is seen that these results are in good qualitative agreement with the simulation results: all effects described in the previous section do, in fact, occur.

In conclusion, the experimental results obtained are in good agreement with the simulation results. Small discrepancies in the experimentally measured FWHM values and the FWHM values obtained in modelling can be explained by the presence of aberrations in the optical system due to the oblique incidence of the initial laser beam on the SLM, and also by the incomplete coincidence of the beam size and the input aperture of the second microobjective. We believe that these results can be useful in the laser trapping of nano- and microparticles, optical microscopy and improvement of high-capacity information storage techniques. For example, the generated 3D-patterns can be used not only for the formation of vector bottle beams with desired polarization distribution [78] but also for creating an experimental setup for polarization sensitive laser trapping and manipulation of different nano-and microparticles [1, 12, 79-82].

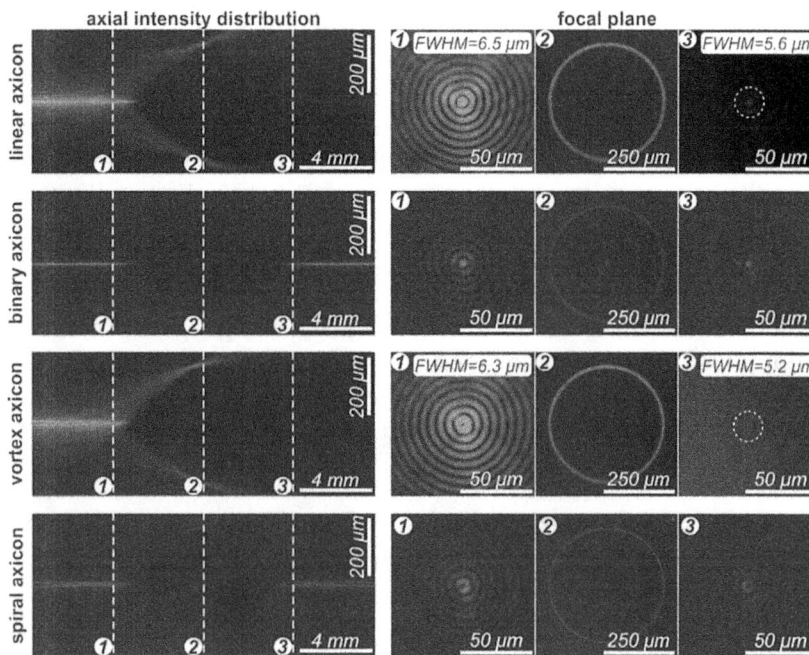

Fig. 2.8. Experimentally obtained results of propagation of light fields generated by various axicons (α_0 = 0.005) in the case of a circularly polarized illuminating laser beam.

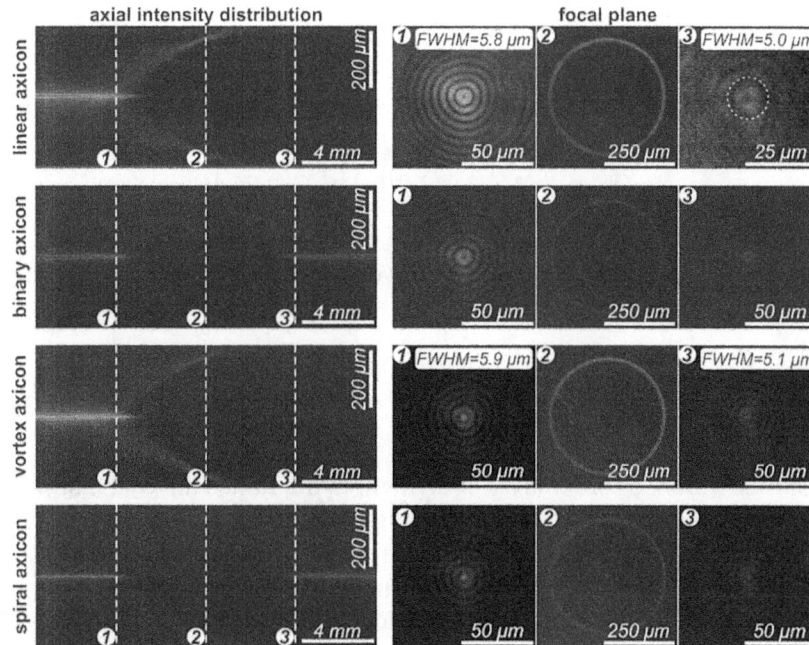

Fig. 2.9. Experimentally obtained results of propagation of light fields generated by various axicons (α_0 = 0.005) in the case of an azimuthally polarized illuminating laser beam.

43

In this case, polarization distribution of the laser radiation is another additional degree of freedom that can be used for the generation of tunable optical traps: changing the incident polarization state of light, it is possible to dynamically reshape the 3D light distribution of the generated optical trap without the use of spatial light modulators or digital micromirror devices. For example, modification of the generated light distribution from needle-like (in the case of a linear axicon with a circularly polarized illuminating laser beam) to ring-shaped in the area before the focal plane (in the case of a linear axicon with an azimuthally polarized illuminating laser beam) allows one to use a single element for manipulation as transparent particles that are attracted to the high-intensity region [83] and opaque particles that are pushed out of the high-intensity region [84]. The control of the polarization distribution of the generated ring-shaped intensity distributions is also crucial for laser material processing because of the sensitivity of orientation and structure of the produced nano- and microelements to the polarization of the radiation [8, 85].

2.4. Polarization Transformations Arising from Focusing of Shifted Vortex Beams of Arbitrary Order with Different Polarization

2.4.1. Theoretical Analysis

Let us consider the input field of the optical system in the form of off-axis optical vortices of arbitrary order within a given axisymmetric beam. First we write it in Cartesian coordinates:

$$
\begin{aligned}
B(x,y) &= R\left(\sqrt{x^2+y^2}\right)\sum_{p}\left(\left(x-x_p\right)+i\left(y-y_p\right)\right)^{m_p} = \\
&= R\left(\sqrt{x^2+y^2}\right)\sum_{p}\left(\left(x+iy\right)-\left(x_p+iy_p\right)\right)^{m_p}
\end{aligned}
\tag{2.21}
$$

For integer values of m_p, Eq. (2.21) can be converted using the binomial formula:

$$
B(x,y) = R\left(\sqrt{x^2+y^2}\right)\sum_{p}\left(\sum_{l=0}^{m_p}(-1)^l C_{m_p}^l \cdot \left(x+iy\right)^{m_p-l}\left(x_p+iy_p\right)^l\right)
\tag{2.22}
$$

That is, the shifted vortex of order m_p is equivalent to the finite sum of the on-axis vortices of orders from 0 to m_p inclusive. So, from an algebraic point of view, we can use already known analytic expressions obtained for an on-axis vortex defined by Eq. (2.21) if we ignore the bulkiness of the resulting expressions. If m_p is not integer, the sum becomes infinite, and it will include on-axis optical vortices of all orders with the same sign.

Further it is more convenient to use polar coordinates:

$$B(r,\phi) = R(r)\sum_{p}\left(\left(r\cos\phi - r_p\cos\phi_p\right) + i\left(r\sin\phi - r_p\sin\phi_p\right)\right)^{m_p} =$$

$$= R(r)\sum_{p}\left(r\exp(i\phi) - r_p\exp(i\phi_p)\right)^{m_p} = \qquad (2.23)$$

$$= R(r)\sum_{p}\left(\sum_{l=0}^{m_p}(-1)^l C_{m_p}^l \cdot \left(r\exp(i\phi)\right)^{m_p-l}\left(r_p\exp(i\phi_p)\right)^l\right)$$

For some values of m_p, we can write the explicit expressions:

$$B_1(r,\phi) = R(r)\left(r\exp(i\phi) - r_0\exp(i\phi_0)\right),$$

$$B_2(r,\phi) = R(r)\left(r\exp(i\phi) - r_0\exp(i\phi_0)\right)^2 =$$

$$= R(r)\left[\left(r\exp(i\phi)\right)^2 - 2r_0\exp(i\phi_0)r\exp(i\phi) + \left(r_0\exp(i\phi_0)\right)^2\right],$$

$$B_3(r,\phi) = R(r)\left(r\exp(i\phi) - r_0\exp(i\phi_0)\right)^3 = \qquad (2.24)$$

$$= R(r)\left[\left(r\exp(i\phi)\right)^3 - 3r_0\exp(i\phi_0)\left(r\exp(i\phi)\right)^2 +\right.$$

$$\left. +3\left(r_0\exp(i\phi_0)\right)^2 r\exp(i\phi) - \left(r_0\exp(i\phi_0)\right)^3\right]$$

As can be clearly seen, the superposition consists of vortex beams with the same direction of rotation. In this case, the pictures, regardless of the order of the vortex and displacement, will look about the same – as a "crescent" [86], reminding also the pictures for fractional orders [87]. This is expected, since the optical vortex of fractional order can be represented as an infinite series on integer orders [88].

To get pictures of another type, for example "camomile-shaped" [89], it is necessary to form a superposition of optical vortices with different directions of rotation. So, if we use a shifted vortex beam of the opposite sign as the illuminating beam, we shall get:

$$B(r,\phi) = \left(r\exp(-i\phi) - r_s\exp(-i\phi_s)\right)^{m_s} \sum_{p}\left(r\exp(i\phi) - r_p\exp(i\phi_p)\right)^{m_p} \qquad (2.25)$$

In particular, for $m_s = 1$ and $p = 1$, $m_l = 2$:

$$B(r,\phi) = \left(r\exp(-i\phi) - r_s\exp(-i\phi_s)\right)\times$$

$$\times\left[\left(r\exp(i\phi)\right)^2 - 2r_0\exp(i\phi_0)r\exp(i\phi) + \left(r_0\exp(i\phi_0)\right)^2\right] =$$

$$= \left[\exp(i\phi)r^3 - 2r_0\exp(i\phi_0)r^2 + \left(r_0\exp(i\phi_0)\right)^2\exp(-i\phi)r\right] - \qquad (2.26)$$

$$-r_s\exp(-i\phi_s)\left(r\exp(i\phi) - r_0\exp(i\phi_0)\right)^2$$

The second term up to the coefficient is similar to the illumination by a plane beam. The first term can be rewritten in the form of:

$$\left[\exp(i\phi)r^3 - 2r_0\exp(i\phi_0)r^2 + \left(r_0\exp(i\phi_0)\right)^2\exp(-i\phi)r\right] =$$

$$= 2\left(r_0\exp(i\phi_0)\right)^2\cos(\phi)r + \tag{2.27}$$

$$+\exp(i\phi)\left[r^3 - r\left(r_0\exp(i\phi_0)\right)^2\right] - 2r_0\exp(i\phi_0)r^2$$

From Eq. (2.27) it is visible that the picture will contain "dumbbell". Obviously, for larger values of m_s the "camomiles" with different numbers of petals will be present.

2.4.2. Simulation Results

Fig. 2.10 shows the results of sharp focusing of a linearly polarized centered first-order vortex beam calculated using Eqs. (2.1) and (2.2). Obviously, linear polarization introduces a certain asymmetry [17] into the beam intensity even in the focal plane ($z = 0$). Outside the focal plane, there are additional changes in the shape of the beam associated not only with the defocusing of the beam, but also with different dynamics of changes in different components of the electromagnetic field.

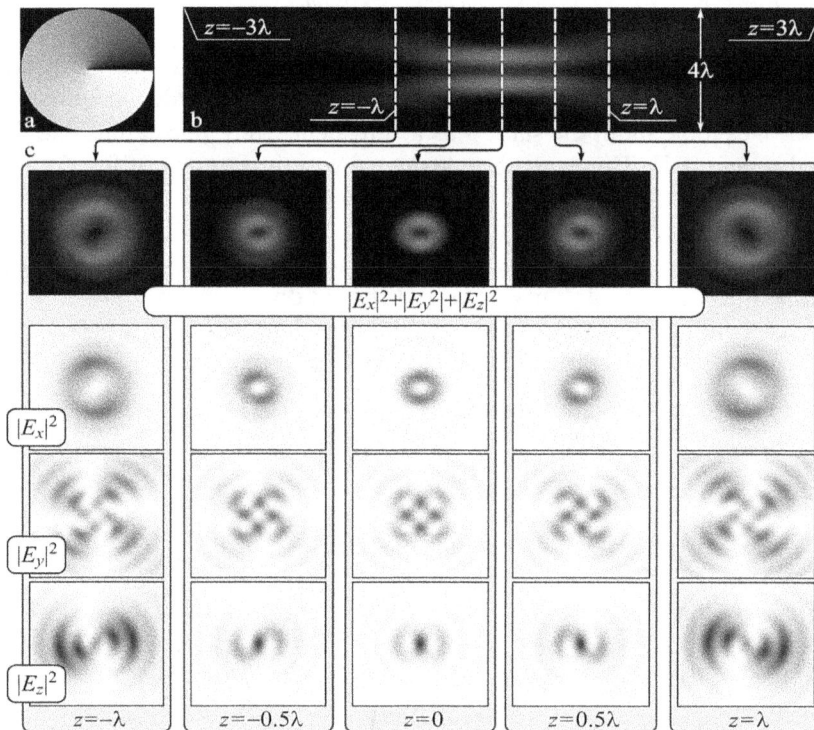

Fig. 2.10. Sharp focusing of linearly polarized centered vortex beam of first order: (a) the input phase; (b) the longitudinal distribution of the total intensity; (c) transverse distributions of total and component's intensities at different distances (red color for x-component, green color for y-component, and blue color for z-component).

Fig. 2.11 shows similar results for a shifted first-order vortex beam. It can be seen from the pattern of the longitudinal distribution (Fig. 2.11b) that the maximum of the beam propagates at an angle to the optical axis. A typical picture of a rotating "crescent" [86] is formed in the focal area at different distances. The main contribution in total intensity is determined by the *x*-component (as initial polarization state). It is noteworthy that in [86] only the first-order vortex beam was considered. To consider the focusing of several vortex beams of different orders at once, we use a multi-order DOE [89, 90].

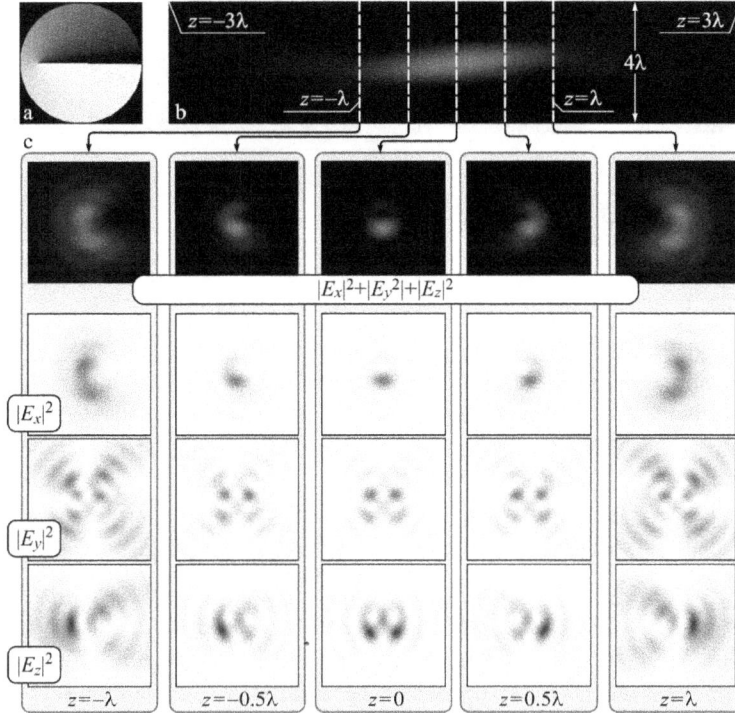

Fig. 2.11. Sharp focusing of a linearly polarized shifted vortex beam of first-order: (a) the input phase; (b) the longitudinal distribution of the total intensity; (c) transverse distributions of total and component's intensities at different distances.

The transmission function of a multi-order DOE is matched with a set of optical vortices:

$$\Psi_m(x,y) = \exp\left(im\tan^{-1}(y/x)\right) = \exp(im\varphi) \tag{2.28}$$

with different spatial carrier frequencies (α, β):

$$\tau(x,y) = \sum_{p=1}^{P} \Psi_{m_p}(x,y)\exp\left[i\left(\alpha_p x + \beta_p y\right)\right] \tag{2.29}$$

Optical vortices defined by Eq. (2.28) are invariant to the passage in free space and lens systems, i.e. invariant to the Fourier transform.

Fig. 2.12 shows the results of sharp focusing of five linearly polarized beams of $m_p = 0, \pm1, \pm2$ orders formed by a multi-channel optical element with a transmission function defined by Eq. (2.29). The pictures of the transverse distribution in the focal plane show that the x-polarized vortex beams of ±1-orders provide the formation of a central focal spot in the z-component (Fig. 2.12f), and x-polarized vortex beams of ±2 orders provide nonzero central intensity in the y-component (Fig. 2.12e). This fact was noted in the work [17].

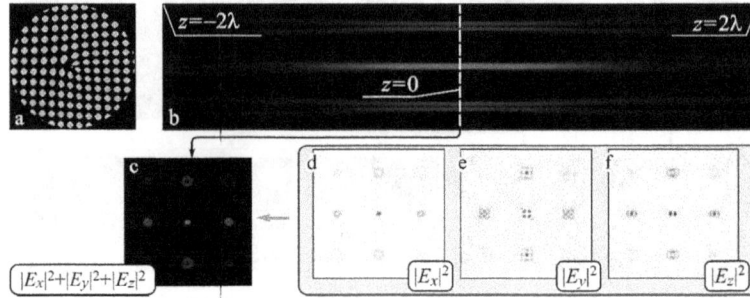

Fig. 2.12. Sharp focusing of five linearly-polarized beams of $m_p = 0, \pm1, \pm2$ orders formed by the multi-channel optical element: (a) the input phase; (b) the longitudinal distribution of the total intensity; (c) transverse distributions in the focal plane for total intensity, as well as the intensity of the (d) x-component; (e) y-component and (f) z-component.

Fig. 2.13 shows similar results for the shifted optical element. It can be seen from the pattern of the longitudinal distribution (Fig. 2.13b) that the maximum of the central beam, which has no phase singularity, propagates along the optical axis, while the maxima of the vortex beams propagate at an angle to the optical axis. Note that in the presence of displacement (as opposed to the previous case) there is a dependence (in the pictures of transverse intensity) not only on the value of the optical vortex, but also its sign (direction of rotation). This is more clearly seen in Fig. 2.13c, which shows the dynamics of changes in the transverse distribution of vortex beams of different orders in the focal area at different distances.

As can be seen from Fig. 2.13c, although there are certain differences of the formed pictures depending on the order of the shifted vortex, they are quite similar and represent a rotating crescent. To complicate the structure of the pictures, it is necessary to embed a vortex of the opposite direction of rotation in the illuminating beam. Fig. 2.14 shows results similar to Fig. 2.13, but with the illuminating beam containing the third order phase vortex. Since the optical element with a transmission function defined by Eq. (2.29) generates the vortex beams of different directions of rotation, then some beams (having a sign opposite to the sign of the illuminating beam vortex) will have pictures of the one type, but others (having the same sign) will have pictures of a different type. The difference depending on the vortex sign can be observed also on the picture of the longitudinal distribution (Fig. 2.14b) – upper and lower beams differ only by sign. A more significant difference is observed in the dynamics of changes in the transverse distribution in the focal area at different distances (Fig. 2.14c). Thus, changing the phase state of the illuminating beam, you can significantly change the form of rotating intensity distributions in the focal area.

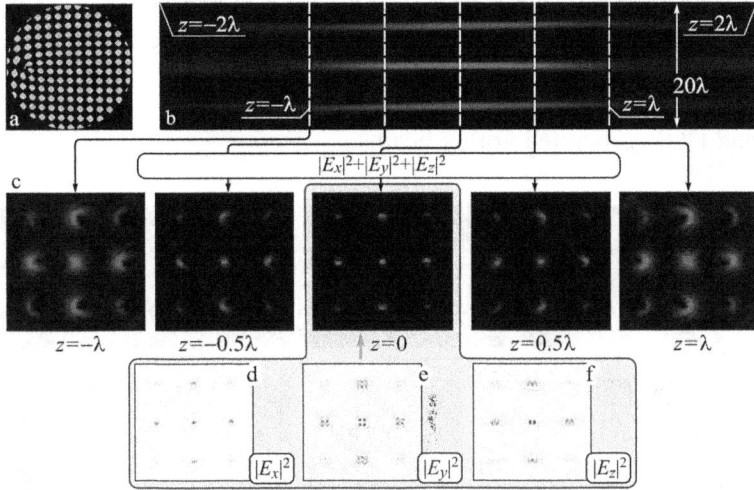

Fig. 2.13. Sharp focusing of five linearly-polarized beams of $m_p = 0, \pm 1, \pm 2$ orders formed by the shifted multi-channel optical element: (a) the input phase, (b) the longitudinal distribution of the total intensity, (c) transverse distributions in the focal plane for total intensity at different distances, as well as the intensity in the focal plane for the (d) x-component, (e) y-component and (f) z-component.

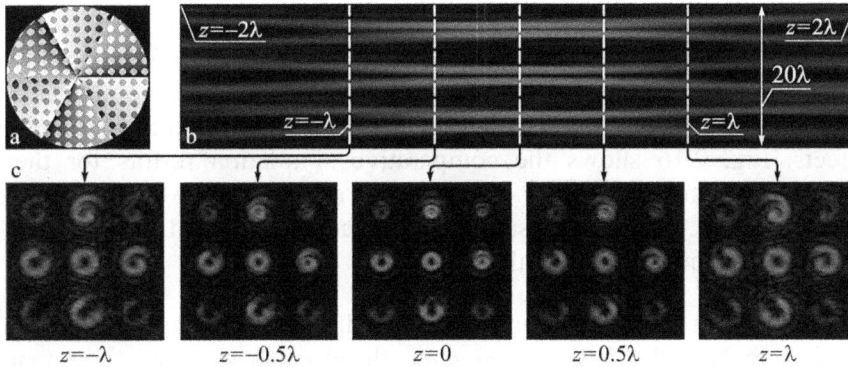

Fig. 2.14. Sharp focusing of five linearly-polarized beams of $m_p = 0, \pm 1, \pm 2$ orders formed by a shifted multi-channel optical element when illuminated by the vortex beam of $m_s = 3$ order: (a) the input phase; (b) the longitudinal distribution of the total intensity, and (c) the dynamics of changes in the transverse distribution at different distances.

In the theoretical analysis it was also shown that the focal patterns will change significantly if the illuminating beam will have a shifted vortex in accordance with Eq. (2.25). In this case, we obtain a product of the superposition of vortex beams analogous to Eq. (2.26). Fig. 2.6 shows the results similar to Fig. 2.14, but with the

49

displacement of the illuminating vortex beam of the third order (m_s = 3). Note that in this case, the torque also acquires a central beam that initially had not vortex phase. As can be seen from Fig. 2.15, the focal patterns for shifted vortex beams have changed significantly: instead of "crescents", they took the form of "dumbbells". At the same time, as can be seen from the dynamics of changes in the transverse distribution at different distances, some beams retain rotational behavior, while others do not rotate, but only change their shape. Obviously, this depends on the ratio of the direction of rotation of illuminating and DOE-generated vortex beams.

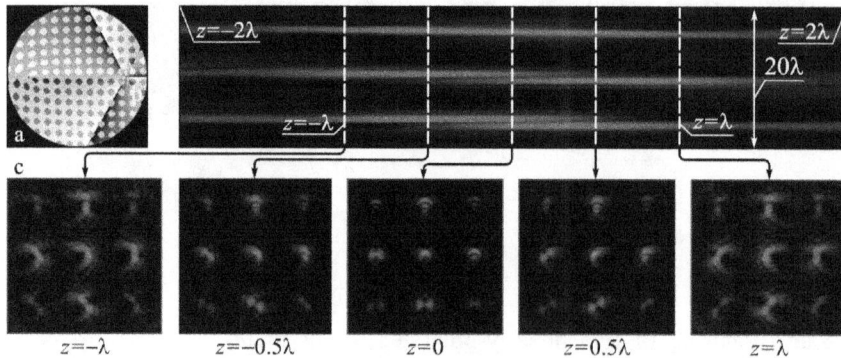

Fig. 2.15. Sharp focusing of five linearly-polarized beams of m_p = 0, ±1, ±2 orders formed by a shifted multi-channel optical element when illuminated by the shifted vortex beam of m_s = 3 order: (a) the input phase; (b) the longitudinal distribution of the total intensity; and (c) the dynamics of changes in the transverse distribution at different distances.

The state of beam polarization is also of great importance in sharp focusing. In this case, there is a significant redistribution of energy between the components of the electromagnetic field [73]. Therefore, let us consider the effect of polarization on the above effects. Fig. 2.16 shows the comparative simulation results for the circular polarization of the illuminating beam. These results are quite close to the results obtained by linear polarization. More interesting results can be obtained by using cylindrical polarization types (azimuthal or radial).

Fig. 2.17 shows the comparative simulation results for azimuthal polarization of the illuminating beam. Note that in this case even in the absence of a vortex component in the illuminating beam (the first line of Fig. 2.17) rather complex pictures are formed. They are similar to the distributions obtained by using vortex illuminated beams with uniform polarization (the second line of Fig. 2.16). The greatest difference between the use of uniform and cylindrical polarization is observed in the illumination of the shifted optical element by the shifted vortex beam (the third line of the Fig. 2.17). Rotating beams in this case do not look like "dumbbells", but have a more complex distribution. The central (initially non-vortex) beam also acquires a rotational moment in this case. The results for radial polarization will be similar. Thus, not only the presence of the vortex phase and the displaced position of the illuminating beam, but also its polarization state affects the pictures formed in the focal area, as well as the dynamics of their changes.

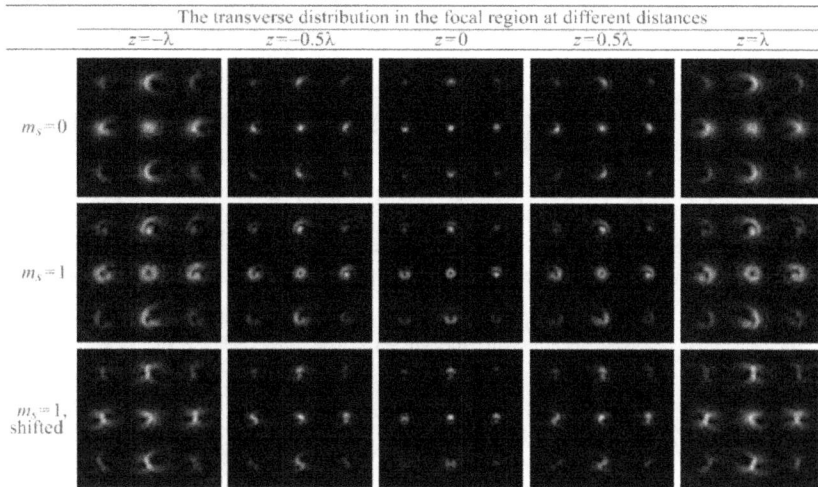

Fig. 2.16. Comparative simulation results of focusing five beams of $m_p = 0, \pm1, \pm2$ orders, formed by a shifted multi-channel optical element when illuminated by different beams with circular polarization.

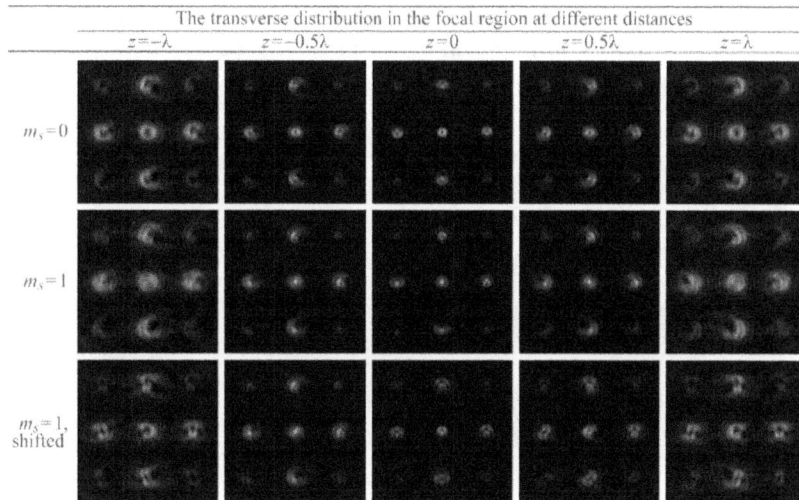

Fig. 2.17. Comparative simulation results of focusing five vortex beams of $0, \pm1, \pm2$ orders, formed by a shifted multi-channel optical element when illuminated by different beams with azimuthal polarization.

In conclusion, the obtained results are relevant for multi-channel communication systems based on OAM (orbital angular momentum)-carrying laser beams division multiplexing, considering distortions and/or walks of vortex beams in turbulent or random medium. Therefore, when a distorted/shifted detecting, the fact that the misaligned higher-order vortex beams contain a set of lower-order vortex beams should be taken into account. The investigated results may also be useful as new tools in optical manipulations.

2.5. Polarization Conversion of Radially and Azimuthally Polarized Vortex Beams

2.5.1. Theoretical Analysis

Using Eq. (2.3) and Eq. (2.5) for tight focusing of CVBs, we can obtain the following expression for the transverse components in the case of focusing an azimuthally polarized beam (in this case, longitudinal component is zero), having a vortex phase of m-th order:

$$
\mathbf{E}_{m,\perp}^{az}(\rho,\varphi,z) = \begin{pmatrix} E_{m,\rho}^{az}(\rho,\varphi,z) \\ E_{m,\varphi}^{az}(\rho,\varphi,z) \end{pmatrix} = \frac{i^{m+1}kf\,e^{im\varphi}}{2} \times
$$
$$
\times \int_0^\alpha R(\theta)T(\theta) \begin{pmatrix} J_{m+1}(k\rho\sin\theta) + J_{m-1}(k\rho\sin\theta) \\ -i\left[J_{m+1}(k\rho\sin\theta) - J_{m-1}(k\rho\sin\theta)\right] \end{pmatrix} \sin\theta \exp(ikz\cos\theta)d\theta
$$

(2.30)

From Eq. (2.30), it follows that the azimuthal polarization is retained only when focusing in the absence of a vortex phase ($m = 0$). In the case of presence of a vortex phase, part of the energy of the azimuthal component will transfer to the orthogonal radial component. Let us determine when this transformation will occur with the greatest level. To simplify the analysis, we consider the limits of integration in Eq. (2.30) in a rather narrow ring $\theta \in [\alpha_1, \alpha_2]$, denoting the central radius of the ring as $\theta_c \in (\alpha_1 + \alpha_2)/2$.

Then the ratio of the intensities of the orthogonal component in the focal region can be approximately estimated by the following expression:

$$
\eta_m^{az\to rad} = \frac{\left| E_{m,\rho}^{az}(\rho,\varphi,z=0)\right|^2}{\left| E_{m,\varphi}^{az}(\rho,\varphi,z=0)\right|^2} \approx \frac{\left| J_{m+1}(k\rho\sin\theta_c) + J_{m-1}(k\rho\sin\theta_c)\right|^2}{\left| J_{m+1}(k\rho\sin\theta_c) - J_{m-1}(k\rho\sin\theta_c)\right|^2}
$$

(2.31)

From Eq. (2.31), it is obvious that the intensity of the radial component will prevail over the azimuthal one when the following is true:

$$
J_{m+1}(k\rho\sin\theta_c)J_{m-1}(k\rho\sin\theta_c) > 0
$$

(2.32)

We can define the boundary of the main dominance area of the radial polarization over the azimuthal one by the following radius:

$$
\rho_m = \frac{j_{m-1,1}}{k\sin\theta_c},
$$

(2.33)

where $j_{v,1}$ is the first zero of the v-th Bessel function of the first kind.

When $|m| \geq 2$ at the central area of the focal plane, the radial polarization will be formed instead of the azimuthal polarization, and with increasing the order of the vortex phase, this region will increase. Note also that the region of conversion decreases with increasing numerical aperture of the focusing system ($\theta_c \rightarrow 90°$); i.e., in the paraxial case (with a weak focus), this effect will be more significant.

Let us now consider the following optical beam as the incident beam:

$$R(\theta) = \exp\left(-\frac{\sin^2 \theta}{\sin^2 \sigma}\right)\frac{\sin \theta}{\sin \alpha},$$ (2.34)

where σ is the angular width of the waist of a Gaussian beam, $\sin \alpha$ is the numerical aperture of the focusing device. Using some approximations, including $T(\theta) \approx \cos \theta / \sqrt{\cos(\alpha/2)}$, we can transform the integral in Eq. (2.30) with a beam described by Eq. (2.34) into the form of tabular integrals [91] and calculate it analytically. Then, using the notation $x = k\rho\sin\sigma$, after mathematical transformations, we obtain:

$$E_{m,\rho}^{az}(x) \approx \frac{\sqrt{\pi} \cdot \sin^3 \sigma}{4\sin \alpha \sqrt{\cos(\alpha/2)}} \exp\left(-\frac{x^2}{8}\right)\left[I_{(m-1)/2}\left(\frac{x^2}{8}\right) - I_{(m+1)/2}\left(\frac{x^2}{8}\right)\right],$$ (2.35)

$$E_{m,\varphi}^{az}(x) \approx \frac{i\sqrt{\pi} \cdot \sin^3 \sigma}{4\sin \alpha \sqrt{\cos(\alpha/2)}} \cdot \exp\left(-\frac{x^2}{8}\right) \times$$

$$\times \left\{ m\left[I_{(m-1)/2}\left(\frac{x^2}{8}\right) + I_{(m+1)/2}\left(\frac{x^2}{8}\right)\right] - \frac{x^2}{2}\left[I_{(m-1)/2}\left(\frac{x^2}{8}\right) - I_{(m+1)/2}\left(\frac{x^2}{8}\right)\right]\right\}$$ (2.36)

Then, in the analysis, we do not take into account the factor that does not depend on x and m, as its module for both components is the same. The radial component $E_{m,\rho}^{az}(x)$ is always positive for $m > 0$, and the azimuthal one $E_{m,\varphi}^{az}(x)$ changes sign once in this area. Since for the constant argument, function $I_m(\cdot)$ decreases with increasing order, then we can show that in the region of positive values of $E_{m,\varphi}^{az}(x)$, the radial component is greater than the azimuthal one. Moreover, the inequality $E_{m,\rho}^{az}(x) > \left|E_{m,\varphi}^{az}(x)\right|$ holds for a wider range. To prove it let us consider another way. In Eq. (2.36), there is a difference of functions $I_\nu(\cdot)$ with orders differ by one. The equation for orders differ by two is known: $I_{\nu-1}(y) - I_{\nu+1}(y) = 2\nu I_\nu(y)/y$. We use the monotonic continuity of a modified Bessel function by index and we approximately write:

$$I_{(m/2)-(1/2)}(\cdot) - I_{(m/2)+(1/2)}(\cdot) \approx \left[I_{(m/2)-1}(\cdot) - I_{(m/2)+1}(\cdot)\right]/2 \approx \frac{4m}{x^2}I_{m/2}(\cdot)$$ (2.37)

Substituting Eq. (2.37) into Eq. (2.36) we get:

$$E_{m,\rho}^{az}(x) \sim m\left[I_{(m-1)/2}\left(\frac{x^2}{8}\right) - I_{(m+1)/2}\left(\frac{x^2}{8}\right)\right],$$

$$E_{m,\varphi}^{az}(x) \sim m\left[I_{(m-1)/2}\left(\frac{x^2}{8}\right) - 2I_{m/2}\left(\frac{x^2}{8}\right) + I_{(m+1)/2}\left(\frac{x^2}{8}\right)\right]$$

(2.38)

The ratio of the intensities similar to Eq. (2.31) will be:

$$\eta_m^{az \to rad} = \frac{\left|E_{m,\rho}^{az}(\rho,\varphi,z)\right|^2}{\left|E_{m,\varphi}^{az}(\rho,\varphi,z)\right|^2} \approx \frac{\left(I_{(m-1)/2} - I_{(m+1)/2}\right)^2}{\left(I_{(m-1)/2} - 2I_{m/2} + I_{(m+1)/2}\right)^2} = $$

$$= \frac{\left[\left(I_{(m-1)/2} - I_{m/2}\right) + \left(I_{m/2} - I_{(m+1)/2}\right)\right]^2}{\left[\left(I_{(m-1)/2} - I_{m/2}\right) - \left(I_{m/2} - I_{(m+1)/2}\right)\right]^2}$$

(2.39)

Since for the constant argument, function $I_\nu(\cdot)$ decreases with increasing order, we get $\eta_m^{az \to rad} > 1$.

When focusing a radially polarized field, having a vortex phase of m-th order we obtain (in the paraxial case, we can consider only the transverse components):

$$\mathbf{E}_{m,\perp}^{rad}(\rho,\varphi,z) = \begin{pmatrix} E_{m,\rho}^{rad}(\rho,\varphi,z) \\ E_{m,\varphi}^{rad}(\rho,\varphi,z) \end{pmatrix} = \frac{i^{m+1}kf\,e^{im\varphi}}{2} \times$$

$$\times \int_0^\alpha R(\theta)T(\theta) \begin{pmatrix} i\left[J_{m+1}(k\rho\sin\theta) - J_{m-1}(k\rho\sin\theta)\right] \\ \left[J_{m+1}(k\rho\sin\theta) + J_{m-1}(k\rho\sin\theta)\right] \end{pmatrix} \cos\theta\sin\theta\exp(ikz\cos\theta)d\theta$$

(2.40)

In this way, we obtain a situation opposite to the one that was discussed above: from the radial polarization the azimuthal polarization is formed with increasing the order of the optical vortex

$$\eta_m^{rad \to az} = \frac{\left|E_{m,\varphi}^{rad}(\rho,\varphi,z=0)\right|^2}{\left|E_{m,\rho}^{rad}(\rho,\varphi,z=0)\right|^2} \xrightarrow[|m|\to\infty]{} \infty$$

(2.41)

2.5.2. Simulation Results

This section presents the results of the simulation of the polarization conversion when we use Gaussian beam of Eq. (2.34) in Eq. (2.30). For simulation by numerical integration of Eq. (2.30), we chose the following parameters: $\sin\alpha = 0.02$, $\sin\sigma = 0.012$. Fig. 2.18 shows the intensity distribution in the focal plane for the focused radially polarized beam described by Eq. (2.40) for $m = \overline{0,4}$. As can be seen, in the absence of vortex phase ($m = 0$) initial radially polarized beam preserves the polarization in the focal plane. If we add a vortex phase of first order (arbitrary sign) in the focused radially polarized beam; in the focal plane, light spot with circular polarization is formed. If we add a vortex phase of higher order, the conversion from radial polarization to azimuthal one is observed.

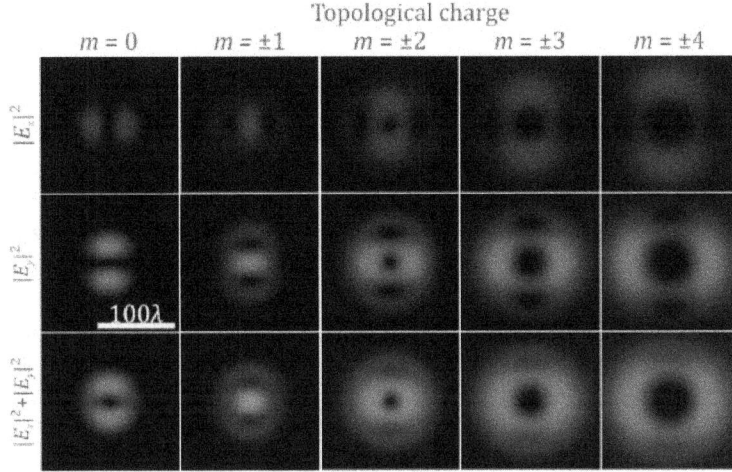

Fig. 2.18. The distribution of the various components of the electric field in the focal plane for a radially polarized incident beam in the absence and in the presence of a vortex phase (the *x*-component is blue and the *y*-component is red).

Fig. 2.19 shows the focal distribution of the azimuthal and radial components of the electric field in this case for quantitative estimation of the conversion degree.

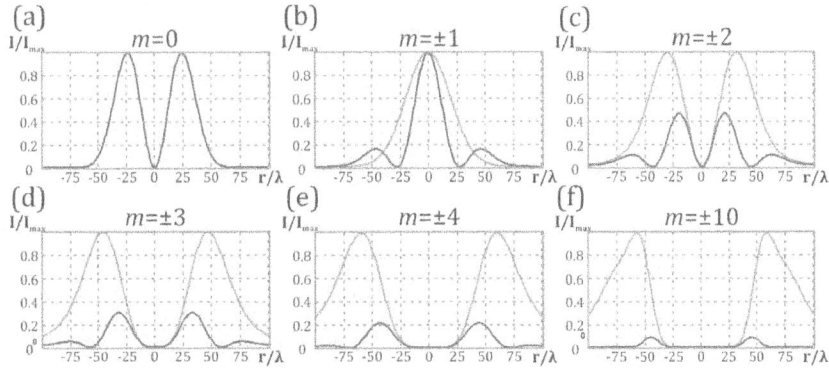

Fig. 2.19. Distribution of the azimuthal and radial components of the electric field: (a) in the absence and (b–f) in the presence of a vortex phase: (b) $m = \pm 1$, (c) $m = \pm 2$, (d) $m = \pm 3$, (e) $m = \pm 4$, (f) $m = \pm 10$. The radial component is denoted with brown color, the azimuthal component is denoted with green color.

We calculated the coefficient $\eta_m^{rad \to az}$ (see Eq. (2.41)) for different m in order to determine the degree of orthogonal polarization conversion. The values are $\eta_{m=0}^{rad \to az} = 0$, $\eta_{m=\pm 1}^{rad \to az} = 1$

, $\eta_{m=\pm2}^{rad \to az} = 2.14$, $\eta_{m=\pm3}^{rad \to az} = 3.33$, $\eta_{m=\pm4}^{rad \to az} = 4.58$, $\eta_{m=\pm10}^{rad \to az} = 11.85$. It can be clearly seen that when the order of vortex increases, the contribution of the azimuthal component increases. However, even in the case of the 10-th vortex, the radial component does not decrease to zero.

2.5.3. Experimental Results

To investigate the polarization conversion experimentally, we utilized the experimental optical setup shown in Fig. 2.20. The output beam from a solid-state laser (λ = 532 nm) passed through a pinhole (100-μm aperture). Then, the laser beam passed through the polarizer P, to obtain linearly polarized light with a predetermined polarization direction. A diaphragm D was used to separate the central spot of the Airy disk resulting from the wave diffraction on the pinhole. The S-waveplate, specially oriented to the direction of linear polarization of the incident laser beam, converted the initial linearly-polarized laser beam into a radially-polarized laser beam. The resulting radially polarized laser beam illuminated the amplitude diffractive optical element DOE, forming a superposition of eight vortex beams with orders ±1, ±2, ±3, and ±4 in different diffraction orders. The lens L (f = 150 mm) focused the laser beam on the camera's sensor.

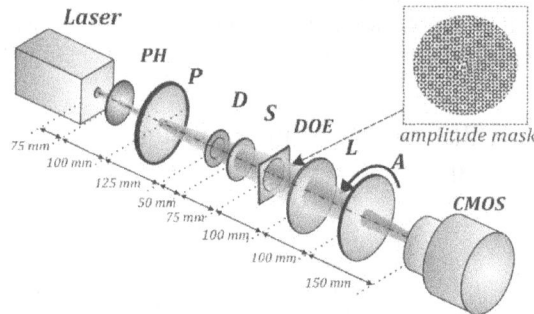

Fig. 2.20. Experimental optical setup: Laser is a solid-state (λ = 532 nm), *PH* is a pinhole (100-μm aperture), *P* is a polarizer, *D* is a diaphragm, *S* is an S-waveplate (radial polarization converter), *DOE* is an amplitude diffractive optical element forming a superposition of eight vortex beams with orders of ±1, ±2, ±3, and ±4, *L* is a lens with a focal length f = 150 mm, *A* is an analyser, and *CMOS* is a CMOS-video camera (LOMO TC-1000, 3664 × 2740 pixel resolution).

The inset in Fig. 2.20 shows the amplitude transmission function of the used diffractive optical element. Amplitude mask is obtained by encoding [92] of the transmission function of the multi-order vortex DOE is

$$\Omega(x, y) = \sum_n \exp(im_n \varphi) \exp\left[i(2\pi u_n x + 2\pi v_n y) \right],$$ (2.42)

where n is an index of diffractive order, m_n is a topological charge of the vortex beam, (u_n, v_n) are the carrier spatial frequencies. Operating principle of such DOE is shown in Fig. 2.21. The action of the DOE is of introducing optical vortices of different order in a radially-polarized Gaussian beam, which initially does not have a vortex phase. Using a multi-order DOE forming multiple optical vortices of different orders allows to demonstrate the degree of conversion of one polarization state to another.

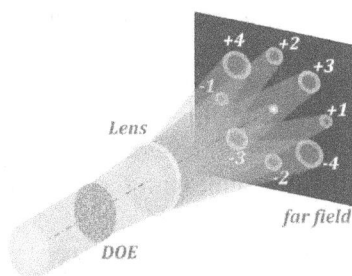

Fig. 2.21. Operating principle of a multi-order vortex diffractive optical element.

The diffraction pattern formed by the amplitude diffractive optical element is shown in Fig. 2.22. Fig. 2.22a shows the simulated intensity distribution generated by DOE. Since we utilized an amplitude DOE, intensity of the zero diffraction order is too high in comparison to intensity of other diffraction orders. In order to see clearly the non-zero diffraction orders we removed the zero diffraction order from the simulated picture. Fig. 2.22b shows the intensity distributions obtained without the analyzer. These experimentally obtained distributions show the simultaneous interaction with eight optical vortices of different orders. Each of them is formed in a separate diffraction order, i.e. at the corresponding location of the focal plane. In the considered paraxial case, the displacement of the diffraction orders from the center of the focal plane has no effect on polarization conversion. To confirm the order of the vortices, we recorded the interference pattern of the vortices with the Gaussian beam, resulting in the characteristic fork fringes, as shown in Fig. 2.22c. The interference patterns confirm the order of the vortices to be $m = \pm 1, \pm 2, \pm 3,$ and ± 4, respectively.

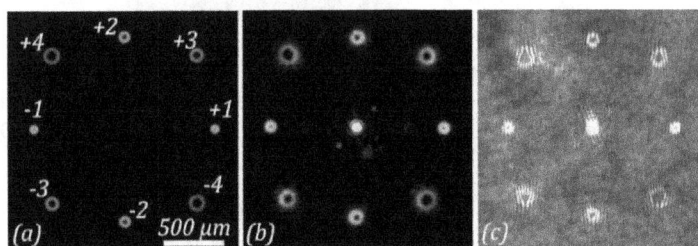

Fig. 2.22. Diffraction pattern formed in the far-field region when passing the laser beam through the amplitude diffractive optical element generated a superposition of eight vortex beams: (a) simulation and (b) experimental result. (c) Experimentally obtained interference patterns of the generated vortex beams.

Intensity distributions obtained with different orientations of the analyzer are shown in Fig. 2.23. The analyzer rotation angles are equal to ± 45 and 90 degrees. The analyzer orientation in these figures represented by white arrows. In addition, the zero diffraction order in which a laser beam with topological charge $m = 0$ is formed gives an indication of the orientation of the analyzer. As was predicted theoretically, in the case of optical vortices with $m = \pm 2, \pm 3$, and ± 4 the conversion of radially polarized light into azimuthally polarized light is observed. For optical vortices with $m = \pm 1$, conversion is not observed. Fig. 2.24 shows the intensity distributions formed in the far-field region using a diffractive optical element forming two vortex beams with topological charge $m = \pm 10$. It can be clearly seen that in this case we obtain a laser beam with a nearly perfect azimuthal polarization. The radial component decreases significantly. Thus, the experimental results are in good agreement with the simulation results presented above.

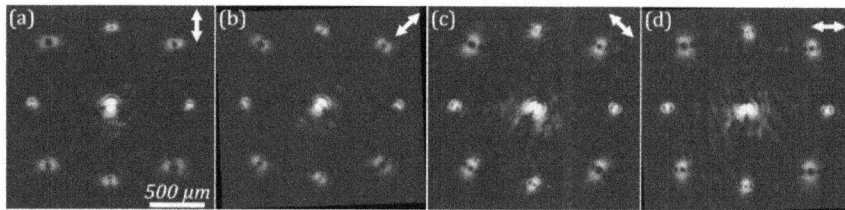

Fig. 2.23. Diffraction pattern formed in the far-field region when passing the laser beam through the amplitude diffractive optical element generated a superposition of eight vortex beams for different orientations of the analyzer: (a) 0, (b) +45, (c) -45, and (d) 90 degrees.

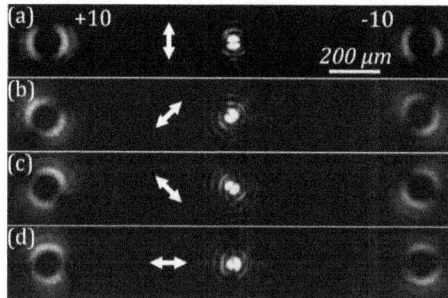

Fig. 2.24. Diffraction pattern formed in the far-field region when passing the laser beam through the amplitude diffractive optical element generated vortex beams with a topological charge $m = \pm 10$ for different orientations of the analyzer: (a) 0, (b) +45, (c) -45, and (d) 90 degrees.

In conclusion, we demonstrated the conversion of cylindrically polarized laser beams in an orthogonal polarization state by introducing a high-order vortex phase singularity. In addition, a numerical study was performed. The experimental results are in good agreement with the simulation results. Our theoretical and experimental investigations show that increasing the order of the phase singularity leads to conversation of the radially polarized laser beam into an azimuthally polarized one. Our results clearly demonstrate tight interrelation of the specific polarization and phase states of electromagnetic beams. Such specific combinations of the polarization and spatial properties of the laser beam are important for various applications, for example, telecommunication or materials processing. So, Bozinovic et al. [93] presented multiplexing techniques that use wavelength, amplitude, phase, and polarization of light to encode information. Taking into account the results presented in the current chapter, it is necessary to cautiously use a combination of orthogonal polarization states and orbital angular momenta. On the other hand, the relationship of the cylindrical types of polarization and phase vortex will allow better understanding of the processes occurring in the interaction of laser radiation with matter, such as ablation [94].

2.6. Formation and Transformation of Higher-order Cylindrical Vector Beams Using Binary Multi-sector Phase Plates

2.6.1. High-order CVBS Focusing

As follows from Eq. (2.5), the z-component of a tightly focused electromagnetic field is not zero at the optical axis only in the case when $p = 1$, and $\phi_0 = 0$ or π, that is, in the case of radial polarization. The z-component completely disappears in the case when $p = 1$, $\phi_0 = \pi/2$ or $3\pi/2$, that is, in the case of azimuthal polarization. For other cases of CVBs, the z-component has cosine or sine dependence on the angle φ: for example, for a CVB with $p = -1$ and $\phi_0 = 0$ $E_z(\rho, \varphi, z) \propto \cos(2\varphi)$, and for a CVB with $p = -1$ and $\phi_0 = \pi/2$ $E_z(\rho, \varphi, z) \propto \sin(2\varphi)$ (see Figs. 2.25 and 2.26). Such negative order CVBs can be obtained by passing the positive order CVB through a half waveplate [95-98]. In the case of negative order CVBs light field patterns generated after passing through a rotating linear polarizer rotates in the direction opposite to the direction of rotation in the case of positive order CVBs [95]. In contrast to the positive order CVBs, the energy contribution of the formed z-component of the tightly focused negative order CVBs is always less than the energy contribution of the x- and y-components (see Fig. 2.26).

2.6.2. Theoretical Analysis

The interrelation of polarization with the phase of light field allows the use of diffractive optical elements for the transformation of first-order CVBs into higher-order $(p > 1)$ CVBs. The easiest way to increase the polarization order is an increase in the multiplicity of the angle in the polarization coefficients of the incident radiation

$$\mathbf{c}(\phi) = \begin{pmatrix} \cos(p\phi + \phi_0) \\ \sin(p\phi + \phi_0) \end{pmatrix}$$

A partial solution of this problem is possible by multiplying the initial cylindrically polarized electromagnetic field by the cosine or sine function of a multiple angle. For example, a combination of the radially polarized beam with the cosine function of the angle φ has the following form:

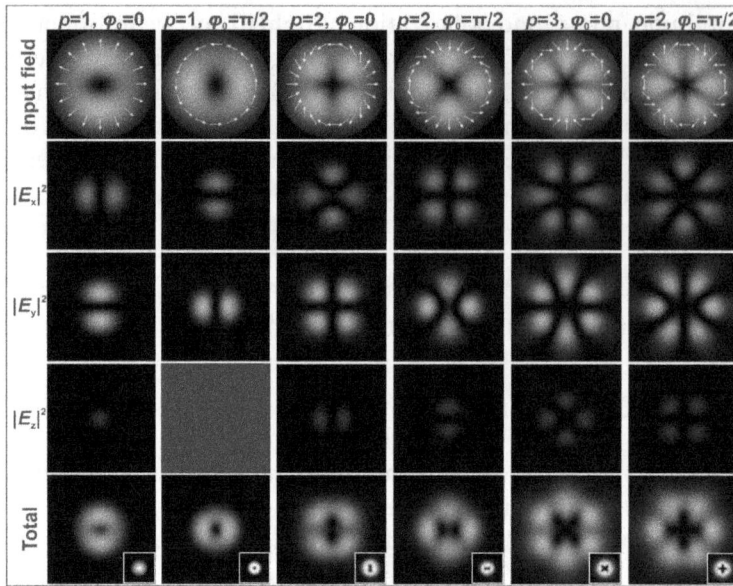

Fig. 2.25. The distribution of the various components of the electric field in the focal plane of a focusing system (NA = 0.99) for an incident CVB with different positive polarization order *p*. The insets in the total field distribution images show the generated grayscale intensity distributions. The white arrows show the schematic polarization distributions.

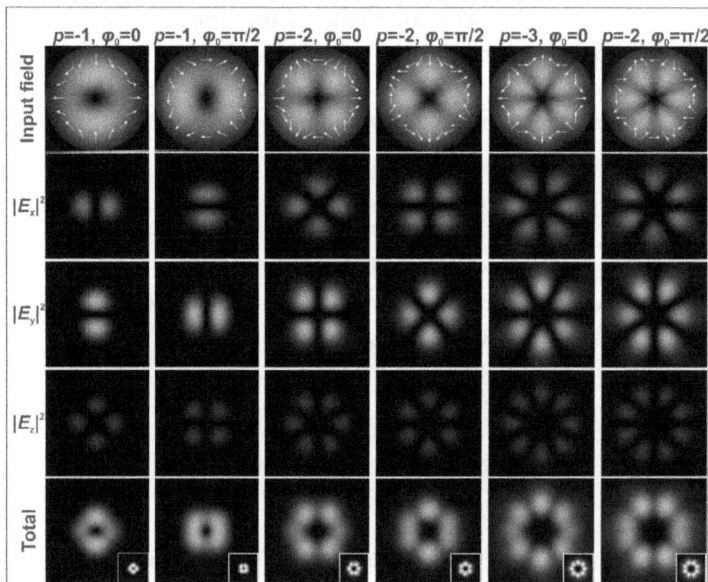

Fig. 2.26. The distribution of the various components of the electric field in the focal plane of a focusing system (NA = 0.99) for an incident CVB with different negative polarization order *p*. The insets in the total field distribution images show the generated grayscale intensity distributions. The white arrows show the schematic polarization distributions.

$$\mathbf{e}_{Rad,1} \cos\varphi = \begin{pmatrix} \cos\varphi \\ \sin\varphi \end{pmatrix} \cos\varphi = \frac{1}{2}\begin{pmatrix} 1+\cos 2\varphi \\ \sin 2\varphi \end{pmatrix} =$$

$$= \frac{1}{2}\begin{pmatrix} 1 \\ 0 \end{pmatrix} + \frac{1}{2}\begin{pmatrix} \cos 2\varphi \\ \sin 2\varphi \end{pmatrix} = \frac{1}{2}\mathbf{e}_{Lin\_x} + \frac{1}{2}\mathbf{e}_{p=2,\phi_0=0}$$

(2.43)

It is clear that the result of Eq. (2.43) is in fact a superposition of two different types of polarization, linear polarization in the x-direction and the second-order cylindrical polarization. For a combination of the radial polarization with the sine function of the angle φ, the following result is obtained:

$$\mathbf{e}_{Rad,1} \sin\varphi = \begin{pmatrix} \cos\varphi \\ \sin\varphi \end{pmatrix} \sin\varphi = \frac{1}{2}\begin{pmatrix} \sin 2\varphi \\ 1-\cos 2\varphi \end{pmatrix} =$$

$$= \frac{1}{2}\begin{pmatrix} 0 \\ 1 \end{pmatrix} - \frac{1}{2}\begin{pmatrix} -\sin 2\varphi \\ \cos 2\varphi \end{pmatrix} = \frac{1}{2}\mathbf{e}_{Lin\_y} - \frac{1}{2}\mathbf{e}_{p=2,\phi_0=\pi/2},$$

(2.44)

that is, a superposition of two polarizations orthogonal to those obtained in Eq. (2.43), namely, the linear polarization in the y-direction and the second-order cylindrical polarization with the inner polarization rotation of $\pi/2$.

In general, an increase in the multiplicity of the angle for the used cosine and sine functions leads to the following transformation:

$$\mathbf{e}_{Rad} \cos m\varphi = \frac{1}{2}\mathbf{e}_{p=-(m-1),\phi_0=0} + \frac{1}{2}\mathbf{e}_{p=m+1,\phi_0=0},$$

$$\mathbf{e}_{Rad} \sin m\varphi = \frac{1}{2}\mathbf{e}_{p=-(m-1),\phi_0=\pi/2} - \frac{1}{2}\mathbf{e}_{p=m+1,\phi_0=\pi/2}$$

(2.45)

Thus, superpositions of higher-order cylindrical polarization are formed.

For a case of using the azimuthal polarization as an initial polarization, the following expressions are obtained:

$$\mathbf{e}_{Az} \cos\varphi = \begin{pmatrix} -\sin\varphi \\ \cos\varphi \end{pmatrix} \cos\varphi = \frac{1}{2}\mathbf{e}_{Lin\_y} + \frac{1}{2}\mathbf{e}_{p=2,\phi_0=\pi/2},$$

$$\mathbf{e}_{Az} \sin\varphi = \begin{pmatrix} -\sin\varphi \\ \cos\varphi \end{pmatrix} \sin\varphi = -\frac{1}{2}\mathbf{e}_{Lin\_x} + \frac{1}{2}\mathbf{e}_{p=2,\phi_0=0},$$

$$\mathbf{e}_{Az} \cos(m\varphi) = \frac{1}{2}\mathbf{e}_{p=-(m-1),\phi_0=\pi/2} + \frac{1}{2}\mathbf{e}_{p=m+1,\phi_0=\pi/2},$$

$$\mathbf{e}_{Az} \sin(m\varphi) = -\frac{1}{2}\mathbf{e}_{p=-(m-1),\phi_0=0} + \frac{1}{2}\mathbf{e}_{p=m+1,\phi_0=0}$$

(2.46)

The well-known binary multi-sector phase plates [99, 100] can be used for generation of functions approximating the above considered trigonometric functions [101, 102]. The expansion in a Fourier series of the transmission function of a phase plate $f(\varphi) = \exp[i\psi(\varphi)]$, where $\psi(\varphi)$ is a phase of the phase plate, has the following form:

$$f(\varphi) = \frac{a_0}{2} + \sum_{n=1}^{\infty} \left[a_n \cos(n\varphi) + b_n \sin(n\varphi) \right],\qquad(2.47)$$

where

$$a_0 = \frac{1}{\pi} \int_0^{2\pi} f(\varphi) \, d\varphi,$$

$$a_n = \frac{1}{\pi} \int_0^{2\pi} f(\varphi) \cos(n\varphi) \, d\varphi,\qquad(2.48)$$

$$b_n = \frac{1}{\pi} \int_0^{2\pi} f(\varphi) \sin(n\varphi) \, d\varphi$$

Two-sector binary phase plate with phase values of φ_1 and φ_2 for different sectors.

In this case, Eq. (2.48) is transformed to the following:

$$a_0 = \exp(i\varphi_1) + \exp(i\varphi_2)$$
$$a_n = 0 \qquad(2.49)$$
$$b_n = \begin{cases} 2(\exp(i\varphi_1) - \exp(i\varphi_2)) / (\pi n), & \text{for odd } n, \\ 0, & \text{for even } n \end{cases}$$

Then, the transmission function of the phase plate has the following form:

$$f(\varphi) = \frac{\exp(i\varphi_1) + \exp(i\varphi_2)}{2} + \frac{2}{\pi} \left(\exp(i\varphi_1) - \exp(i\varphi_2) \right) \sum_{n=0}^{\infty} \frac{\sin\left[(2n+1)\varphi\right]}{2n+1}\qquad(2.50)$$

From Eq. (2.50), it is evident that such phase plate simultaneously generates a set of trigonometric functions with decreasing weight factors. In this case, if $\varphi_1 - \varphi_2 = \pi$, then the free term a_0 is absent. In particular, when $\varphi_1 = 0$ and $\varphi_2 = \pi$, Eq. (2.50) is transformed to the following:

$$f(\varphi) = \frac{4}{\pi} \cdot \sum_{n=0}^{\infty} \frac{\sin\left[(2n+1)\varphi\right]}{2n+1} = \frac{4}{\pi} \left(\sin(\varphi) + \frac{\sin(3\varphi)}{3} + \frac{\sin(5\varphi)}{5} + \ldots \right),\qquad(2.51)$$

that is, the phase plate corresponds to a sum of the sine functions of the odd orders.

Taking into account the decreasing weight factors, as well as the concentration of the energy in the focal region near the optical axis, we can assume that such a phase plate substantially corresponds to the $\sin(\varphi)$ function and can be used for transformations described by Eqs. (2.43)–(2.46). It is evident, when the phase plate rotates by 90 degrees, we get an analogous sum of cosine functions with $\cos(\varphi)$ as the main term.

N-sector binary phase plate with phase values of φ_1 and φ_2 for different sectors.

The general formula for an N-sector binary phase plate has the following form:

$$f(\varphi) = \frac{\exp(i\varphi_1) + \exp(i\varphi_2)}{2} +$$
$$+ \frac{2N}{\pi}(\exp(i\varphi_1) - \exp(i\varphi_2)) \sum_{n=0}^{\infty} \frac{\sin((2n+1)N\varphi)}{(2n+1)N} \tag{2.52}$$

As follows from Eq. (2.52), the first term of the series corresponds to the $\sin(N\varphi)$ function, while the other harmonics have orders changing by $2N$ and proportionally decreasing weight factors. The presence of the free term (for $\varphi_1 - \varphi_2 \neq \pi$) leads to the presence of an additional term corresponding to the initial polarization state in the polarization superpositions described by Eqs. (2.43)–(2.46). In order to estimate the energy contribution of the free term and the first (main) term in the series, we assume that $\varphi_1 = 0$, then the energy contribution of the free term equals b, and the energy contribution of the main term is $|b_1|^2 = 16/\pi^2 \sin^2(\varphi_2/2)$. Thus, it is possible to change the ratio between the initial polarization and the formed one by varying the value φ_2. In particular, the equal energy contributions of these two components is when $\varphi_2 = 2\text{arctg}(\pi/2\sqrt{2}) \approx 96°$. It is evident that the average value of $\sin^2 N\varphi$ equal to 0.5 does not change when N is replaced with a multiple of it, so the total fraction of the total sum in Eq. (2.52) is $16/\pi^2 \sin^2(\varphi_2/2) \cdot (1/2) \cdot (1 + 1/3^2 + 1/5^2 + ...) = \sin^2(\varphi_2/2)$, taking into account that the sum of the series in brackets is $\pi^2/8$. Thus, the energy contributions of the free term and the total sum of the series are the same when $\varphi_2 = 90° = \pi/2$.

2.6.3. Simulation Results

Fig. 2.27 shows the modelling results for tight focusing (NA = 0.99) of a laser beam with the radial polarization passing through binary multi-sector phase plates with $\varphi_1 = 0$, and $\varphi_2 = \pi$ or $\pi/2$. It is clear that the modelling results are in good agreement with the presented theoretical analysis. In the case of a two-sector phase plate with $\varphi_2 = \pi$ (the first column of Fig. 2.27), whose action analogous to the action of the $\sin(\varphi)$ function, an initial radially-polarized CVB is transformed into a sum of linearly polarized light in the y direction $\mathbf{e}_{Lin\_y}$ and light with the second order cylindrical polarization $\mathbf{e}_{p=2,\phi_0=\pi/2}$. When comparing the individual field components, it is evident that the x- and z-components correspond to the second-order cylindrical polarization $\mathbf{e}_{p=2,\phi_0=\pi/2}$ (see Fig. 2.25), and the y-component substantially consists of linear y-polarization, because it is much larger than the corresponding component $\mathbf{e}_{p=2,\phi_0=\pi/2}$. When this phase plate is rotated 90 degrees (the second column of Fig. 2.27), then its action is similar to the action of the $\cos(\varphi)$ function, that is, a superposition of different polarizations (a linear x-polarization $\mathbf{e}_{Lin\_x}$ and the second-order cylindrical polarization $\mathbf{e}_{p=2,\phi_0=0}$) described by Eq. (2.43). As follows from Eq. (2.52) and the modelling results, the presence of the two-sector phase plate with $\varphi_2 = \pi/2$ (the third column of Fig. 2.27) leads to a situation when half of the initial light energy does not change its initial polarization state (the radial polarization $\mathbf{e}_{Rad}$) and half of the initial light energy is transformed to a superposition of a linear y-polarization and the second-order cylindrical polarization $\mathbf{e}_{p=2,\phi_0=0}$.

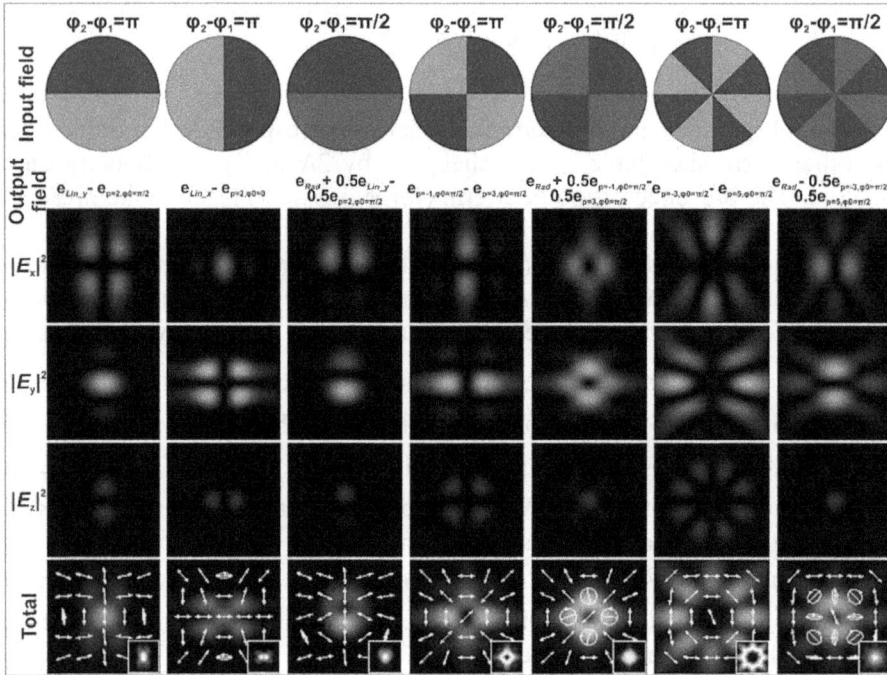

Fig. 2.27. Transformation of the initial radially-polarized laser beam passed through different binary multi-sector phase plates. In fact, the generated output fields are superpositions of different CVBs. The distribution of the various components of the electric field in the focal plane of a focusing system (NA = 0.99) are shown. The insets in the total field distribution images show the generated grayscale intensity distributions. The white arrows show the schematic polarization distributions.

A 4-sector phase plate with $\varphi_2 = \pi$ (the 4th column of Fig. 2.27), whose action is analogous to the $\sin(2\varphi)$ function, allows the superposition of cylindrical polarization of the -1st and 3rd order: $0.5\mathbf{e}_{p=-1,\phi_0=\pi/2} - 0.5\mathbf{e}_{p=3,\phi_0=\pi/2}$. In this case, the longitudinal component of the generated superposition coincides with the CVB with $p = -1$ and $\phi_0 = \pi/2$. The presence of a 4-sector phase plate with $\varphi_2 = \pi/2$ (the 5th column of Fig. 2.27) leads to the superposition of the three CVBs $0.5\mathbf{e}_{p=1,\phi_0=0} + 0.25\mathbf{e}_{p=-1,\phi_0=\pi/2} + 0.25\mathbf{e}_{p=3,\phi_0=\pi/2}$. In this case, the central part of the generated light pattern has circular polarization and the peripheral part of the pattern has the polarization analogous to the polarization formed by the 4-sector phase plate with $\varphi_2 = \pi$. Finally, the presence of a 8-sector phase plate with $\varphi_2 = \pi$ or $\pi/2$ (the 6th and 7th columns of Fig. 2.27) leads to the superpositions of polarization in the form of $0.25\mathbf{e}_{p=-3,\phi_0=\pi/2} - 0.25\mathbf{e}_{p=5,\phi_0=\pi/2}$ or $0.5\mathbf{e}_{Rad} + 0.25\mathbf{e}_{p=-3,\phi_0=\pi/2} - 0.25\mathbf{e}_{p=5,\phi_0=\pi/2}$, respectively. In the latter case, the peripheral part of the generated light pattern is azimuthally polarized and the central part of the pattern has a hybrid polarization state, partially radial and partially circular. When $\varphi_2 = \pi$, the azimuthal polarization is substantial.

Analogous modelling results obtained when using a light field with azimuthal polarization as the initial field are shown in Fig. 2.28. The results are similar to those obtained in the case of the radial polarization with the exception of a 90 degrees rotation of vectors and the absence of the longitudinal z-component. In these cases, it is necessary to use Eq. (2.46), from which it follows that multi-sector phase plates correspond to the sine functions of multiple angles and lead to a superposition in the form of $c_1\mathbf{e}_{Az} + c_2\left(\mathbf{e}_{p=m+1,\phi_0=0} - \mathbf{e}_{p=-(m-1),\phi_0=0}\right)$, where coefficients c_1 and c_2 depend on the phase difference $\varphi_2 - \varphi_1$ ($c_1 = 0$ for $\varphi_2 - \varphi_1 = \pi$). For $\mathbf{e}_{p=m+1,\phi_0=0}$ and $\mathbf{e}_{p=-(m-1),\phi_0=0}$, the longitudinal components of the generated light field in the focal plane are equal, therefore this component is absent in the superposition. At small numerical apertures (for example, for NA = 0.3 used for further experimental verification), the contribution of the longitudinal component is also insignificant in other cases, and the difference between the results obtained for an initial radial or azimuthal polarization is only in the rotation of the light pattern in the focal plane.

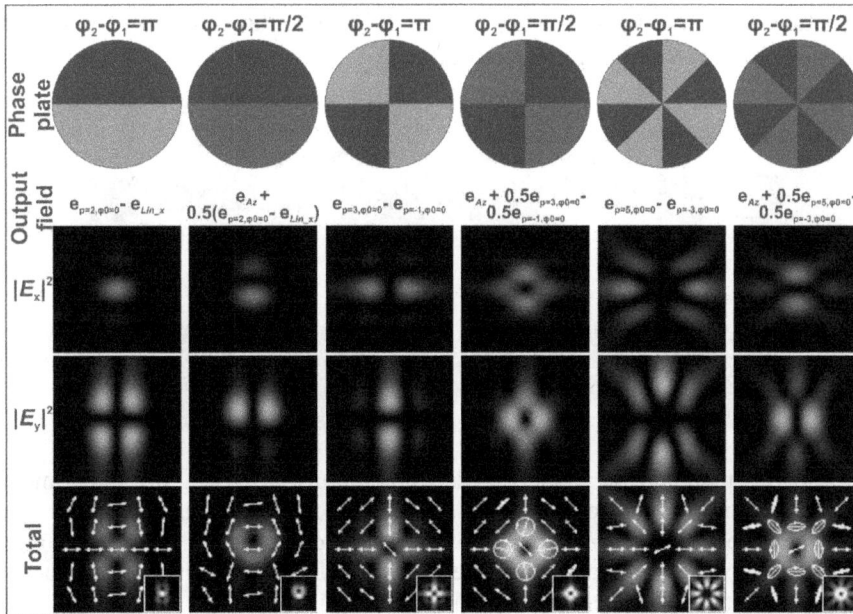

Fig. 2.28. Transformation of the initial azimuthally polarized laser beam passed through different binary multi-sector phase plates. In fact, the generated output fields are superpositions of different

CVBs. The distribution of the various components of the electric field in the focal plane of a focusing system (NA = 0.99) are shown. The insets in the total field distribution images show the generated grayscale intensity distributions. The white arrows show the schematic polarization distributions.

2.6.4. Experimental Results

The optical setup for the experimental investigation of the higher-order hybrid CVBs generation and polarization transformation is shown in Fig. 2.29a. The input laser beam was extended and spatially filtered by a system composed of a microobjective MO1 (10×, NA = 0.2), a pinhole PH (aperture size of 40 μm), and a lens Lens1 (focal length f_1 = 150 mm). The collimated linearly polarized laser beam with a Gaussian profile of intensity distribution (waist diameter is approximately 3 mm) was transformed into a "donut"-shaped radially/azimuthally polarized laser beam using a commercially available S-waveplate (Altechna, clear aperture diameter of 4 mm). Then, a wavefront of the formed laser beam was modulated using the fabricated 2, 4, or 8-sector phase plate. The 2, 4, and 8-sector phase plates with a diameter of 4 mm were manufactured on surfaces of 2-mm thick fused silica plates. Two variants of each of the sector plates, with a height difference between neighbouring sectors of approximately 290 ± 20 and 580 ± 20 nm corresponding to π/2 and π-phase shift at 532 nm, were manufactured (see an example of the manufactured 8-sector phase plate with relief steps with the height h = 580 ± 20 nm and side-wall inclination angle of 5 ± 1 degrees in Fig. 2.29b). A combination of two lenses, Lens2 (f_2 = 250 mm) and Lens3 (f_3 = 150 mm), and a diaphragm was used for spatial filtering of the modulated laser beam. Finally, the generated higher-order hybrid CVB was focused by microobjective MO1 (40×, effective numerical aperture NA_{eff} = 0.3) and imaged by microobjective MO2 (100×, NA = 0.8) onto the sensor of the CMOS-video camera (ToupCam, 3328×2548 pixel resolution).

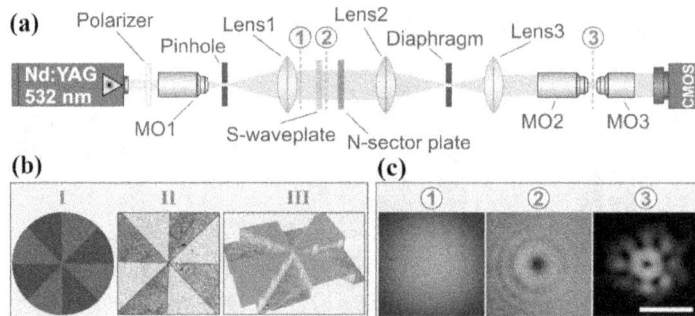

Fig. 2.29. Experimental investigation of polarization transformation. (a) Experimental setup utilized for generation of higher-order CVBs; (b) Example of the utilized 8-sector phase plate: (I) designed phase profile with φ_2 - φ_1 = π/2, (II) optical microscopy and (III) optical profilometry image of the central part of the manufactured element; (c) Beam intensity distributions measured in different planes of the experimental setup. The scale bar is 500 μm.

Fig. 2.30 shows intensity distributions for different components of the generated hybrid CVBs. A polarizer was utilized for analysing the separated transverse electromagnetic field components. Despite the fact the obtained experimental pictures completely coincide with the transverse component distributions obtained in the numerical modelling, the total intensity distributions were different, which may be explained by the difference in the NA values used in the modelling and experiment. It is obvious that the longitudinal component of the electromagnetic field has a noticeable contribution to the formed intensity distribution only under sharp focusing conditions (NA > 0.7). Thus, the obtained experimental results have confirmed the possibility of generating and transforming hybrid CVBs with the help of binary phase elements, even in the case of weak focusing.

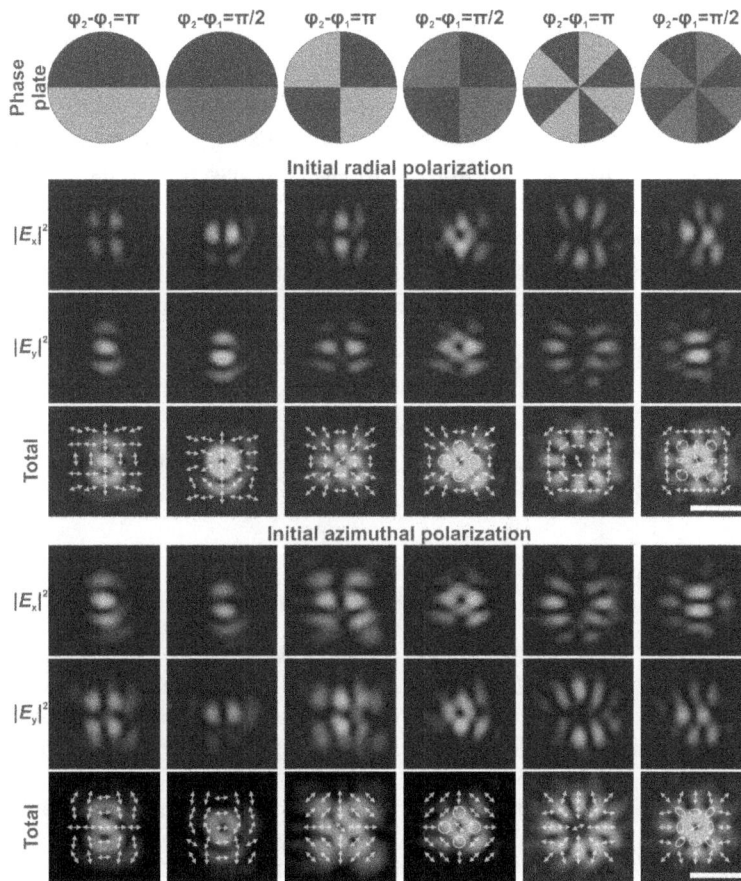

Fig. 2.30. Experimentally obtained transverse component distributions of the focused radially/azimuthally polarized laser beam passed through different binary multi-sector phase plates. The intensity distributions were formed in the focal plane of a microobjective with $NA_{eff} = 0.3$. The scale bar is 500 µm.

In conclusion, we conducted a theoretical analysis of the generation of hybrid cylindrical polarization states of the light field using optical elements having a transmission function defined by the cosine or sine function of the azimuthal angle in the polar coordinate system. Easy-to-manufacture binary phase elements in the form of the multi-sector phase plate allowed the approximation of these functions and to experimentally realize the polarization conversion from radially/azimuthally polarized beams into higher-order superpositions. Such transformation does not depend on the numerical aperture of the focusing optical system and can also be performed under conditions of weak focusing (NA < 0.7). In the latter case, the difference between the results obtained for the transformation of radial or azimuthal polarization is only in the rotation of the light pattern in the focal plane. The experimental results obtained with the help of multi-sector phase plates manufactured on surfaces of fused silica substrates are in good agreement with the numerical modelling results, providing evidence of the proposed technique for the generation of hybrid cylindrical vector beams.

The utilized multi-sector phase plates have a high-damage threshold which allows the use of these plates with a high power laser. However, in our opinion, the main advantage of the described polarization transformation approach in comparison with well-known techniques for the generation of higher-order CVBs using single or double spatial light modulators supporting implementation of multi-level phase functions, is the simplicity of the transmission function of the binary multi-sector phase elements. Due to this, such elements can be realized not only as 'static' phase plates, but also with the help of low cost binary spatial light modulators with low resolution for a higher frame rate for fast switching between different polarization states.

Acknowledgements

This chapter contains research results financially supported by the Russian Scientific Foundation (RSF) grant No. 17-12-01258.

References

[1]   V. G. Shvedov, C. Hnatovsky, N. Eckerskorn, A.V. Rode, W. Krolikowski, Polarization-sensitive photophoresis, *Appl. Phys Lett.*, Vol. 1, 2012, 051106.

[2]   A. Turpin, V. Shvedov, C. Hnatovsky, Y. V. Loiko, J. Mompart, W. Krolikowski, Optical vault: a reconfigurable bottle beam based on conical refraction of light, *Opt. Express*, Vol. 21, 2013, pp. 26335-26340.

[3]   M. Dienerowitz, M. Mazilu, K. Dholakia, Optical manipulation of nanoparticles: A review, *J. Nanophotonics*, Vol. 2, 2008, 021875.

[4]   M. Kraus, M.A. Ahmed, A. Michalowski, A. Voss, R. Weber, T. Graf, Microdrilling in steel using ultrashort pulsed laser beams with radial and azimuthal polarization, *Opt. Express*, Vol. 18, 2010, pp. 22305-22313.

[5]   C. Hnatovsky, V. G. Shvedov, W. Krolikowski, The role of light-induced nanostructures in femtosecond laser micromachining with vector and scalar pulses, *Opt. Express*, Vol. 21, 2013, pp. 12651-12656.

[6]   K. Lou, S.-X. Qian, Z.-C. Ren, C. Tu, Y. Li, H.-T. Wang, Femtosecond laser processing by using patterned vector optical fields, *Sci. Rep.*, Vol. 3, 2013, 2281.

[7]   S. Hasegawa, Y. Hayasaki, Holographic femtosecond laser manipulation for advanced material processing, *Advanced Optical Technologies*, Vol. 5, 2015, pp. 1-16.

[8]   J. J. J. Nivas, F. Cardano, Z. Song, A. Rubano, R. Fittipaldi, A. Vecchione, D. Paparo, L. Marrucci, R. Bruzzese, S. Amoruso, Surface structuring with polarization-singular femtosecond laser beams generated by a q-plate, *Sci. Rep.*, Vol. 7, 2017, 42142.

[9]   Q. Zhan, Cylindrical vector beams: From mathematical concepts to applications, *Adv. Opt. Photon.*, Vol. 1, 2009, pp. 1-57.

[10]   M.V. Berry, M.R. Dennis, Polarization singularities in isotropic random vector waves, *Proc. R. Soc. London, Ser. A*, Vol. 457, 2001, pp. 141-155.

[11]   M.R. Dennis, K. O'Holleran, M. J. Padgett, Singular optics: Optical vortices and polarization singularities, *Prog. Opt.*, Vol. 53, 2009, pp. 293-363.

[12]   V. Shvedov, A. R. Davoyan, C. Hnatovsky, N. Engheta, W. Krolikowski, A long-range polarization-controlled optical tractor beam, *Nat. Photonics*, Vol. 8, 2014, pp. 846-850.

[13]   G. Milione, M. P. J. Lavery, H. Huang, Y. Ren, G. Xie, T.A. Nguyen, E. Karimi, L. Marrucci, D. A. Nolan, R. R. Alfano, A. E. Willner, 4×20 Gbit/s mode division multiplexing over free space using vector modes and a q-plate mode (de)multiplexer, *Opt. Lett.*, Vol. 40, 2015, pp. 1980-1983.

[14]   J. Wang, Advances in communications using optical vortices, *Photon. Res.*, Vol. 4, 2016, pp. B14-B28.

[15]   M. Soskin, S. V. Boriskina, Y. Chong, M.R. Dennis, A. Desyatnikov, Singular optics and topological photonics, *J. Opt.*, Vol. 19, 2017, 010401.

[16]   J. Pu, Z. Zhang, Tight focusing of spirally polarized vortex beams, *Opt. Laser Technol.*, Vol. 42, 2010, pp. 186-191.

[17]   S. N. Khonina, N. L. Kazanskiy, S. G. Volotovsky, Vortex phase transmission function as a factor to reduce the focal spot of high-aperture focusing system, *J. Mod. Opt.*, Vol. 58, 2011, pp. 748-760.

[18]   M. Rashid, O. M. Marago, P. H. Jones, Focusing of high order cylindrical vector beams, *J. Opt. A: Pure Appl. Opt.*, Vol. 11, 2009, 065204.

[19]   L. Rao, J. Pu, Z. Chen, P. Yei, Focus shaping of cylindrically polarized vortex beams by a high numerical-aperture lens, *Opt. Laser Technol.*, Vol. 41, 2009, pp. 241-246.

[20]   S. N. Khonina, A. V. Ustinov, S. G. Volotovsky, Shaping of spherical light intensity based on the interference of tightly focused beams with different polarizations, *Opt. Laser Technol.*, Vol. 60, 2014, pp. 99-106.

[21]   M. Martínez-Corral, G. Saavedra, The resolution challenge in 3D optical microscopy, *Prog. Opt.*, Vol. 53, 2009, pp. 1-67.

[22]   E. Walker, A. Dvornikov, K. Coblentz, S. Esener, P. Rentzepis, Toward terabyte two-photon 3D disk, *Opt. Express*, Vol. 15, 2007, pp. 12264-12276.

[23]   S.C. Tidwell, D.H. Ford, W.D. Kimura, Generating radially polarized beams interferometrically, *Appl. Opt.*, Vol. 29, 1990, pp. 2234-2239.

[24]   K. Yonezawa, Y. Kozawa, S. Sato, Compact laser with radial polarization using birefringent laser medium, *Japan. J. Appl. Phys.*, Vol. 46, 2007, pp. 5160-5163.

[25]   T. Fadeyeva, V. Shvedov, N. Shostka, C. Alexeyev, A. Volyar, Natural shaping of the cylindrically polarized beams, *Opt. Lett.*, Vol. 35, 2010, pp. 3787-3789.

[26]   S. N. Khonina, S. V. Karpeev, S. V. Alferov, V. A. Soifer, Generation of cylindrical vector beams of high orders using uniaxial crystals, *J. Opt.*, Vol. 17, 2015, 065001.

[27]   Z. Bomzon, G. Biener, V. Kleiner, E. Hasman, Radially and azimuthally polarized beams generated by space variant dielectric subwavelength gratings, *Opt. Lett.*, Vol. 27, 2002, pp. 285-287.

[28]   L. Marrucci, C. Manzo, D. Paparo, Pancharatnam-Berry phase optical elements for wave front shaping in the visible domain: switchable helical mode generation, *Appl. Phys. Lett.*, Vol. 88, 2006, 221102.

[29] M. Beresna, M. Gecevičius, P. G. Kazansky, T. Gertus, Radially polarized optical vortex converter created by femtosecond laser nanostructuring of glass, *Appl. Phys. Lett.*, Vol. 98, 2011, 201101.

[30] Y. Zhang, X. Guo, L. Han, P. Li, S. Liu, H. Cheng, J. Zhao, Gouy phase induced polarization transition of focused vector vortex beams, *Opt. Express*, Vol. 25, 2017, pp. 25725-25733.

[31] X. Yi, X. Ling, Z. Zhang, Y. Li, X. Zhou, Y. Liu, S. Chen, H. Luo, S. Wen, Generation of cylindrical vector vortex beams by two cascaded metasurfaces, *Opt. Express*, Vol. 22, 2014, pp. 17207-17215.

[32] W. Shu, Y. Liu, Y. Ke, X. Ling, Z. Liu, B. Huang, H. Luo, X. Yin, Propagation model for vector beams generated by metasurfaces, *Opt. Express*, Vol. 24, 2016, pp. 21177-21189.

[33] D. Naidoo, F.S. Roux, A. Dudley, I. Litvin, B. Piccirillo, L. Marrucci, A. Forbes, Controlled generation of higher-order Poincaré sphere beams from a laser, *Nat. Photonics*, Vol. 10, 2016, pp. 327-332.

[34] E. Karimi, B. Piccirillo, E. Nagali, L. Marrucci, E. Santamato, Efficient generation and sorting of orbital angular momentum eigenmodes of light by thermally tuned q-plates, *Appl. Phys. Lett.*, Vol. 94, 2009, 231124.

[35] B. Piccirillo, V. D'Ambrosio, S. Slussarenko, L. Marrucci, E. Santamato, Photon spin-to-orbital angular momentum conversion via an electrically tunable q-plate, *Appl. Phys. Lett.*, Vol. 97, 2010, 241104.

[36] S. Slussarenko, A. Murauski, T. Du, V. Chigrinov, L. Marrucci, E. Santamato, Tunable liquid crystal q-plates with arbitrary topological charge, *Opt. Express*, Vol. 19, 2011, pp. 4085-4090.

[37] C. Alpmann, C. Schlickriede, E. Otte, C. Denz, Dynamic modulation of Poincaré beams, *Sci. Rep.*, Vol. 7, 2017, 8076.

[38] W. Jue, W. Lin, X. Yu, Generation of full Poincare beams on arbitrary order Poincare sphere, *Curr. Opt. Photon.*, Vol. 1, 2017, pp. 631-636.

[39] J.E. Holland, I. Moreno, J.A. Davis, M.M. Sánchez-López, D.M. Cottrell, Q-plates with a nonlinear azimuthal distribution of the principal axis: application to encoding binary data, *Appl. Opt.*, Vol. 57, 2018, pp. 1005-1010.

[40] S. Fu, C. Gao, T. Wang, Y. Zhai, C. Yin, Anisotropic polarization modulation for the production of arbitrary Poincaré beams, *J. Opt. Soc. Am. B*, Vol. 35, 2018, pp. 1-7.

[41] T. H. Lu, T. D. Huang, J. G. Wang, L. W. Wang, R. R. Alfano, Generation of flower high-order Poincaré sphere laser beams from a spatial light modulator, *Sci. Rep.*, Vol. 6, 2016, 39657.

[42] R. Dorn, S. Quabis, G. Leuchs, Sharper focus for a radially polarized light beam, *Phys. Rev. Lett.*, Vol. 91, 2003, 233901.

[43] N. Davidson, N. Bokor, High-numerical-aperture focusing of radially polarized doughnut beams with a parabolic mirror and a flat diffractive lens, *Opt. Lett.*, Vol. 29, 2004, pp. 1318-1320.

[44] G. M. Lerman, U. Levy, Effect of radial polarization and apodization on spot size under tight focusing conditions, Opt. Express, Vol. 16, 2008, pp. 4567-4581.

[45] H. Wang, L. Shi, B. Lukyanchuk, C. Sheppard, C.T. Chong, Creation of a needle of longitudinally polarized light in vacuum using binary optics, *Nat. Photonics*, Vol. 2, 2008, pp. 501-505.

[46] S. N. Khonina, D. A. Savelyev, High-aperture binary axicons for the formation of the longitudinal electric field component on the optical axis for linear and circular polarizations of the illuminating beam, *J. Exp. Theor. Phys.*, Vol. 117, 2013, pp. 623-630.

[47] A. P. Porfirev, A. V. Ustinov, S. N. Khonina, Polarization conversion when focusing cylindrically polarized vortex beams, *Sci. Rep.*, Vol. 6, 2016, 6.

[48] A. H. S. Holbourn, Angular momentum of circularly polarized light, *Nature*, Vol. 137, 1936, p. 31.

[49] N. B. Simpson, K. Dholakia, L. Allen, M. J. Padgett, Mechanical equivalence of spin and orbital angular momentum of light: an optical spanner, *Opt. Lett.*, Vol. 22, 1997, pp. 52-54.

[50] A. M. Yao, M. J. Padgett, Orbital angular momentum: origins, behavior and applications, *Adv. Opt. Photon.*, Vol. 3, 2011, pp. 161-204.

[51] D. L. Andrews, M. Babiker, The Angular Momentum of Light, *Cambridge University Press*, Cambridge, 2012.

[52] J. Poynting, The wave motion of a revolving shaft, and a suggestion as to the angular momentum in a beam of circularly polarized light, *Proc. R. Soc. London, Ser. A*, Vol. 82, 1909, pp. 560-567.

[53] R. Beth, Mechanical detection and measurement of the angular momentum of light, *Phys. Rev.*, Vol. 50, 1936, pp. 115-125.

[54] L. Allen, S. M. Barnett, M. J. Padgett, Optical Angular Momentum, *Institute of Physics Publishing*, London, 2003.

[55] E. Hasman, G. Biener, A. Niv, V. Kleiner, Space-variant polarization manipulation, *Prog. Opt.*, Vol. 47, 2005, pp. 215-289.

[56] G. Molina-Terriza, J. P. Torres, L. Torner, Twisted Photons, *Nat. Phys.*, Vol. 3, 2007, pp. 305-310.

[57] S. Franke-Arnold, L. Allen, M. Padgett, Advances in optical angular momentum, *Laser Photonics Rev.*, Vol. 2, 2008, pp. 299-313.

[58] Y. Zhao, J.S. Edgar, G.D.M. Jeffries, D. McGloin, D.T. Chiu, Spin-to-orbital angular momentum conversion in a strongly focused optical beam, *Phys. Rev. Lett.*, Vol. 99, 2007, 073901.

[59] C.J.R. Sheppard, W. Gong, K. Si, Polarization effects in 4Pi microscopy, *Micron*, Vol. 42, 2011, pp. 353 - 359.

[60] A. Ciattoni, G. Cincotti, C. Palma, Circularly polarized beams and vortex generation in uniaxial media, *J. Opt. Soc. Am. A*, Vol. 20, 2003, pp. 163-171.

[61] L. Marrucci, C. Manzo, D. Paparo, Optical spin-to-orbital angular momentum conversion in inhomogeneous anisotropic media, *Phys. Rev. Lett.*, Vol. 96, 2006, 163905.

[62] T. A. Fadeyeva, V. G. Shvedov, Y. V. Izdebskaya, A. V. Volyar, E. Brasselet, D. N. Neshev, A. S. Desyatnikov, W. Krolikowski, Y. S. Kivshar, Spatially engineered polarization states and optical vortices in uniaxial crystals, *Opt. Express*, Vol. 18, 2010, pp. 10848-10863.

[63] I. Moreno, J. A. Davis, I. Ruiz, D. M. Cottrell, Decomposition of radially and azimuthally polarized beams using a circular-polarization and vortex-sensing diffraction grating, *Opt. Express*, Vol. 18, 2007, pp. 7173-7183.

[64] S. N. Khonina, D. A. Savelyev, N. L. Kazanskiy, Vortex phase elements as detectors of polarization state, *Opt. Express*, Vol. 23, 2015, pp. 17845-17859.

[65] B. Richards, E. Wolf, Electromagnetic diffraction in optical systems II. Structure of the image field in an aplanatic system, *Proc. R. Soc. A*, Vol. 253, 1959, pp. 358-379.

[66] S. N. Khonina, A. V. Ustinov, E. A. Pelevina, Analysis of wave aberration influence on reducing focal spot size in a high-aperture focusing system, *J. Opt.*, Vol. 13, 2011, 095702.

[67] S. N. Khonina, S. G. Volotovsky, Application axicons in a large-aperture focusing system, *Opt. Mem. Neural Network*, Vol. 23, 2014, pp. 201-217.

[68] P. Belanger, M. Rioux, Ring pattern of a lens-axicon doublet illuminated by a Gaussian beam, *Appl. Opt.*, Vol. 17, 1978, pp. 1080-1086.

[69] S. N. Khonina, A. P. Porfirev, A.V. Ustinov, Diffractive axicon with tunable fill factor for focal ring splitting, *Proceedings of SPIE*, Vol. 10233, 2017, 102331P.

[70] A.V. Ustinov, A. P. Porfirev, S. N. Khonina, Effect of the fill factor of an annular diffraction grating on the energy distribution in the focal plane, *J. Opt. Technol.*, Vol. 84, 2017, pp. 580-587.

[71] C. J. R. Sheppard, Jones and Stokes parameters for polarization in three dimensions, *Phys. Rev. A*, Vol. 90, 2014, 023809.

[72]   S. N. Khonina, V. V. Kotlyar, V. A. Soifer, M. V. Shinkaryev, G. V. Uspleniev, Trochoson, *Opt. Commun.*, Vol. 91, 1992, pp. 158-162.

[73]   S. N. Khonina, S. G. Volotovsky, Controlling the contribution of the electric field components to the focus of a high-aperture lens using binary phase structures, *J. Opt. Soc. Am. A*, Vol. 27, 2010, pp. 2188-2197.

[74]   H. McLeod, The axicon: A new type of optical element, *J. Opt. Soc. Am.*, Vol. 44, 1954, pp. 592-597.

[75]   M. Abramowitz, I. A. Stegun, Handbook of Mathematical Functions, *Dover*, 1972.

[76]   J. Stamnes, Waves, rays, and the method of stationary phase, *Opt. Express*, Vol. 10, 2002, pp. 740-751.

[77]   S. N. Khonina, V. V. Kotlyar, V. A. Soifer, J. Lautanen, M. Honkanen, J. Turunen, Generation of Gauss-Hermite modes using binary DOEs, *Proceedings of SPIE*, Vol. 4016, 1999, pp. 234-239.

[78]   V. G. Shvedov, C. Hnatovsky, N. Shostka, W. Krolikowski, Generation of vector bottle beams with a uniaxial crystal, *J. Opt. Soc. Am. B*, Vol. 30, 2013, pp. 1-6.

[79]   M.-C. Zhong, L. Gong, D. Li, J.-H. Zhou, Z.-Q. Wang, Y.-M. Li, Optical trapping of core-shell magnetic microparticles by cylindrical vector beams, *Appl. Phys. Lett.*, Vol. 105, 2014, 181112.

[80]   F. P. Wu, B. Zhang, Z. L. Liu, Y. Tang, N. Zhang, Optical trapping forces of a focused azimuthally polarized Bessel-Gaussian beam on a double-layered sphere, *Opt. Commun.*, Vol. 405, 2017, pp. 96-100.

[81]   T. A. Nieminen, N. R. Heckenberg, H. Rubinsztein-Dunlop, Forces in optical tweezers with radially and azimuthally polarized trapping beams, *Opt. Lett.*, Vol. 33, 2008, pp. 122-124.

[82]   B.-L. Yao, S.-H. Yan, T. Ye, W. Zhao, Optical trapping of double-ring radially polarized beam with improved axial trapping efficiency, *Chin. Phys. Lett.*, Vol. 27, 2010, 108701.

[83]   A. Ashkin, J. M. Dziedzic, J. E. Bjorkholm, S. Chu, Observation of a single-beam gradient force optical trap for dielectric particles, *Opt. Lett.*, Vol. 11, 1986, pp. 288-290.

[84]   A. P. Porfirev, R. V. Skidanov, Generation of an array of optical bottle beams using a superposition of Bessel beams, *Appl. Opt.*, Vol. 52, 2013, pp. 6230-6238.

[85]   Z. Kuang, W. Perrie, S. P. Edwardson, E. Fearon, G. Dearden, Ultrafast laser parallel microdrilling using multiple annular beams generated by a spatial light modulator, *J. Phys. D: Appl. Phys.*, Vol. 47, 2014, 115501.

[86]   X. Zhao, J. Zhang, X. Pang, G. Wan, Properties of a strongly focused Gaussian beam with an off-axis vortex, *Opt. Commun.*, Vol. 389, 2017, pp. 275-282.

[87]   N. Zhang, J. A. Davis, I. Moreno, J. Lin, K.-J. Moh, D. M. Cottrell, X. Yuan, Analysis of fractional vortex beams using a vortex grating spectrum analyzer, *Appl. Opt.*, Vol. 49, 2010, pp. 2456-2462.

[88]   S. N. Khonina, A. P. Porfirev, A. V. Ustinov, Diffraction patterns with m-th order symmetry generated by sectional spiral phase plates, *J. Opt.*, Vol. 17, 2015, 125607.

[89]   S. N. Khonina, V. V. Kotlyar, V. A. Soifer, K. Jefimovs, J. Turunen, Generation and selection of laser beams represented by a superposition of two angular harmonics, *J. Mod. Opt.*, Vol. 51, 2004, pp. 761-773.

[90]   V. V. Kotlyar, S. N. Khonina, V. A. Soifer, Light field decomposition in angular harmonics by means of diffractive optics, *J. Mod. Opt.*, Vol. 45, 1998, pp. 1495-1506.

[91]   A. P. Prudnikov, Y. A. Brychkov, O. I. Marichev, Integrals and Series: Special Functions, Vol.2, *Nauka Publisher*, Moscow, 1983.

[92]   V. A. Soifer, Methods for Computer Design of Diffractive Optical Elements, *Wiley-Interscience*, New Jersey, 2002.

[93]   N. Bozinovic, Y. Yue, Y. Ren, M. Tur, P. Kristensen, H. Huang, A. E. Willner, S. Ramachandran, Terabit-scale orbital angular momentum mode division multiplexing in fibers, *Science*, Vol. 340, 2013, pp. 1545-1548.

[94]   C. Hnatovsky, V. G. Shvedov, N. Shostka, A.V. Rode, W. Krolikowski, Polarization-dependent ablation of silicon using tightly focused femtosecond laser vortex pulses, *Opt. Lett.*, Vol. 37, 2012, pp. 226-228.

[95]   M. Stalder, M. Schadt, Linearly polarized light with axial symmetry generated by liquid-crystal polarization converters, *Opt. Lett.*, Vol. 21, 1996, pp. 1948-1950.

[96]   X. Zhang, L. Qiu, Generation of high-order polarized vortex beam by achromatic meniscus axicon doublet, *Opt. Eng.*, Vol. 52, 2013, 080503.

[97]   X. Weng, L. Du, Z. Yang, C. Min, X. Yuan, Generating arbitrary order cylindrical vector beams with inherent transform mechanism, *IEEE Photonics J.*, Vol. 9, 2017, pp. 1-8.

[98]   I. Moreno, J. A. Davis, K. Badham, M. M. Sánchez-López, J. E. Holland, D. M. Cottrell, Vector beam polarization state spectrum analyzer, *Sci. Rep.*, Vol. 7, 2017, 2216.

[99]   L. Zhou, Y. Jiang, P. Zhang, W. Fan, X. Li, Directly writing binary multi-sector phase plates on fused silica using femtosecond laser, *High Power Laser Science and Engineering*, Vol. 6, 2018, e6.

[100]   A. A. Kuchmizhak, A. P. Porfirev, S. A. Syubaev, P. A. Danilov, A. A. Ionin, O. B. Vitrik, Yu. N. Kulchin, S. N. Khonina, S. I. Kudryashov, Multi-beam pulsed-laser patterning of plasmonic films using broadband diffractive optical elements, *Opt. Lett.*, Vol. 42, 2017, pp. 2838-2841.

[101]   L. Novotny, E. J. Sanchez, X. S. Xie, Near-field imaging using metal tips illuminated by higher-order Hermite-Gaussian beams, *Ultramicroscopy*, Vol. 71, 1998, pp. 21-29.

[102]   S. N. Khonina, I. Golub, Optimization of focusing of linearly polarized light, *Opt. Lett.*, Vol. 36, 2011, pp. 352-354.

Chapter 3
Dispersion as a Noise Source in Direct and Coherent Detection Optical Channels

Er'el Granot

3.1. Introduction

It is well known that dispersion is one of the main obstacles in transmitting high data rates in optical fiber- based communications channels. Dispersion distorts the signal; it decreases the eye-opening and ultimately (if not compensated) it prevents data decoding.

In principle, the dispersion effects can be almost completely compensated by moving to a zero-dispersion working point (such as 1310 nm wavelength in most common optical fibers). However, in practice it is uncommon to work in these working points due to the relatively costly infrastructure (mostly amplifiers), and in addition, dispersion effect decreases nonlinear effects, and therefore can be useful (see, for example Section 4.2 in [1]). As a consequence the most ubiquitous approaches are either ignoring the dispersion effects (in short distances or low frequencies networks), using less sensitive methods (like DuoBinary [2-9], OFDM [10-12] etc.) or to compensate them partially or completely afterward. The compensations methods are quite diverse; however, they are all based on the fact that dispersion is a linear distortion, and therefore can be mitigated by an equivalent inverse linear operation. The problem is, of course, that the distortion occurs in the field domain, and ordinary detectors measure power rather than field. Therefore, mitigation should take place in the optical domain, where the inverse system operates directly on the field (like Dispersion Compensating fibers or modules [13-21] and alike), or in the electrical domain, but then the detection must be coherent to keep the linearity between the signal and the distortion [22-24]. Otherwise, the distortion can be regarded as an additional source of noise.

Usually, the dispersion effects are measured in power penalty, eye-opening shrinkage, or a corresponding increase in the Bit-Error Rate (BER). In this chapter the dispersion effect is analyzed as a source of noise, and it is shown that its Signal-to-Noise Ratio (SNR) obeys simple formulae. We calculate the SNR for both rectangular and sinc-shaped

Er'el Granot
Department of Electrical and Electronics Engineering, Ariel Univesity, Ariel, Israel

Nyquist pulses, both for direct and coherent detections. The result *analytical* expressions are surprisingly simpler.

3.2. General Theory

In the first approximation, dispersion is governed by the Schrödinger equation [25]

$$i\frac{\partial A(t,z)}{\partial z} = -\frac{\beta_2}{2}\frac{\partial^2 A(t,z)}{\partial t^2}, \tag{3.1}$$

where $A(t,z)$ is the electromagnetic (EM) pulse envelope, z is the propagation direction in the dispersive medium (the fiber), β_2 is the dispersion coefficient of the medium, and $t = t'-z/v$ is the time in a frame of reference that travels at the speed of light in the medium (v), i.e., t' is the real time.

The relation between the transmitted signal $A(t,0)$ at one end of the medium, and the detected one at its other end $A(t,z)$ can be formulated with a simple convolution

$$A(t,z) = \int_{-\infty}^{\infty} K(t-t',z)A(t',0)\,dt' \tag{3.2}$$

with the dispersion Kernel [25]

$$K(t-t',z) = (2\pi i\beta_2 z)^{-1/2}\exp\left(-i\frac{(t-t')^2}{2\beta_2 z}\right), \tag{3.3}$$

and therefore, it is a linear deterministic effect. It is deterministic in the sense that the dispersion distortions are completely determined by the initial conditions, which are the initial signal's profile, which enters the dispersive medium (i.e. the fiber).

However, since the Schrödinger equation has a nonlocal effect, then the distortions of a specific bit/symbol are affected by its adjacent neighbors. Consequently, when the initial signal carries pseudo random data (as in most networks), the distortions are pseudo random as well. In fact, when the incoming signal is completely random (for distances, which exceed the sampling interval) then the local distortions behave like an additional stochastic noise. Moreover, when the distortions are location sensitive (as in the rectangular pulse case, which will be investigated in what follows) the presence of a temporal jitter is an additional source of noise.

This Dispersion Generated Noise (DGN) shows a pattern that depends not only on the data statistics but more importantly on the data protocol, i.e., on the shape of the single bit/symbol-pulse.

Since in coherent-detection based networks dispersion can be mitigated relatively easily by using digitial FIR filters (see, for example, [22-24]), we will focus in this chapter on direct detection technologies, where the dispersion's distortions cannot be compensated completely after detection (for partial compensation see [25-26]). These networks are affordable, and are common in Passive Optical Networks (PONs). Such a system is illustrated in Fig. 3.1. The system consists of a simple transmitter, a diffusive medium (usually an optical fiber), and a simple receiver. The transmitter consists of a coherent source (a laser), which may be modulated directly or externally (usually by a Mach-Zehnder modulator). The finite bandwidth of the RF channel is expressed by the low-pass (LP) filter. The receiver part of the channel consists of a simple power detector and a DC blocker, i.e., a capacitor. For comparison we will introduce the coherent system analog at the end of the chapter.

As will be shown below, the DGN is extremely sensitive on the shape of the single bit pulse. Therefore, we will investigate the two extreme cases: rectangular and sinc pulses. Clearly, the one is the Fourier counterpart of the other: while the rectangular pulse is bounded in its allocation *time*, the sinc pulse is *spectrally* bounded. They are both orthogonal pulses, i.e., initially there is no Inter Symbol Interference (ISI); however, the dispersion increases the ISI (see, for example, [27] where this ISI is used to measure dispersion).

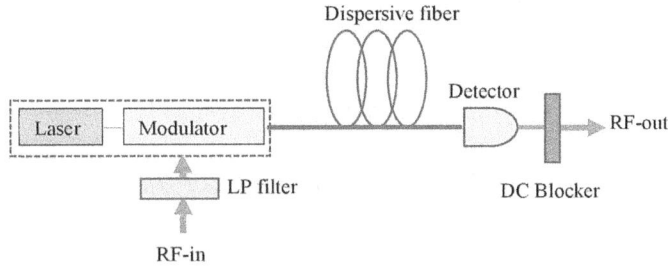

Fig. 3.1. System schematic. The laser power can be directly or externally modulated. After the dispersive medium the signal is directly detected, and its DC part is removed.

3.3. Rectangular Pulses

When the transmitted signal consists of a stream of rectangular pulses, then its field amplitude at the transmitting end of the medium (at $z = 0$) can be written as

$$A(t, z = 0) = A_0 + \sum_n a_n \operatorname{rect}_{\xi T}(t - nT), \tag{3.4}$$

where T is the pulses' period, ξ is the duty cycle of the pulses, and

$$\operatorname{rect}_{\xi T}(t) \equiv \begin{cases} 1 & |t| \leq \xi T / 2 \\ 0 & |t| > \xi T / 2 \end{cases} \tag{3.5}$$

is the rectangular pulse function (whose width is ξT). In particular $\xi = 1$ stands for Non-Return-to-Zero (NRZ) sequence, while $\xi < 1$ represents Return-to-Zero (RZ) one. A_0 is the dc part of the field, which is required for intensity modulated signals. The data is decoded in the coefficients a_n's. In particular, in the On-Off-Keying (OOK) method, $a_n = \pm A_0$, in which case the intensity levels, which represent the binary "1" and "0" are

$$I("1") = |A("1")|^2 = |A_0 + A_0|^2 = 4|A_0|^2 \text{ and } I("0") = |A("0")|^2 = |A_0 - A_0|^2 = 0 \text{ respectively.}$$

Clearly, in real networks the pulses are not perfect rectangles, and in fact, have a finite rise/fall time. This fact can be implemented by applying a Low Pass Filter (LPF) on the signal (3.4). For simplicity, we apply the following Gaussian LPF

$$H_{LPF}(f) = \exp\left[-2\ln 2 (f/\Delta)^2\right], \tag{3.6}$$

since this case has an exact analytical solution [28, 29], and yet it shows high agreement with laboratory experimental studies [27].

Furthermore, the dispersion effect of the dispersive medium (with a length L and a dispersion coefficient β_2) is manifested by the following spectral filter

$$H(f) = \exp\left[i\beta_2 L (2\pi f)^2 / 2\right] \tag{3.7}$$

The resultant signal, at the end of the fiber, is

$$A(t, z > 0) = A_0 + \sum_n a_n \text{ srect}_\xi \left(t/T - n, \Delta/T, \beta_2 z/T^2\right), \tag{3.8}$$

where (see [27-29])

$$\text{srect}_\xi(t, \delta, \zeta) \equiv \left\{\text{erfc}\left[(t - \xi/2)/\sqrt{i2\zeta + 2\ln(2)/\pi^2\delta^2}\right] - \right.$$
$$\left. -\text{erfc}\left[(t + \xi/2)/\sqrt{i2\zeta + 2\ln(2)/\pi^2\delta^2}\right]\right\} / 2, \tag{3.9}$$

and *erfc* is the complementary error function [30].

Therefore, since the current is proportional to the detected intensity, then the dc current is

$$i_0(t, z) = r_D \left\langle |A(t, z)|^2 \right\rangle, \tag{3.10}$$

and the ac current (after the dc reduction) is

$$\Delta i(t, z) = r_D \left[|A(t, z)|^2 - \left\langle |A(t, z)|^2 \right\rangle\right], \tag{3.11}$$

where r_D is the normalized detector's responsivity, which takes account of the propagating mode's cross-section, the coupling to the detector and its quantum efficiency. The triangular brackets stand for temporal averaging.

Then, the noise variance, or the power noise is usually defined as [31, 32]

$$N \equiv R\left\langle \left[i(t,z) - i(t,0) \right]^2 \right\rangle = Rr_D^2 \left[\left| A(t,z) \right|^2 - \left\langle \left| A(t,0) \right|^2 \right\rangle \right]^2, \tag{3.12}$$

where R is the detector load resistance.

To simplify the notations, we will adopt dimensionless units: $\tau \equiv t/T$, $\delta \equiv \Delta/T$ and $\zeta \equiv \beta_2 z/T^2$, in which case the field amplitude reads

$$A(\tau, \zeta > 0) = A_0 + \sum_n a_n \, \text{srect}_\xi (\tau - n, \delta, \zeta) \tag{3.13}$$

Since the measurements are taken approximately at the centers of the bits, i.e., at $t = nT$, then unless $2\beta_2 z > 2\ln(2)/\pi^2\Delta^2$ the noise is negligible. Therefore, in what follows we assume that $\Delta^2 \gg 1/\beta_2 z$. In this case (following Eq. (3.9))

$$\text{srect}_{\xi T} (t, \Delta, z) \cong \frac{1}{2}\left\{ \text{erf}\left[(\tau + \xi/2)/\sqrt{2i\zeta} \right] - \text{erf}\left[(\tau - \xi/2)/\sqrt{2i\zeta} \right] \right\} \tag{3.14}$$

Then, following Appendix A, around the measurements points, i.e., at $t = nT$ (or, in the normalized units, $\tau = n$), Eq. (3.13) can be rewritten in a simple form

$$A(\tau \cong m, \zeta > 0) = A_0 + \sum_n a_n h(m - n), \tag{3.15}$$

where

$$h(n) \equiv \delta(n) + \delta h(n), \tag{3.16}$$

and

$$\delta h(n) \equiv -\sqrt{\frac{i\zeta}{2\pi}} \exp\left[i\left(\frac{n^2 + \xi^2/4}{2\zeta} \right) \right] \left\{ \frac{\exp(in\xi/2\zeta)}{(n + \xi/2)} - \frac{\exp(-in\xi/2\zeta)}{(n - \xi/2)} \right\}, \tag{3.17}$$

$h(n)$ is the digital impulse response of a single "1" bit.

Similarly, in terms of the original signal $A(\tau \cong m, 0)$, the measured signal at the end of the fiber is

$$A(\tau \cong m, \zeta > 0) = A(\tau \cong m, 0) + \sum_n a_n \delta h(m - n). \qquad (3.18)$$

Without loss of generality one can assume that both $A(\tau \cong m, 0)$ and a_n are real (otherwise this is not amplitude modulation), therefore, after taking only the first order correction (in terms of the small perturbation $\delta h(n)$

$$|A(m, \zeta > 0)|^2 - |A(\tau \cong m, 0)|^2 \cong 2(A_0 + a_m) \sum_n a_{m-n} \Re \delta h(n), \qquad (3.19)$$

where the convolution's commutativity property was used.

Then (see a detailed derivation in Appendix B) the power noise (3.12) is

$$N \cong 4 R r_D^2 \left\{ \left(A_0^2 + \langle a_m^2 \rangle \right) \langle a_m^2 \rangle \sum_n \Re \delta h^2(n) + \Re \delta h^2(0) \left(\langle a_m^4 \rangle - \langle a_m^2 \rangle^2 \right) \right\} \qquad (3.20)$$

The dc level of the original signal, i.e., the transmitted signal, is

$$i_0(m, \zeta = 0) = r_D \left(A_0^2 + \langle a_m^2 \rangle \right), \qquad (3.21)$$

and its amplitude is

$$\Delta i(m, \zeta = 0) = r_D 2 A_0 a_m. \qquad (3.22)$$

Therefore, by denoting

$$\overline{i} \equiv r_d A_0^2, \qquad (3.23)$$

and using $\langle \Delta i^2(0) \rangle = 4 r_D^2 A_0^2 \langle a_m^2 \rangle$, Eq. (3.20) can be rewritten

$$N \cong \frac{i_0}{\overline{i}} R \langle \Delta i^2(0) \rangle \sum_n [\Re \delta h(n)]^2 + 4 R r_d^2 [\Re \delta h(0)]^2 \left(\langle a_m^4 \rangle - \langle a_m^2 \rangle^2 \right) \qquad (3.24)$$

Since

$$\Re \delta h(n) \cong -\sqrt{\frac{\zeta}{2\pi}} \left\{ \frac{\cos\left(\frac{(n + \xi/2)^2}{2\zeta} + \frac{\pi}{4} \right)}{n + \xi/2} - \frac{\cos\left(\frac{(n - \xi/2)^2}{2\zeta} + \frac{\pi}{4} \right)}{n - \xi/2} \right\}, \qquad (3.25)$$

then, due to jitter in the measurement location Δn, which for small ζ obeys $\Delta n \gg \zeta$,

$$\left[\Re\delta h(0)\right]^2 \cong 4\frac{\zeta}{\pi\xi^2}, \tag{3.26}$$

and

$$\sum_n\left(\Re\delta h(p)\right)^2 \cong \frac{\zeta}{4\pi}\sum_n\left(\frac{1}{\left(n+\xi/2\right)^2}+\frac{1}{\left(n-\xi/2\right)^2}\right)=\frac{\zeta}{2}\pi\csc^2\left(\pi\xi/2\right) \tag{3.27}$$

Therefore, the noise power is

$$N \cong R\frac{i_0}{\overline{i}}\left\langle\Delta i^2(0)\right\rangle\frac{\zeta}{2}\pi\csc^2\left(\pi\xi/2\right)+16Rr_d^2\frac{\zeta}{\pi\xi^2}\left(\left\langle a_m^4\right\rangle-\left\langle a_m^2\right\rangle^2\right) \tag{3.28}$$

In the case of OOK, for example, the second term vanishes, and

$$N \cong R\frac{i_0}{\overline{i}}\left\langle\Delta i^2(0)\right\rangle\frac{\zeta}{2}\pi\csc^2\left(\pi\xi/2\right) \tag{3.29}$$

As was expected, for narrow RZ pulses the noise is inversely proportional to the square of ξ.

In the case of NRZ, i.e., $\xi=1$, then

$$N \cong R\frac{i_0}{\overline{i}}\left\langle\Delta i^2(0)\right\rangle\frac{\zeta}{2}\pi \tag{3.30}$$

Furthermore, in the case of weak modulation (as in RF-over-Fiber optical links) $i_0 \cong \overline{i}$ and therefore

$$N \cong R\left\langle\Delta i^2(0)\right\rangle\frac{\zeta}{2}\pi, \tag{3.31}$$

the power SNR reduces to the simple analytical expression

$$SNR = \frac{\left\langle\Delta i^2(0)\right\rangle}{N} \cong \frac{2}{\pi\zeta} \tag{3.32}$$

On the other hand, in the case of deep modulation (as in most digital channels), $i_0 \cong 2\overline{i}$ and then

$$N \cong R\left\langle\Delta i^2(0)\right\rangle\zeta\pi, \tag{3.33}$$

and similarly the power SNR reduces to a simple analytical expression

$$SNR = \frac{\left\langle \Delta i^2(0) \right\rangle}{N} \cong \frac{1}{\pi \zeta} \qquad (3.34)$$

It is interesting to note that the distance, for which the SNR is equal to 1, i.e., $\zeta = \pi^{-1}$ is exactly equal to the fundamental upper limit in decoding the data without dispersion mitigation (see [33]).

Using real physical units, i.e., not dimensionless, the noise power and the SNR are

$$N \cong R \left\langle \Delta i^2(0) \right\rangle z \beta_2 B^2 \pi, \qquad (3.35)$$

and

$$SNR \cong \frac{T^2}{\pi z \beta_2} = \frac{1}{\pi z \beta_2 B^2} \qquad (3.36)$$

respectively.

In Fig. 3.2 Eq. (3.36) was compared to a simulation result. The simulation consisted of a Pseudo-Random-Bit-Sequence (PRBS) of $2^{11} - 1$ binary OOK sequence $a_m = \pm A_0$, which was substituted in Eq. (3.13) along with a bandwidth $\Delta = 10^4 T^{-1}$ for different normalized distances ζ. As can be seen from this figure Eq. (3.36) is a good approximation for short distances.

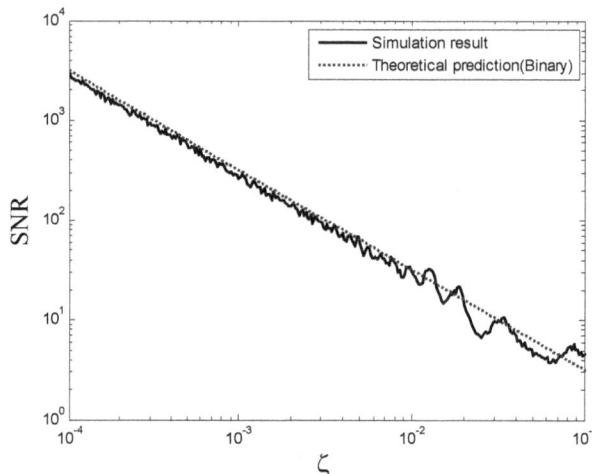

Fig. 3.2. A comparison between the theoretical SNR (solid curve), i.e., Eq. (3.34) and the numerical simulation (dotted curve).

In general (when the data is not necessarily binary-decoded), the SNR is a little lower

$$SNR \cong \frac{1}{\frac{i_0}{\bar{i}} \frac{\zeta}{2} \pi \csc^2 \left(\pi \xi / 2 \right) + 4 \frac{\zeta}{\pi \xi^2} \left(\langle a_m^4 \rangle / \langle a_m^2 \rangle - \langle a_m^2 \rangle \right) / A_0^2} \qquad (3.37)$$

In particular, for NRZ in the deep modulation case, i.e., $i_0 \cong 2\bar{i}$, and a_m is distributed uniformly, i.e., the probability density is constant over the range $a_m \in \left[-A_0, A_0 \right]$, then

$$SNR \cong \frac{\zeta^{-1}}{\pi + 16/15\pi}, \qquad (3.38)$$

and when they have a normal distribution (with zero mean, $\mu = 0$, and standard deviation equal to $\sigma = A_0$) then

$$SNR \cong \frac{\zeta^{-1}}{\pi + 8/\pi} \qquad (3.39)$$

A comparison between the theoretical power SNR (Eq. (3.39)) and the numerical simulation results for Gaussian distribution is presented in Fig. 3.3. As can be seen, there is a good agreement especially in the short distances limit. The binary case (Eq. (3.34)) is plotted as well for comparison.

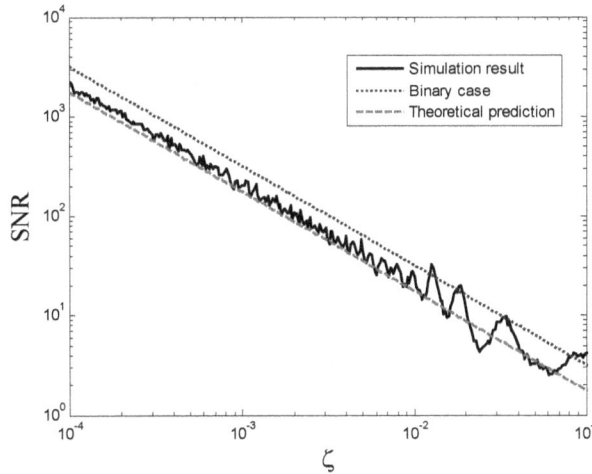

Fig. 3.3. Comparison between the SNR calculated by a simulation where the a_m's are distributed normally (solid curve), Eq. (3.39) (dashed line) and Eq. (3.34) (dotted line).

3.4. Sinc-shaped Nyquist Pulses: Spectrally Bounded Signals

Another important pulse shape, which received a lot of attention recently [34, 35], is the sinc-shaped Nyquist pulse, which is important in optical communication because it has the highest spectral efficiency (its spectrum is confined within a bandwidth, which is exactly equal to the reciprocal of the bit-rate, i.e. $B = 1/T$) with zero ISI. Moreover, sinc-shaped Nyquist pulses have additional advantages over rival high spectral efficiency technologies (such as OFDM) by requiring simpler receivers [36, 37], it is less sensitive to nonlinearities [37], it has lower peak-to-average power ratio [38], and yet its spectral efficiency is higher [39].

Due to their narrower (and bounded) spectrum these pulses are less affected by dispersion. Therefore, in this case, for the same sequence a_n and the same usage of dimensionless parameters $\zeta \equiv \beta_2 z / T^2$ and $\tau \equiv t/T$, the transmitted signal's field envelope is

$$A(\tau, \zeta = 0) = A_0 + \sum_n a_n \operatorname{sinc}(\tau - n), \tag{3.40}$$

where $\operatorname{sinc}(\tau) \equiv \dfrac{\sin(\pi\tau)}{\pi\tau}$ is the well-known sinc function.

Then, it is straightforward to validate that the received field, i.e., at the end of the dispersive medium, is

$$A(\tau, \zeta) = A_0 + \sum_n a_n \operatorname{dsinc}(\tau - n, \zeta) = A_0 + \sum_n a_n \operatorname{dsinc}(t/T - n, \beta_2 z / T^2), \tag{3.41}$$

where we define a new dynamic-sinc (dsinc) function [40, 41]

$$\operatorname{dsinc}(\tau, \zeta) \equiv \frac{1}{2}\sqrt{\frac{i}{2\pi\zeta}} \exp\left(-i\frac{\tau^2}{2\zeta}\right)\left[\operatorname{erf}\left(-\frac{\tau - \pi\zeta}{\sqrt{i2\zeta}}\right) - \operatorname{erf}\left(-\frac{\tau + \pi\zeta}{\sqrt{i2\zeta}}\right)\right] \tag{3.42}$$

Note that $\operatorname{dsinc}(\tau, \zeta = 0) = \operatorname{sinc}(\tau)$.

Therefore, the field at the center of the mth bit is

$$A(m, \zeta) = A_0 + \sum_n a_n g(m - n) = A(m, 0) + \sum_n a_n \delta g(m - n), \tag{3.43}$$

where

$$g(n) \equiv \operatorname{dsinc}(n, \zeta) \text{ and } \delta g(n) \equiv \operatorname{dsinc}(n, \zeta) - \delta(n) \tag{3.44}$$

For simplicity we focus on weak modulation, but unlike the previous section, the final result holds for the deep modulation case as well. Therefore, the second order distortion can be neglected, and

$$\left|A(m,\zeta)\right|^2 - \left|A(m,0)\right|^2 = \left|A(m,0) + \sum_n a_n \delta g(m-n)\right|^2 - \left|A(m,0)\right|^2 \cong$$

$$\cong 2A(m,0)\sum_n a_n \Re \delta g(m-n) \tag{3.45}$$

Thus, we can follow the same derivation that led to Eq. (3.20) when $\Re \delta g(n)$ replaces $\Re \delta h(n)$, and $i_0 \cong \overline{i}$. The result is

$$N \cong R\langle \Delta i^2(0)\rangle \sum_n \left[\Re \delta g(n)\right]^2 \tag{3.46}$$

Following Appendix C,

$$\Re \delta g(n) = -\frac{1}{2}\zeta^2 w(n) * w(n) = -\frac{1}{2}\zeta^2 \sum_m w(n-m)w(m), \tag{3.47}$$

where $w(m) = \begin{cases} (-1)^{m+1}/m^2 & m \neq 0 \\ -\pi^2/6 & m = 0 \end{cases}$.

After some tedious, albeit straightforward calculations

$$\left[\Re \delta g(0)\right]^2 = \frac{1}{4}\zeta^4 \sum_m w^2(m) = \frac{\pi^4}{80}\zeta^4, \tag{3.48}$$

and

$$\sum_m w(n-m)w(m) = \begin{cases} (-1)^n \left\{\dfrac{-6}{n^4} + \dfrac{\pi^2}{n^2}\right\} & n \neq 0 \\ \dfrac{\pi^4}{20} & n = 0 \end{cases} \tag{3.49}$$

When substituting (3.49) in (3.47) it follows

$$\sum_n \left[\Re \delta g(n)\right]^2 = \sum_n \frac{1}{4}\zeta^4 \left[\sum_m w(n-m)w(m)\right]^2 = \frac{1}{4}\zeta^4 \frac{\pi^8}{144}, \tag{3.50}$$

which can be substituted in (3.46) to finally give

$$N \cong R\langle \Delta i^2(0)\rangle \zeta^4 \frac{\pi^8}{576}, \tag{3.51}$$

and the power SNR is simply the analytical expression

$$SNR = \frac{R\langle \Delta i^2(0) \rangle}{N} \cong \frac{576}{\pi^8 \zeta^4}, \tag{3.52}$$

and in real physical units

$$SNR \cong \frac{576}{\pi^8 B^8 (\beta_2 z)^4}, \tag{3.53}$$

High agreement between equation (3.52) and the simulation results (similar simulation to the one in the previous section where the rectangular pulses were replaced by the sinc ones), is presented in Fig. 3.4.

It should be stressed that Eq. (3.52) is universal in the sense that it is valid *for any protocol and any signal's values distribution.*

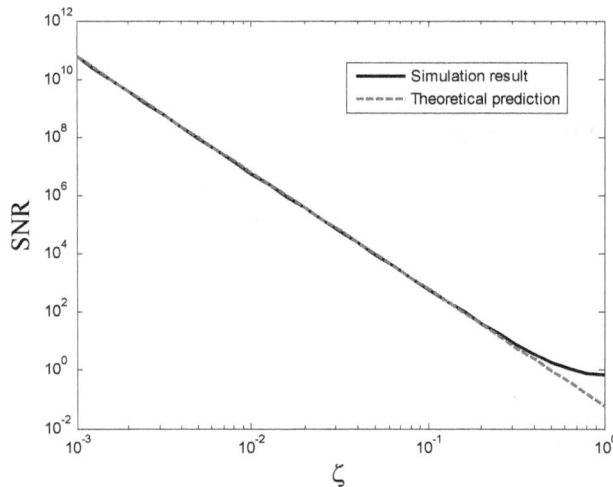

Fig. 3.4. A comparison between Eq. (3.52) (dashed line) and the simulation results (solid curve).

3.5. Dispersion Noises in Coherent Channel Based on Nyquist Pulses

As was explained above, Nyquist pulses became recently common in coherent optical communication channels [34-39]. Moreover, these pulses can be incorporated in any QAM protocol, and therefore seems to be a promising technology. In coherent system, every carrier can carry four orthogonal channels: two components (real and imaginary) on two polarizations (see Fig. 3.5). In the transmitter side there is a complex coherent modulator, which consists of several cascaded interferometers, while in the receiver side there is a coherent detector, which interferes the detected signal with a local oscillator and extracts the four components of the field (for details about the inner structure of these elements see, for example, [42]).

Therefore, to transmit the random complex sequence $a_n = x_n + iy_n$ for $n = -\infty, \ldots -1, 0, 1, 2, \ldots \infty$, the signal can be modulated as overlapping sinc pulses

$$A(\tau, \zeta = 0) = \sum_n a_n \operatorname{sinc}(\tau - n) \tag{3.54}$$

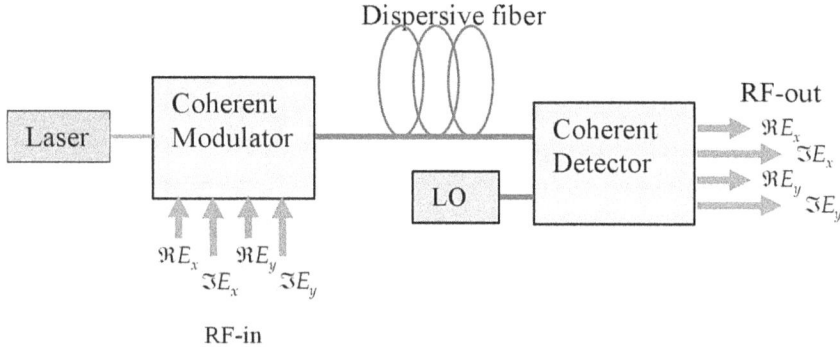

Fig. 3.5. Coherent system schematic. Every carrier can carry four orthogonal channels: two components (real and imaginary) on two polarizations. LO stands for local oscillator.

Then, clearly, at the received end of the optical channel, the field is (following (3.41) and (3.42) but without the dc term)

$$A(\tau, \zeta) = \sum_n a_n \operatorname{dsinc}(\tau - n, \zeta) = \sum_n a_n \operatorname{dsinc}\left(t / T - n, \beta_2 z / T^2\right) \tag{3.55}$$

In coherent channels both components of a_n (real and imaginary) carry information, and both of them are detected (for both polarizations).

The noise level of the signal (the variance) can be evaluated using Eq. (3.C6)

$$\left\langle \left[\Delta \Im A(m, \zeta)\right]^2 \right\rangle \cong \zeta^2 \left\langle \left[\sum_n w(m - n) \Re A(n, 0)\right]^2 \right\rangle, \tag{3.56}$$

and in an analogy to (3.B3),

$$\left\langle \Re A(n, 0) \Re A(p, 0)\right\rangle = \delta(n - p)\left\langle \left[\Re A(n, 0)\right]^2 \right\rangle, \tag{3.57}$$

and therefore

$$\left\langle \left[\Delta \Im A\left(m,\zeta\right)\right]^{2} \right\rangle \cong \zeta^{2} \sum_{n} w\left(m-n\right)^{2} \left\langle \left[\Re A\left(n,0\right)\right]^{2} \right\rangle \qquad (3.58)$$

Finally, since on average there is no difference between the real and imaginary parts, i.e.,

$$\left\langle \left[\Im A\left(m,\zeta\right)\right]^{2} \right\rangle = \left\langle \left[\Re A\left(n,0\right)\right]^{2} \right\rangle, \qquad (3.59)$$

it follows that the SNR has a simple analytical expression

$$SNR = \frac{\left\langle \left| A\left(m,\zeta\right)\right|^{2} \right\rangle}{\left\langle \left| \Delta A\left(m,\zeta\right)\right|^{2} \right\rangle} = \frac{\left\langle \left[\Im A\left(m,\zeta\right)\right]^{2} \right\rangle}{\left\langle \left[\Delta \Im A\left(m,\zeta\right)\right]^{2} \right\rangle} \cong \frac{1}{\zeta^{2} \sum_{n} w\left(n\right)^{2}} = \frac{20}{\zeta^{2} \pi^{4}} \qquad (3.60)$$

Again, it should be stressed that this is a generic result, which holds true for any data protocol and any data distribution.

In Fig. 3.6 the SNR of a coherent channel simulation with a very high QAM (1024), was compared to Eq. (3.60) for different dispersion values (ζ). As can be shown, Eq. (3.60) is an excellent approximation.

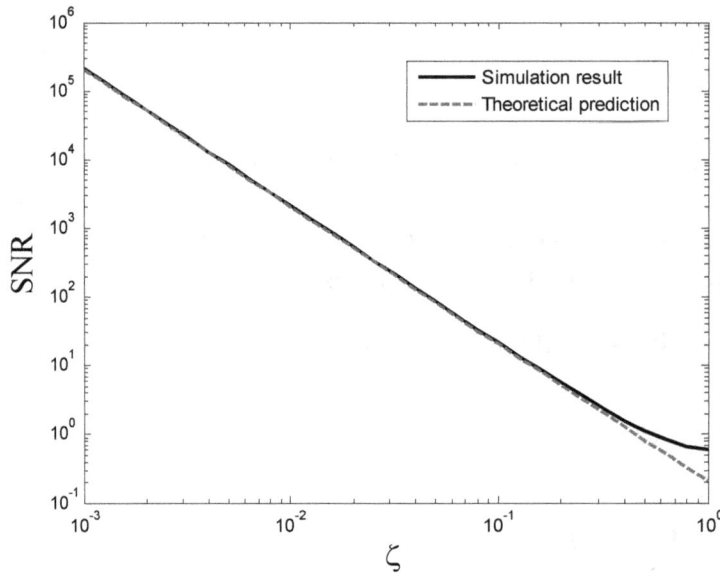

Fig. 3.6. The SNR plot as a function of the normalized distance ζ for QAM 1024 which is a good approximation for a uniform distribution. The solid line stands for numerical simulation while the dashed line stands for Eq. (3.60).

3.6. Conclusions

Dispersion's distortions are very common in optical communications. Since dispersion is a linear deterministic effect, its distortions are commonly regarded as deterministic ones. However, due to the nonlocality nature of the dispersion equation and the randomness of the transmitted data, the dispersion's distortions are effectively random, and can be regarded as a random noise source.

In this chapter the noise and the SNR of the signal, which was caused by dispersion, are calculated for several scenarios. It is shown that in all these cases the SNR has a simple analytical expression, which depends only on the value of the dimensionless dispersion parameter $\zeta \equiv \beta_2 z / T^2$.

We focus on rectangular and sinc-shaped Nyquist pulses. The rectangular pulses, due to their sharp boundaries are responsible for relatively low SNR. In the case of binary data protocol in the weak modulation regime the result is $SNR = 2/\pi\zeta$, while in the deep modulation case the SNR is reduced by half to $SNR = 1/\pi\zeta$. In the case of a uniform distribution $SNR \cong [(\pi + 16/15\pi)\zeta]^{-1}$ and in a normal Gaussian one $SNR \cong [(\pi + 8/\pi)\zeta]^{-1}$. When the pulses have a sinc shape the SNR is much larger and it is independent of the data's distribution. It is always $SNR = 576\pi^{-8}\zeta^{-4}$ in direct modulation systems and $SNR = 20\pi^{-4}\zeta^{-2}$ in coherent ones.

References

[1]. G. P. Agrawal, Nonlinear Fiber Optics, *Academic Press*, 2001.
[2]. X. Gu, L. C. Blank, 10 Gbit/s unrepeated three-level optical transmission over 100 km of standard fibre, *Electron. Lett.*, Vol. 29, 1993, pp. 2209-2210.
[3]. K. Yonenaga, S. Kuwanto, S. Norimatsu, N. Shibata, Optical duobinary transmission system with no receiver sensitivity degradation, *Electron. Lett.*, Vol. 31, 1995, pp. 302-304.
[4]. D. Penninckx, M. Chbat, L. Pierre, J.-P. Thiery, The Phase-Shaped Binary Transmission (PSBT): A new technique to transmit far beyond the chromatic dispersion limit, *Photon. Technol. Lett.*, Vol. 9, 1997, pp. 259-261.
[5]. H. Bissessur, G. Charlet, C. Simonneau, S. Borne, L. Pierre, C. De Barros, P. Tran, W. Idler, R. Dischler, 3.2 Tb/s 80×40 Gb/s C-band transmission over 3×100 km with 0.8 bit/s/Hz efficiency, in *Proceedings of the European Conference on Optical Communication (ECOC'01)*, Amsterdam, The Netherlands, Sep. 2001, Vol. 6, pp. 22-23.
[6]. T. Ono, Y. Yano, K. Fukuchi, T. Ito, H. Yamazaki, M. Yamaguchi, K. Emura, Characteristics of optical duobinary signals in terabit/s capacity, high-spectral efficiency WDM systems, *J. Lightwave Technol.*, Vol. 16, 1998, pp. 788-797.
[7]. R. Kaur, S. Dewra, Duobinary modulation format for optical system – A review, *International Journal of Advanced Research in Electrical, Electronics and Instrumentation Engineering*, Vol. 3, 2014, pp. 11039-11045.
[8]. M. Joindot, G. Bosco, A. Carena, V. Curri, P. Poggiolini, Fundamental performance limits of optical duobinary, *Opt. Exp.*, Vol. 16, 2008, pp. 19600-19614.

[9]. I. N. Cano, A. Lern, M. Presi, V. Polo, E. Ciaramella, J. Prat, 6.25 Gb/s differential duobinary transmission in 2 GHz BW limited direct phase modulated DFB for udWDM-PONs, in *Proceedings of the European Conference on Optical Communication (ECOC'14)*, Cannes, 2014, pp. 1-3.

[10]. W. Shieh, I. Djordjevic, OFDM for Optical Communications, *Academic Press*, 2009.

[11]. W. Shieh, H. Bao, Y. Tang, Coherent optical OFDM: Theory and design, *Optics Express*, Vol. 16, 2008, pp. 841-859.

[12]. J. Armstrong, OFDM for optical communications, *J. Lightwave Technol.*, Vol. 27, 2009, pp. 189-204.

[13]. C. Lin, H. Kogelnik, L. G. Cohen, Optical-pulse equalization of low-dispersion transmission in single-mode fibers in the 1.3-1.7-µm spectral region, *Opt. Lett.*, Vol. 5, 1980, pp. 476-478.

[14]. D. S. Larner, V. A. Bhagavatula, Dispersion reduction in single-mode fiber links, *Electron. Lett.*, Vol. 21, 1985, pp. 1171-1172.

[15]. A. M. Vengsarkar, W. A. Reed, Dispersion-compensating single-mode fibers: Efficient designs for first- and second-order compensation, *Opt. Lett.*, Vol. 18, 1993, pp. 924-926.

[16]. C. D. Poole, J. M. Wiesenfeld, A. R. McCormick, K. T. Nelson, Broadband dispersion compensation by using the higher-order spatial mode in a two-mode fiber, *Opt. Lett.*, Vol. 17, 1992, pp. 985-987.

[17]. C. D. Poole, J. M. Wiesenfeld, D. J. DiGiovanni, Elliptical-core dual mode fiber dispersion compensator, *IEEE Photon. Technol. Lett.*, Vol. 5, 1993, pp. 194-197.

[18]. C. D. Poole, J. M. Wiesenfeld, D. J. DiGiovanni, A. M. Vengsarkar, Optical fiber-based dispersion compensation using higher order modes near cutoff, *J. Lightwave Technol.*, Vol. 12, 1994, pp. 1746-1758.

[19]. R. C. Youngquist J. L. Brooks, H. J. Shaw, Two-mode fiber modal coupler, *Opt. Lett.*, Vol. 9, 1984, pp. 177-179.

[20]. J. N. Blake, B. Y. Kim, H. J. Shaw, Fiber-optic modal coupler using periodic microbending, *Opt. Lett.*, Vol. 11, 1986, pp. 177-179.

[21]. C. D. Poole, C. D. Townsend, K. T. Nelson, Helical-grating two-mode fiber spatial-mode coupler, *J. Lightwave Technol.*, Vol. 9, 1991, pp. 598-604.

[22]. S. J. Savory, Digital filters for coherent optical receivers, *Opt. Express*, Vol. 16, 2008, pp. 804-807.

[23]. S. J. Savory, Compensation of fibre impairments in digital coherent systems, in *Proceedings of the European Conference on Optical Communication (ECOC'08)*, Brussels, Belgium, 2008, pp. 1-4.

[24]. T. Xu, G. Jacobsen, S. Popov, J. Li, E. Vanin, K. Wang, A. T. Friberg, Y. Zhang, Chromatic dispersion compensation in coherent transmission system using digital filters, *Opt. Express*, Vol. 18, 2010, pp. 16243-16257.

[25]. E. Granot, S. Bloch, S. Sternklar, Affordable dispersion mitigation with an analog electrical filter, *Applied Optics*, Vol. 55, 2016, pp. 7956-7963.

[26]. S. Bloch, S. Sternklar, E. Granot, Affordable dispersion mitigation method for the next generation RF-over-Fiber optical channels, *Applied Optics*, Vol. 56, 2017, pp. 6777-6784.

[27]. S. Marciano, S. Ben-Ezra, E. Granot, Eavesdropping and network analyzing using network dispersion, *Appl. Phys. Res.*, Vol. 7, 2015, 27.

[28]. E. Granot, A. Marchewka, Emergence of currents as a transient quantum effect in nonequilibrium systems, *Phys. Rev. A*, Vol. 84, 2011, 032110.

[29]. E. Granot, E. Luz, A. Marchewka, Generic pattern formation of sharp-boundaries pulses propagation in dispersive media, *J. Opt. Soc. Am. B*, Vol. 29, 2012, pp. 763-768.

[30]. M. Abramowitz, A. Stegun, Handbook of Mathematical Functions, *Dover Publication*, New-York, 1965.

[31]. C. H. Cox III, Analog Optical Links, Theory and Practice, *Cambridge University Press*, 2004.

[32]. S. Haykin, M. Moher, Introduction to Analog and Digital Communications, *John Wiley & Sons Inc.*, 2007.

[33]. E. Granot, Fundamental dispersion limit for spectrally bounded On-Off-Keying communication channels and its implications to Quantum Mechanics and the Paraxial Approximation, *Europhys. Lett.*, Vol. 100, 2012, 44004.

[34]. M. A. Soto, M. Alem, M. A. Shoaie, A. Vedadi, C-S Brès, L. Thévenaz, T. Schneider, Optical sinc-shaped Nyquist pulses of exceptional quality, *Nature Communications*, Vol. 4, 2013, 2898.

[35]. R. Schmogrow, R. Bouziane, M. Meyer, P. A. Milder, P. C. Schindler, R. I. Killey, P. Bayvel, C. Koos, W. Freude, J. Leuthold, Real-time OFDM or Nyquist pulse generation – Which performs better with limited resources?, *Optics Express*, Vol. 20, 2012, pp. B543-B551.

[36]. T. Hirooka, P. Ruan, P. Guan, M. Nakazawa, Highly dispersion-tolerant 160 Gbaud optical Nyquist pulse TDM transmission over 525 km, *Opt. Express*, Vol. 20, 2012, pp. 15001-15007.

[37]. T. Hirooka, M. Nakazawa, Linear and nonlinear propagation of optical Nyquist pulses in fibers, *Opt. Express*, Vol. 20, 2012, pp. 19836-19849.

[38]. R. Schmogrow, et al., 512QAM Nyquist sinc-pulse transmission at 54 Gbit/s in an optical bandwidth of 3 GHz, *Opt. Express*, Vol. 20, 2012, pp. 6439-6447.

[39]. G. Bosco, A. Carena, V. Curri, P. Poggiolini, F. Forghieri, Performance limits of Nyquist-WDM and CO-OFDM in high-speed PM-QPSK systems, *IEEE Phot. Technol. Lett.*, Vol. 22, 2010, pp. 1129-1131.

[40]. E. Granot, Information loss in quantum dynamics, in Advanced Technologies of Quantum Key Distribution, *INTECH*, Rijeka, 2017.

[41]. E. Granot, Analytical Solutions for the Propagation of UltraShort and UltraSharp Pulses in Dispersive Media, *Applied Science,* Vol. 9, 2019, 527.

[42]. S. Betti, G. De Marchis, E. Iannone, Coherent Optical Communication Systems, *Wiley-Interscience*, 1995.

Appendix A. Approximations of the Complex ERF and the SRECT Functions

Following [30], one can use a more useful approximation: for $|z| \to \infty$ the complex error-function can be written in a single term, which can apply for both $z \to \pm\infty$, namely

$$erf\left(z/\sqrt{i}\right) \to \left[1 - \frac{1}{|z|}\sqrt{\frac{i}{\pi}}\exp\left(iz^2\right)\right]\operatorname{sgn}\left(z\right) = \operatorname{sgn}\left(z\right) - \frac{1}{z}\sqrt{\frac{i}{\pi}}\exp\left(iz^2\right) \qquad (3.A1)$$

Then, for any τ that holds $\left|\tau - |\xi/2|\right|^2 >> \zeta$, i.e., it is not too close to the pulse's edges, the following approximation can be used to the distorted pulse

$$\operatorname{srect}_\xi\left(\tau,\delta,\zeta\right) \cong \operatorname{rect}_{\xi T}\left(t,\Delta\right) -$$
$$-\sqrt{\frac{i\zeta}{2\pi}}\exp\left[i\left(\frac{\tau^2 + \xi^2/4}{2\zeta}\right)\right]\left\{\frac{\exp\left(i\tau\xi/2\zeta\right)}{\left(\tau + \xi/2\right)} - \frac{\exp\left(-i\tau\xi/2\zeta\right)}{\left(\tau - \xi/2\right)}\right\} \qquad (3.A2)$$

Since $\zeta \ll 1$ it is a weak distortion from the original pulse $\text{rect}_{\xi T}(t, \Delta)$.

Appendix B. Derivation of Eq. (3.20)

$$\left\langle \left(\left| A(m, \zeta > 0) \right|^2 - \left| A(\tau \cong m, 0) \right|^2 \right)^2 \right\rangle \cong$$

$$\cong 4 \left\langle (A_0 + a_m)^2 \sum_n a_{m-n} \Re \delta h(n) \sum_p a_{m-p} \Re \delta h(p) \right\rangle = \qquad (3.B1)$$

$$= 4 A_0^2 \sum_{n,p} \Re \delta h(n) \Re \delta h(p) \left\langle a_{m-n} a_{m-p} \right\rangle + 4 \sum_{n,p} \Re \delta h(n) \Re \delta h(p) \left\langle a_m^2 a_{m-n} a_{m-p} \right\rangle$$

Therefore, the noise power is equal to

$$N \cong 4 r_d^2 \left\{ A_0^2 \sum_{n,p} \Re \delta h(n) \Re \delta h(p) \left\langle a_{m-n} a_{m-p} \right\rangle + \sum_{n,p} \Re \delta h(n) \Re \delta h(p) \left\langle a_m^2 a_{m-n} a_{m-p} \right\rangle \right\} \quad (3.B2)$$

Due to the fact that there is no correlation between adjacent symbols, i.e.,

$$\left\langle a_{m-n} a_{m-p} \right\rangle = \delta(n-p) \left\langle a_m^2 \right\rangle, \qquad (3.B3)$$

then

$$N \cong 4 r_d^2 \left\{ A_0^2 \left\langle a_m^2 \right\rangle \sum_n \Re \delta h^2(n) + \left[\sum_n \Re \delta h^2(n) - \Re \delta h^2(0) \right] \left\langle a_m^2 \right\rangle^2 + \Re \delta h^2(0) \left\langle a_m^4 \right\rangle \right\} =$$

$$= 4 r_d^2 \left\{ \left(A_0^2 + \left\langle a_m^2 \right\rangle \right) \left\langle a_m^2 \right\rangle \sum_n \Re \delta h^2(n) + \Re \delta h^2(0) \left(\left\langle a_m^4 \right\rangle - \left\langle a_m^2 \right\rangle^2 \right) \right\} \qquad (3.B4)$$

Appendix C. Derivation of Eq. (3.47)

$\Re \delta g(n)$ is the real part of the change, or the distortion, that a sinc function experiences. Since

$$\left. \frac{\partial^2 \operatorname{sinc}(\tau)}{\partial \tau^2} \right|_{\tau = n \neq 0} = \frac{2}{n^2} (-1)^{n+1} \quad \text{and} \quad \left. \frac{\partial^2 \operatorname{sinc}(\tau)}{\partial \tau^2} \right|_{\tau = 0} = -\frac{\pi^2}{3} \qquad (3.C1)$$

then for $\tau = m$ Eq. (3.1) can be replaced with a discrete convolution

$$\frac{dA(m, \zeta)}{d\zeta} = i w(m) * A(m, \zeta) = i \sum_n w(m-n) A(n, \zeta), \qquad (3.C2)$$

where

$$w(m) = \begin{cases} (-1)^{m+1} / m^2 & m \neq 0 \\ -\pi^2 / 6 & m = 0 \end{cases} \tag{3.C3}$$

Alternatively, it can be written as a set of two coupled equations

$$\frac{d\Re A(m,\zeta)}{d\zeta} = -\sum_n w(m-n)\Im A(n,\zeta), \tag{3.C4}$$

$$\frac{d\Im A(m,\zeta)}{d\zeta} = \sum_n w(m-n)\Re A(n,\zeta) \tag{3.C5}$$

If initially the field was real, i.e. $A(n,0) = \Re A(n,0)$, then after a short distance

$$\Im A(m,\zeta) \cong \zeta \sum_n w(m-n)\Re A(n,0), \tag{3.C6}$$

and after substituting (3.C6) in (3.C4)

$$\Re A(m,\zeta) - \Re A(m,0) \cong -\int_0^\zeta \sum_n w(m-n)\Im A(n,\zeta')d\zeta', \tag{3.C7}$$

or

$$\Re \delta A(m,\zeta) \cong -\frac{\zeta^2}{2} \sum_n \sum_m w(n-m)w(m)A(n,0), \tag{3.C8}$$

and therefore

$$\Re \delta g(n) \cong -\frac{1}{2}\zeta^2 w(n) * w(n) = -\frac{1}{2}\zeta^2 \sum_m w(n-m)w(m) \tag{3.C9}$$

Chapter 4
AOLS Technique Survey and an All Fiber Realization

E. N. Lallas

4.1. Introduction

The rapid growth of packet based Internet traffic and big data, associated with interconnected Petascale Data Centers (DC) and High Performance Computing (HPC) systems have imposed the need for ultrahigh link capacities and ultrahigh packet switching speeds, at network nodes. Multiprotocol label switching (MPLS) technology is a promising solution to the problem of the imposing need for ultrahigh link capacities and packet switching speeds at network nodes. Instead of reading huge route lookups, a single label is read on each packet [1]. MPLS protocol is supported by Generalized MPLS (GMPLS), when IP packets are bypassed directly over the wavelength division multiplexing (WDM) layer, avoiding the two intermediate layers, asynchronous transfer mode (ATM) and synchronous digital hierarchy (SDH) respectively. In this case, the label switched paths of MPLS protocol have been replaced by the wavelength switched paths of GMPLS protocol [2]. Apart from the wavelength labeling, additional label information can be encoded and attached to the IP packet, via various label encoding methods, at the edge nodes before entering the WDM core network layer. All optical label swapping (AOLS) is the method of coding the optical label onto the packet, after having removed the old one, for all optical packet routing and forwarding. Many label encoding techniques for label swapping have been proposed so far [3], such as subcarrier modulated (SCM) header [4], bit serial header and orthogonal modulation schemes [5]. In this chapter, at first, a short survey attempts, apart from conventional label swapping techniques updates also to point out new trends and solutions to overcome major label swapping limitations related to speed and capacity matters. Moreover, an orthogonal four wave mixing (FWM) based, intensity modulated (IM) payload, frequency shift keying (FSK) label, IM/FSK high bitrate scheme is studied. Its implementation is based on a dispersion shifted fiber (DSF) fiber, which is proven to be more promising at high bit rates, compared to the

E. N. Lallas
Technical Educational Institute of Sterea Ellada, Lamia, Greece

semiconductor optical amplifier (SOA) based IM/FSK scheme [6]. The principle of a typical FSK modulated label, combined with an intensity modulated IM payload on the same optical carrier, is implemented via a DSF fiber based FWM scheme.

AOLS takes place at intermediate nodes, inside the core WDM optical network and requires wavelength conversion and probably regeneration of the payload, and at the same time old header removal and insertion of a new one, in order to support the dominant routing and forwarding functions. To achieve all optical wavelength conversion and regeneration, various strategies have been investigated. There are two kinds of media: one medium is based on SOAs and possible methods are cross gain modulation (XGM) and cross phase modulation (XPM) individually [7], or combined together [6], and FWM [8]. The other medium is based on DSF fiber and methods are FWM [9] and XPM alone [10] or combined together with self-phase modulation (SPM) [11]. Many efforts have been made on DSF fiber based wavelength conversion schemes supporting label swapping schemes, via XPM, but they concern mostly return to zero (RZ) pulses [12]. The basic idea involves combination of incoming RZ data pulses with continuous wave (CW) signal through DSF fiber. The data imposes phase modulation onto CW via XPM and generates optical sidebands on the CW spectra, which can be converted to amplitude modulation by suppressing the CW carrier and filtering one sideband. Rise and fall time of the RZ pulses determine the conversion efficiency, hence non return to zero (NRZ) data cannot be converted, as only the edges of a signal are converted. In the solutions proposed in the literature, NRZ pulse wavelength conversion has been reported and demonstrated, using nonlinear optical loop mirrors (NOLM) [13, 14]. By proper adjustment of a polarization controller (PC), the relative state of polarization (SOP) between probe and drive signals changes, a phase shift is added and the disturbing counter-propagating nonlinear phase shift is cancelled out, while the co-propagating phase shift is maintained. However this adjustment is not easy to obtain, as it requires tuning of the PC while observing the output signal, and maximizing the output power and extinction ratio as well. As a solution here for the label swapping module, we adopt an already proposed interferometer DSF based wavelength conversion scheme [15]. This architecture is not limited to special formats and presents high insensitivity against slow varying environmental noise, due to its common mode configuration as in NOLMs. On the other hand, it is not a completely common path structure, allowing a simple operation point regulation, by means of a piezo-ceramic stretcher, to overcome the undesirable counter-propagating nonlinear phase shift, thus avoiding complicated SOP regulations [15].

In this chapter, apart from the label swapping technique survey, we report a 40 Gbps payload / 2.5 Gbps header AOLS scheme based on all fiber realization for the implementation of NRZ payload and header coupling via a FWM mechanism, as well as for the payload wavelength conversion and header removal via an interferometric mechanism. Our chapter is structured as follows: in Section 4.2 there is a survey related to recent label swapping techniques including newcomer techniques with high operating speed potential, appropriate for big data traffic management among DCs and HPCs. In Section 4.3, there is a description of the all fiber based system architecture and its parameters. In Section 4.4, there is a description of the unidirectional and bidirectional DSF fiber modules, used in our simulations along with a component investigation.

Simulation results for the 40 Gbs NRZ IM payload with the 2.5 Gbs NRZ, FSK header is included in Section 4.5. Finally, Section 4.6 concludes the chapter.

4.2. Label Swapping Technique Survey

It is widely known that MPLS protocol is an IP packet forwarding mechanism based on label swapping techniques. Traditionally, the routing and forwarding of the packet were based on the exhaustive search and matching in large routing tables of the destination IP address, and this procedure was enforced for each intermediate node of the route. With MPLS, forwarding a packet to the next node is only done by reading and exchanging a label included in the packet header, without the need for the knowledge and the tedious search of the destination IP address. Thus, the promotion of the packet to the final destination of the route is simpler and faster, and it is no more related to routing decision procedure.

For the implementation of MPLS protocol and its critical routing and forwarding functions, directly in the optical layer, AOLS mechanism is suggested. AOLS routing and forwarding functions at the intermediate nodes are usually implemented via wavelength conversion and regeneration of the payload, along with old label removal and insertion of a new one, onto the forwarding packet. An ideal AOLS technique should combine simple and feasible label replacement, operation at the highest possible rates, high signal quality combined with low interference between the label and the payload signals, as packet moves on to its destination, and with a minimum overhead occupation percentage compared to the total data capacity.

There are several techniques to encode and attach a label into the IP packet, along with the payload:

a) Time domain techniques also known as bit serial labeling techniques, where the label is placed bit serial to the payload in the time domain [16].

b) Parallel labeling techniques where the label is processed in parallel with the payload, either multiplexed on a separate wavelength [17] or on a band of lambdas (out of band) [18], or on some microwave subcarrier frequency maintaining the same wavelength channel with the data (in band, subcarrier modulation) [19].

c) Parallel labeling techniques where the label is processed in parallel with the payload, but at different modulation format, the so called orthogonal modulation schemes (orthogonal modulation label swapping). Most popular orthogonal schemes are IM / FSK, and IM / DPSK (differential phase shift keying) where the payload is intensity modulated while the label is angular modulated [6].

d) Other parallel labeling techniques where the label is processed in parallel with the payload, using different code, also called optical code (OC) label swapping [20].

e) Hybrid techniques, which are mainly combinations of the abovementioned techniques (e.g. 2-D and 3-D labeling that combines labeling of separate lambdas and separate codes, and polarization states) [21].

f) New label swapping techniques, such as spectral amplitude codes (SAC) with label stack, combined with advanced fully coherent labeling techniques.

4.2.1. Bit Serial Label Swapping

In bit serial label swapping, the label, is placed in front of the IP packet in time domain, [4] and there may be a guard time between them, delaying packet data as long as necessary in order for the old label to be extracted, processed and replaced by a new one. Both label and payload are driven by clock pulses to facilitate synchronization recovery. This label swapping technique is particularly attracted to time slotted, packet switched networks, as they require synchronous and fixed-length packet switching with unavoidably strict timing control.

The first AOLS demonstrations were based on bit serial technique. Specifically, in [22], one of the first high bit rate serial labeling demonstrations, was based on a 3 cascaded SOA-Mach Zender Interferometric switches (MZI) configuration, that each one was used as a wavelength converter, a packet clock recovery (CR) unit, and a logical AND gate between the input data packet and its properly delayed clock signal respectively, for a 40 Gbps data rate signal.

Apart from SOAs, other media may be used for AOLS functioning as well, such as XPM based DSFs, followed by appropriate optical filters. In [23], the basic serial label swapping demonstration was relied on the use of a XPM based DSF fiber, followed by a loop mirror notch filter. The incoming data consisted of 40 Gbps RZ pulses with 2.5 Gbps NRZ label that is combined with a CW signal and sent through the DSF. Pulses impose phase modulation that generates optical sidebands on the CW signal, which can be easily converted to amplitude modulation by suppressing the original CW carrier and selecting one of the side bands using an optical notch filter.

Strict timing synchronization which is considered the main drawback of bit serial AOLS schemes, may be avoided when using angular modulated labels, such as FSK or DPSK labels [24]. The DPSK label can be easily and effectively erased via a delay interferometer based on SOA-XPM wavelength converter. Moreover this method offers 2R regeneration with high extinction ratio of the new data packet.

Apart from single wavelength carrier bit serial labeling demonstrations, simultaneous AOLSimplementation on multiple independent WDM channels was feasible via the use of periodicallypoledlithiumniobate (PPLN) waveguides. Various PPLN waveguides were used as simultaneous wavelength converters for incoming WDM data channels, which are coming out as their mirror images with respect to a pump reference signal. Specifically, in [25], the related demonstration involved two WDM channels at 10 Gbps, implemented with two PPLN waveguides that were used for the removal of the original label from the incoming packets and the insertion of a new label.

4.2.2. Parallel InBand/Out of Band Label Swapping

4.2.2.1. InBand (Subcarrier Modulation) Labeling

In subcarrier multiplexing (SCM) technique, the label is placed in the same wavelength with the data payload on a microwave (RF) subcarrier frequency that is outside the spectra limits of the baseband data signal. SCM spectrum consists of two sidebands lying symmetrically outside the optical carrier, and that is why, it is called double side band (DSB) multiplexing.

One of the first popular AOLS implementations used for the support of AOLS for direct IP addressing in the WDM layer, was based on a cascaded combination of a SOA-XGM stage, followed by a SOA-XPM second stage, for packet data rates of 2.5 Gbps, and SCM labels of 100 Mbps at 16 GHz frequency [26]. The second stage which is a SOA-MZI, was used as a 2R regenerator as it enhances the payload extinction ratio due to the inherent non-linearity of the MZI transfer function. That popular SCM labeling technique was also used in other major network demonstrations for control and routing at those times, such as OPERA (Optical Packet Experimental Routing Architecture) [27], and MOSAIC (multiwavelength optical subcarrier multiplexed controlled network) [28].

Subcarrier AOLS techniques mainly utilize steep notch filters in order to extract the desired spectral sideband (SCM label), from the data baseband signal. Specifically, in [29], a fiber loop mirror was used as a filter for separating label from data, while in [30] alternatively, a Fiber Bragg Grating (FBG) filter was used for separating baseband data signal from the double RF frequency sidebands, by simply adjust its peak reflectivity to match the baseband signal [30].

4.2.2.2. Out of Band Labeling (Single Wavelength or Band of Wavelengths)

As far as concerns wavelength multiplexing label swapping techniques, a separate wavelength or a band of wavelengths may be used for label encoding process, which is different from the wavelength of the data. Eventually, this method has the advantage of the easy and separate label processing and recovery, without interleaving with the payload data at all, using though, separate transponders for the different wavelengths, which may be quite costly.

A remarkable, high data rate, multi wavelength (multi-λ) band label, experimental demonstration is described in [31], for 40 Gbps optical time multiplexed (OTDM) data that propagate over three fiber spans of 40 km each, to end up to 10 Gbps data streams, via demultiplexing unit. Multiwavelength (Mλ) labels are processed via FBG filter series, which act as label bank correlators, for routing and forwarding the OTDM data. Critical modules here, are the multiwavelength edge node (Mλ-EN) transmitter that generates and transmits the 40 Gbps OTDM data with Mλ labels attached on it, intermediate nodes such as, the multiwavelength label switched node (Mλ-LSN), consisted of FBG label bank correlators for label swapping, and the wavelength band optical cross connect (λB-OXC) nodes, consisted of array waveguide gratings (AWG) arrays, implemented via planar lightwave circuit (PLC) technology, for packet routing and forwarding, and finally the detectors for payload and labels.

Unfortunately, as the number of wavelength data channels increases, and hence the wavelengths used for labeling, there will be an increasing difficulty in managing all of them all optically with existing state of the art optical correlators, and optical gates.

4.2.3. Orthogonal Modulated Label Swapping

Orthogonal modulation labelling techniques are characterised as one of the most innovative approaches. They are based on label and data information modulated in orthogonal modulation formats of each other (amplitude, phase or frequency domains), in every possible combination. The most popular demonstrated orthogonal technique involves ON-OFF keying, intensity modulated (OOK-IM) data payloads with FSK [32], or DPSK labels [33]. The proposed scheme which is described in detail in next paragraph is an all fiber based orthogonal IM/FSK label swapping technique.

4.2.4. Optical Code Label Swapping

According to the code label swapping technique, the encoding of a label bit sequence is associated with the fragmentation of each bit into a sequence of smaller pulses (named chips), in a unique pattern that also represents its label code. Optical code labeling is an alternate parallel labeling method, being characterized for its all optical correlation layout consisted of massive integrated optical circuits working in parallel, consuming far less time than electronic processing, for identifying the match in a forwarding table lookup, at intermediate nodes. It is a method which is mainly targeted for packet and even burst switched networks when labeling can be adapted to various packet rates.

Critical parameters of OC labeling is the proper choice of code family with its desired characteristics namely be, long word codebook for generation of as many encoded labels as possible, high signal to interference ratio (SIR), high processing gain, and high autocorrelation function for instant discrimination at the detector.

All optical label encoding and decoding can be implemented with correlation devices, such as gates accompanied with phase shifters, interferometers and FBG filter banks. When the encoded label enters into an ordered optical correlation bank, which is actually the table that contains all the possible code words, many label copies are created at its inputs as much as the code word count, and there can be only one code word matching that produces at the device output a high autocorrelation peak, while all the other outputs lead to low cross-correlation peaks where there will be no code word matching at all. Various popular all optical label code correlator architectures are illustrated in Fig. 4.1 [34].

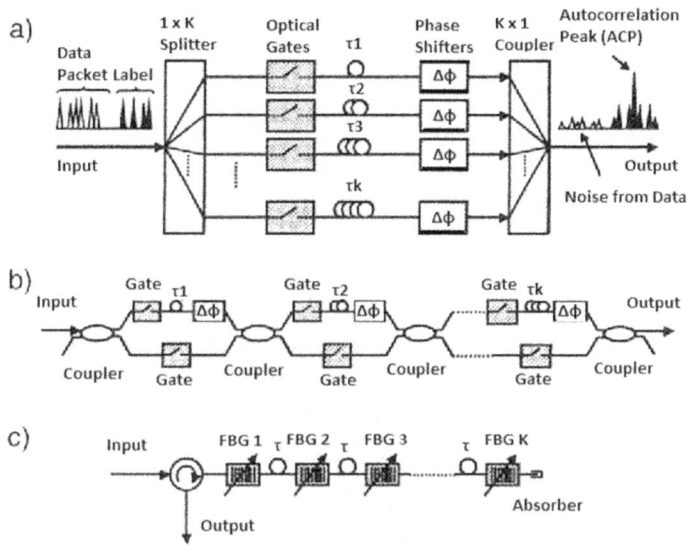

Fig. 4.1. All optical label code correlators.

4.2.5. Hybrid 2-D and 3-D Label Swapping Techniques

Nowadays, as number of network nodes tremendously scales up, and much more labels than ever before, are required at the same time, hybrid label swapping schemes, that combine code and wavelength label swapping techniques, allowing the encoded label to be used more than once, carried on different wavelengths may offer a solution to this label scalability problem. This is called hybrid WDM/OC or 2-D label swapping, in which both the wavelength and the optical code (λ, C) are the critical parameters to fully characterize the identifying label.

In [35], a remarkable WDM/OC testbed is presented for 40 Gbps data rates and 1.12 Gbps direct sequence optical code division multiplexed (DS-OCDM) labels. The router layout is fully equipped by all optical OCDM devices, such as fiber Bragg grating correlators and MZI interferometers, acting as optical gates, and dispersion compensating fibers (DCFs) acting as delay lines for synchronization purposes. Each OCDM module is held responsible for producing L different optical codes that correspond to L different node addresses, and can be assigned to more than one of the available WDM wavelengths. The incoming packets are split over an array of optical gates, and routed towards a predetermined output. The label extraction can be done either by FBGs or an AWG demultiplexer.

3-D labeling is an extension of 2-D labeling, which is simply achieved by adding polarization as a third parameter, into predefined wavelength and optical code (λ, C) critical parameters of the original 2-D label scheme [36]. The 3-D labeling system fabrication, easily comes out of the original 2-D labeling infrastructure, simply by addition of an optical coupler and polarization controllers (PCs), so as to split the original signal into two branches, at two orthogonal polarization states respectively (X-POL and Y-POL).

101

The experimental demonstration of a 4-code × 2-wavelength × 2-polarization increases the code labels up to more than 1.000 labels.

4.2.6. SAC Labeling

Spectral amplitude codes with optical label stacking is another sub category of OC labeling, based on the encoding of incoherent broadband light sources in the frequency domain. As far as concerns SAC labeling, the wavelengths (single or a band of wavelengths) that used to define the payload and the labels do not overlap with each other. Hence, SAC labeling inherently separates the processing between the label and payload, thus accelerating the intermediate node forwarding process. There are two techniques of creating SAC optical packets, one with separable SAC labels and the other with SAC-encoded payloads. In the first technique as mentioned, the SAC label is a collection of spectral tones modulated at the packet rate and lasts for the entire packet time frame, and the payload is on a separate wavelength modulated at the data rate (Fig. 4.2 a). Hence, the payload and the label have different wavelengths and independent modulation rates, and the typical AOLS functions such as label extraction and label detection can be performed all optically, by simply using optical filters without affecting the payload at all. In the second technique, the payload data modulates a band of wavelengths that constitute the code of the SAC label, and the label now is implicit in payload bits (Fig. 4.2 b). One of the first SAC labeling experimental demonstrations was described in ref [37], and involved two SAC labels in the label stack over a fiber span of 80 km. SAC labels are generated by the encoding of incoherent multilaser broadband sources (BBS) in the frequency domain. In multilaser sources, the laser emission is spectrally formed as bins, placed at short distances. The BBS encoding and decoding which is identical with each other, is usually done via a bank of chirped FBG's and determine the spectral bins that SAC labels consist of. Optical filters, couplers and single or balanced photodetectors are also required in order to complete the AOLS process.

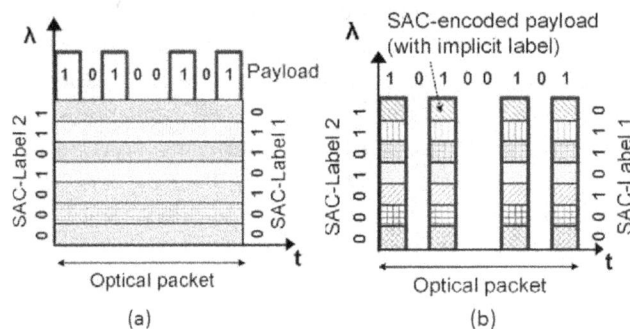

Fig. 4.2. (a) Separable payload SAC labels, (b) SAC encoded payload with implicit label.

However, a drawback of SAC labeling technique, is the appearance of the so called phase-induced intensity noise (PIIN) resulting from the multiple label interference (MLI), during encoding of broadband light source, due to its incoherent origin. PIIN produces a frequent change in intensity of the broadband light source thus degrading system performance. It has been found that MLI can be effectively eliminated by the use of a balanced photodetector [38].

Composite M-labels encoded with small dimension AWGs with small number of input-output ports is a promising solution for decreasing hardware complexity requirements [39]. Composite M-labels, are constructed by the combination of two single M-sequences originating from two separate AWGs into a composite code implementing a modulo-2 operation. Hence, long code length labels can be implemented without large-dimension AWG, and composite M-labels have been proved more flexible and robust for AOLS applications, compared with conventional M-sequence labels.

Advanced coherent modulation techniques such as quadrature amplitude modulation (QAM) or differential quadrature phase shift keying (DQPSK) combined with polarization division multiplexing (PDM), may be used in order to improve the data speed rates. In [40] a simulation setup of a PDM-DQPSK modulated SAC labeling scheme for a 100 GHz channel spacing WDM system at 112×4 Gbps rates is presented.

4.3. System Description

A block diagram of the proposed scheme is shown in Fig. 4.3 where the starting node, the end node and the intermediate node are shown. Between successive nodes there is also the transmission module, consisting of a 50 km single mode fiber (SMF) dispersion compensated. Dispersion compensation is required for proper propagation of the IM modulated payload signal along the fiber, as well as for proper FSK label demodulation, due to the walk off effect, when the tone spacing is sufficiently big. The starting node includes the transmitter modules for payload and header and the unit where initially the two signals are combined together onto a common optical carrier, which is the FWM based DSF fiber module.

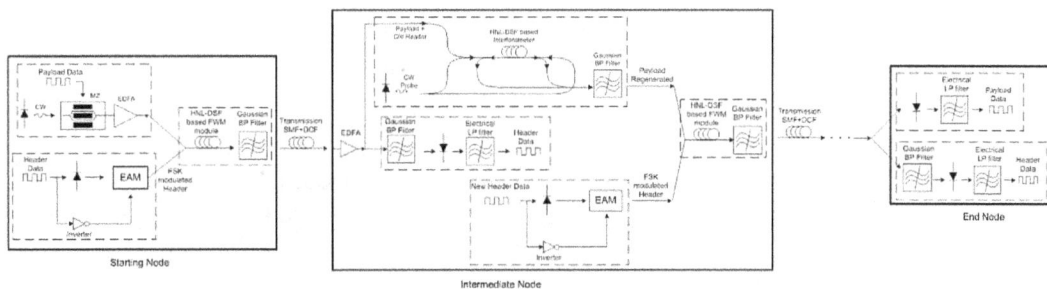

Fig. 4.3. Detailed block diagram of the proposed scheme.

IM 40 Gbps modulated packet payload is generated by externally modulating a Mach Zender (MZ) amplitude modulator at a low extinction ratio (around 3 dB). Header 2.5 Gbps information is generated by chirping through direct modulation the laser transmitter at a low modulation index, according to a typical FSK scheme. An already proposed FSK compensation scheme [41], has been applied on the FSK transmitter. According to this scheme, an external electroabsorbsion (EA) modulator, accepts the optical FSK data, while at the same time is driven with the inverse electrical data, thus removing the label intensity variations at the output of this transmitter (see Fig. 4.4) and minimizing the residual intensity modulation effect on the payload signal when they are coupled together. Both signals enter the DSF fiber after being amplified in such a way that payload is the pump and label is the signal according to a typical FWM scheme. A 500 m long Highly Nonlinear (HNL) DSF fiber, with a zero dispersion wavelength of 1554 nm is used. The nonlinear index coefficient of the fiber (γ) is 12 (1/W/km) and its dispersion coefficient β_2 is 0.1 ps^2/km.

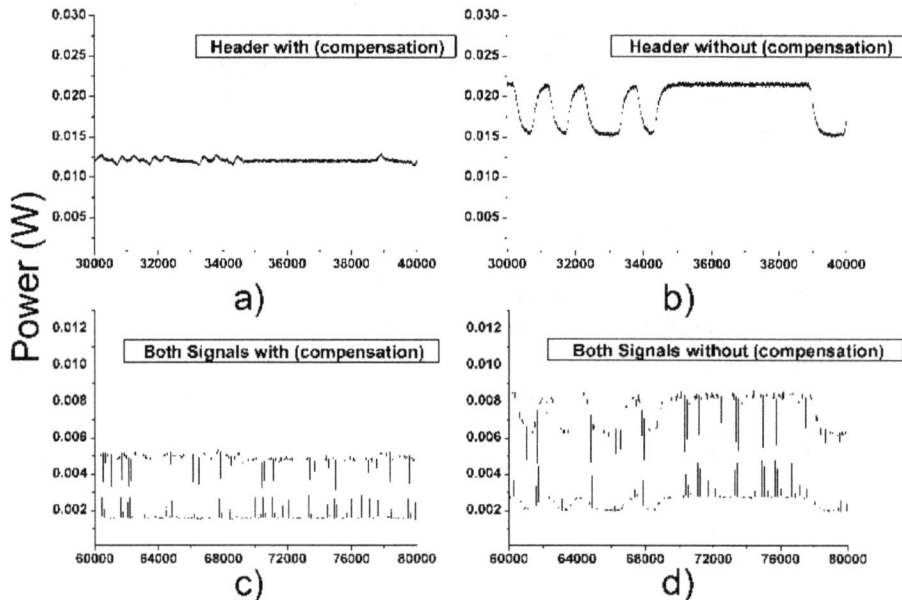

Fig. 4.4. FSK compensation. a) Compensated header; b) Uncompensated header; c) Payload after FWM with compensated header; d) Payload after FWM with uncompensated header.

In Fig. 4.5 individual header spectra and total spectra at the FWM output are shown. The signal after having propagating over a multispan route, according to label swapping mechanism, removes its final label information at the end node, so that the pure IP packet remains. The end node simply consists of individual receivers for header and payload demodulation respectively.

According to Fig. 4.3, the intermediate node consists of three units: the label extraction unit, the label removal and payload wavelength conversion unit and the label insertion unit. The label extraction unit is a typical optical FSK receiver and is composed by a Gaussian 10 GHz optical bandpass filter (BPF), a p-i-n (PIN) photodiode and the appropriate electrical lowpass filter (LPF). The same module also applies to the end node, for the final label extraction.

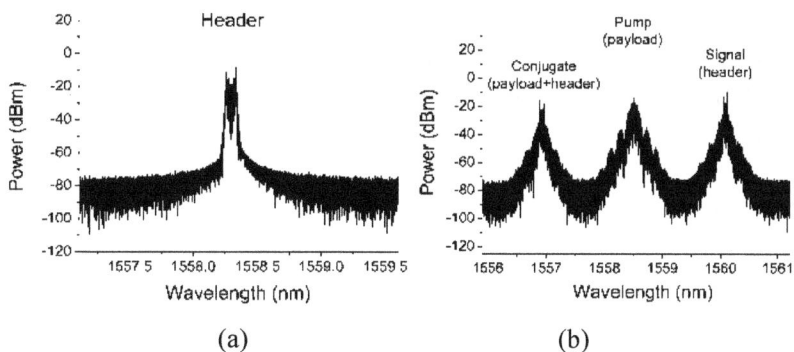

Fig. 4.5. a) Header spectrum; b) FWM spectrum.

The dominant element of the label removal and payload wavelength conversion and regeneration as already mentioned [15], is a bidirectional interferometric HNL DSF fiber. A 2.5 km long HNL DSF fiber, with a zero dispersion wavelength of 1550 nm is used. The effective core area of the fiber is 70 μm^2, the nonlinear index coefficient γ is 11 (1/W/km) and its dispersion coefficient and dispersion slope are 0.2 ps/nm/km and 0.08 $ps/nm^2/km$ respectively. The CW probe signal is fed into the interferometer via an input coupler and split into two portions. The combined IM modulated 40 Gbs NRZ payload signal with the 2.5 Gbs FSK modulated NRZ header, after being amplified by an erbium doped fiber amplifier (EDFA) is coupled into the DSF with the probe. Part of the probe, co-propagates with the drive signal experiencing a nonlinear phase modulation proportional to the power. XPM and fiber nonlinearity are enhanced, due to the combined action of the two nonlinear Kerr and Raman effects respectively. The remaining part of the probe enters the fiber from the opposite side and counter-propagates with respect to the drive beam. The nonlinear phase shift imposed on counter-propagating CW probe, by the drive signal, is considered by our simulation model static and constant, due to the short interaction time and thus can be neglected. In the literature [15], the counter-propagating phase shift is also overcome by means of a piezo-ceramic stretcher. Optical interference between the two probe fractions is obtained at the output coupler, so phase modulation is transformed into intensity modulation and thus wavelength conversion has been achieved. Due to the XPM process, all the coherently encoded FSK label information is not transferred to the output of the coupler and thus header removal has also been achieved. A new label can be inserted into the bare payload by means of a label insertion unit. It is actually an FSK transmitter combined with an FSK compensation scheme, as in the starting node. As the payload is combined with the new header via FWM based DSF fiber, it propagates over the next transmission span, where it meets the next intermediate node for a new label swapping. Simulations have proven the feasibility of the method for five successive spans.

4.4. DSF Module Theory and Investigation

As already mentioned, the two basic functions in an AOLS high bitrate (40 Gbps for the payload and 2.5 Gbps for the header) signaling scheme are the combination of FSK

105

modulated header with the IM modulated payload via FWM and the label (header) erasure and payload 2R regeneration via XPM. In order to perform these functions in such a high bitrate system, DSF fiber is used as the dominant element for both FWM and XPM, due to its inherent ultrafast nonlinear response.

4.4.1. FWM Fiber Module

Concerning the FWM fiber module the propagation of the IM/FSK signal in the fiber, is governed by the nonlinear Schrodinger equation:

$$\frac{\partial A}{\partial z} = -\frac{i}{2} \beta_{2(\lambda_i)} \frac{\partial^2 A}{\partial t^2} + \frac{1}{6} \beta_{3(\lambda_i)} \frac{\partial^3 A}{\partial t^3} + i\gamma |A|^2 A - \frac{\alpha}{2} A \qquad (4.1)$$

The first two terms account for the dispersion of the fiber, the third for its non-linearity, and the last one for its losses. A is the propagating optical field, a is the loss of the fiber (in dB/km) and γ is the nonlinear coefficient (in 1/W/km) defined by:

$$\gamma = \frac{n_2 \cdot \omega}{c \cdot A_{EFF}} = \frac{2\pi n_2}{A_{EFF}\lambda} , \qquad (4.2)$$

where n_2 is the nonlinear index coefficient, A_{EFF} is the effective fiber core area, and λ is the wavelength of the optical carrier of the signal. The term β_2 (ps^2/km) and β_3 (ps^3/km), account for the first and second order group velocity dispersion (GVD), and they are derived by:

$$\beta_2 = -\frac{\lambda^2 D}{2\pi c} , \qquad (4.3)$$

$$\beta_3 = \frac{dD}{d\lambda} = \frac{\lambda^3 (3\lambda - \lambda_0)}{(2\pi c)^2} , \qquad (4.4)$$

$$D = \frac{dD}{d\lambda}(\lambda - \lambda_0), \qquad (4.5)$$

where D is the dispersion parameter (ps/nm/km), c is the speed of light, dD/dλ the dispersion slope (ps/nm^2/km), and λ_0 the zero dispersion wavelength. Typical values for the β_2 and β_3 and dispersion slope dD/dλ, may sort fibers into classic SMF, DSF or HNL-DSF fibers.

In order to evaluate the behavior of the aforementioned module, and to clarify the importance of some critical parameters that characterize it, a thorough investigation via numerical simulations has been conducted. At first one of the crucial parameters of the HNL DSF fiber module is its length. In Fig. 4.6 we illustrate the performance of the module using Q-factor values of payload and header. As shown in this figure, when the length of the FWM fiber module is increasing its performance is getting better at first, but when the length of the device exceeds 500 m then its performance deteriorates. This is attributed to the increased influence of pump depletion effect to our signal, despite the FWM efficiency increase.

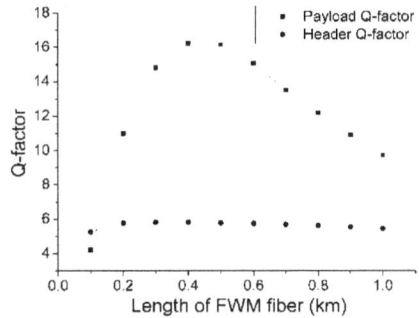

Fig. 4.6. Payload Q-factor vs length of HNL DSF fiber used as FWM module.

Another crucial parameter that determines the effectiveness of the FWM is the detuning of the signals. Knowing that we use a high bitrate system, the detuning of pump and signal should be adequately large in order to avoid interference between adjacent channels. This can be clearly illustrated at Fig. 4.7, where we observe that the Q-factor values of the payload (our high bitrate signal) is getting better and better as detuning increases from 100 GHz to 200 GHz. When the spacing between the channels increases above 200 GHz the quality of the payload remains almost constant at adequately good levels and the best performance of the signal is achieved when the detuning is around the value of 400 GHz. Finally when the detuning between pump and signal increases to larger spacings (around 1 THz or even greater values) the efficiency of the FWM and the quality of the signals deteriorate.

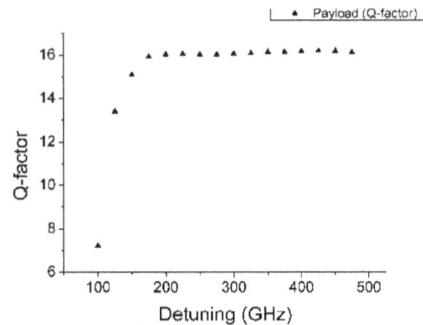

Fig. 4.7. Regenerated payload Q-factor vs signals detuning.

4.4.2. XPM Fiber Module

Concerning the interferometric wavelength converter model and due to its resemblance with NOLMs, the label swapping module requires a bidirectional HNL-DSF fiber part, where the CW signal, after splitting into two arms, is co-propagating and

107

counter-propagating with the NRZ payload signal, over forward and backward inputs of the fiber. The NRZ drive signal imposes a nonlinear phase shift on the co-propagating CW probe signal due to Kerr induced XPM properties. On the other hand, when counter-propagating the drive signal also imposes a nonlinear phase shift on the CW counter-propagating signal.

The expressions for the two nonlinear phase shifts are alike, except for a plus sign in the latter case, taking into account the different directions of probe and drive signals. At high bit rates and when the nonlinear medium length is long, the bit time is much lower than the fiber transit time and the phase shift for the counter-propagating case is constant but not minimum. This phase shift is overcome by means of a piezo-ceramic stretcher [15]. In our simulation model, dynamic effects induced by interaction of the counter-propagating waves are neglected. Signals walk through each other with an extremely high velocity, therefore the interaction time between counter-propagating pulses is very short and will lead to some constant phase shift, so such static effects are neglected as well.

Moreover, in the bidirectional DSF fiber module, in addition to Kerr effect nonlinearity, Raman nonlinearity is taken into account too and thus phase response will be a joint action of the above two phenomena. Raman gain in combination with the bidirectional modes greatly enhances the fiber nonlinearity and XPM, by changing the power evolution along the fiber, thus giving better performances.

The above same equations are also valid for the bidirectional module, except the nonlinear coefficient that is now attributed to both Kerr and Raman effects. The Raman gain experienced by the signal at f_i frequency due to stimulated Raman scattering of the wave with frequency f_k, is determined by an imaginary part of the Fourier transformation, of the time domain response function $h(t)$:

$$g(f_k, f_i) = 2 * \rho * \gamma(f_k, f_i) \text{Im}[H(f_k - f_i)],$$
(4.6)

while its real part determines the Raman contribution to the nonlinear coefficient:

$$\gamma_R = \rho * \gamma(f_k, f_i) \text{Re}[H(f_k - f_i)]$$
(4.7)

An approximated response function is used:

$$h(t) = \frac{\tau_1^2 + \tau_2^2}{\tau_1 * \tau_2^2} \exp\left(-\frac{t}{\tau_2}\right) * \sin\left(\frac{t}{\tau_1}\right),$$
(4.8)

where τ_1, τ_2 are the characteristic times defining the frequency offset of the Raman peak and FWHM width of the Raman gain peak.

Finally, there is an interaction length, dependent on bit rate and dispersion magnitude, after which, probe and drive waveforms are separated by more than a bit time due to dispersion distortion. Propagation for distance longer than the interaction length, results in severe modulation distortion. Operating lengths in our simulations are well below the interaction length limits.

In Fig. 4.8, the Q-factor of the regenerated payload is illustrated versus the length of the HNL DSF bidirectional fiber acting as XPM module. When the interaction length of the fiber increases the quality of the signal is getting better at first but when its length exceeds 2.5 km the signal deteriorates due to the increased influence that nonlinear Kerr and Raman effects have on it.

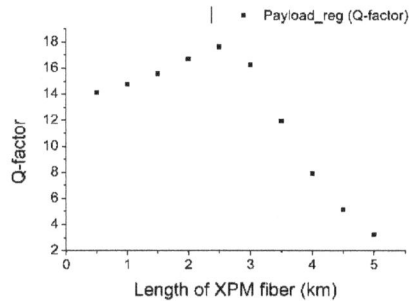

Fig. 4.8. Regenerated payload Q-factor vs the length of fiber XPM module.

Another very important parameter that determines the quality of the regenerated signal is its power level, when it enters to the XPM fiber module. The power level of the input signal must be high enough, in order to stimulate Kerr effect in the HNL DSF fiber. In Fig. 4.9, it is illustrated that the power of the input signal must be more than 10 dBm and the optimum performance is achieved with input signal power of 16 dBm.

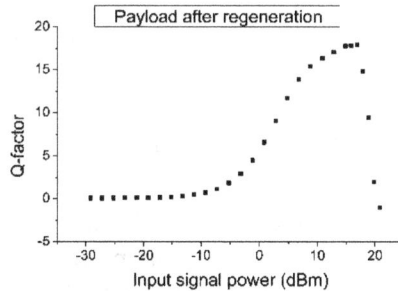

Fig. 4.9. Regenerated performance vs input power level.

4.5. Numerical Results and Discussion

Due to the ultrafast nonlinear response of the fiber, and being the only dominant element of our system, the all fiber based system is applicable at high bit rates. In this section, simulation results have been carried out successfully for a 40 Gbs NRZ IM modulated payload signal with a 2.5 Gbs NRZ FSK modulated header, onto a common optical carrier.

The conjugate signal after propagating 50 km dispersion compensated SMF, bypasses the label swapping node for header removal and payload wavelength conversion and the same procedure is repeated for four spans. Many crucial operating parameters of the system are examined via extensive simulations, such as the extinction ratio of the IM modulated signal, the modulation index of the FSK label which varies the FSK tone spacing and the existing trade off of the above two parameters, and the dynamic range of the DSF based FWM module. Moreover, the system's limitations are tested according to the propagation distance without intermediate label swapping processing, and with intermediate label swapping nodes along with their successive propagation spans.

The IM modulated payload extinction ratio tradeoff is one of the most crucial parameters for the orthogonal IM/FSK encoding method. High extinction ratio ensures high performance for the payload, while at the same time low extinction ratio continues to support optical FSK even during the zeros of the payload. Fig. 4.10 shows the 40 Gbs payload and the 2.5 Gbs header BER performances, for various extinction ratio values. As shown, payload performance is getting better as extinction ratio increases while, header performance is decreasing, as expected. Its performance is very poor for extinction ratios higher than 3.5 dB, emphasizing the low extinction required for the implementation of IM/FSK and any other orthogonal modulation scheme.

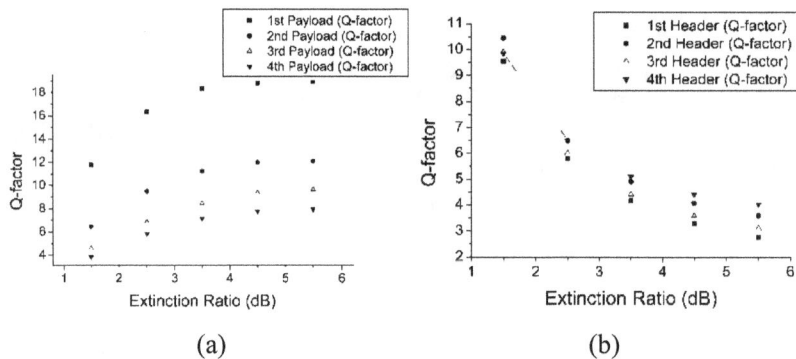

Fig. 4.10. Q-factor values vs Extinction ratio a) payload, b) header.

Fig. 4.10 a-b show IM payload and FSK header performance for various spectral distances between the two FSK tones. It is obvious that the more we modulate the FSK label, the greater frequency deviations we achieve and the better Q factor results we get at the FSK receiver, since the two tone spectra would be more easily separated and filtered. Simulation measurements have been carried out for 10, 15, 20, 25 and 30 GHz FSK tone deviations. Fig. 4.11 a shows that frequency spacing of 20 GHz, which is the operational value of our system, is the optimum choice. For larger FSK spacings the header BER performance is only slightly improved. On the contrary, the more we modulate the FSK label, the more residual intensity modulation affects IM payload, thus deteriorating its performance. Indeed, as shown in Fig. 4.11 b, for FSK spacings greater than 20 GHz, the payload performance is heavily degraded, despite the FSK compensation scheme, at the FSK transmitter.

Another benefit of the large frequency spacing is related to the chirp characteristics of the total signal (IM combined with FSK). Chirping generally results in a broadening of the signal spectrum, so if this broadening is too large, the FSK modulated label will be influenced. Luckily, there's no degradation of the FSK signal as long as chirp falls within the bandwidth of the filters used for direct detection of the FSK tone [42], which is our case. Indeed, Fig. 4.12 shows chirp measurements at the output of the wavelength converter module, for various ER values of the 40 Gbs IM signal. As shown, chirp spectral broadening at the ER operating area of 3 dB, is around 4 GHz zero to peak, and 8 GHz peak to peak value respectively, which are much smaller than the operating FSK tone spacing of our system and within the full width half maximum (FWHM) bandwidth of the Gaussian optical BPF, used for direct detection of the tone at the FSK receiver.

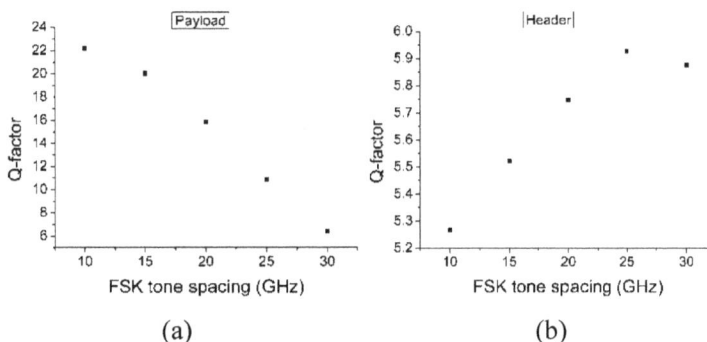

(a)   (b)

Fig. 4.11. Q-factor evolution with FSK tone spacing a) payload, b) header.

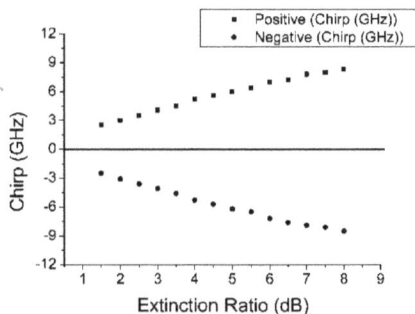

Fig. 4.12. Chirp measurement at the output of the wavelength converter vs. extinction ratio.

One of the most important operating parameters to characterize, is the dynamic range of DSF based FWM module. Dependent on the propagation distance, signals do not always have the same power level, when they reach the FWM module. Fig. 4.13 shows IM signal's performance, for various average pump power levels at the input of the FWM module and having propagated over a 50 km DCF fiber span afterwards. As shown, the optimum pump power value, for optimum performance, is around 20 dBm. It is worth noticing that the performance of the signal remains at very high levels, for a wide pump power range.

Simulation measurements have also been carried out, for testing the system's limitations, as far as propagation distance and cascadability of intermediate label swapping nodes followed by their corresponding propagation spans. Fig. 4.14 shows system performance for a range of propagation distances from 30 km to 70 km DCF fiber, before the signal meets the label swapping module. As shown, IM payload performance is adequately good even after 50 km propagation distance, while the header preserves an almost constant, moderate performance, for a variety of distance ranges. It is worth mentioning that, apart from an EDFA placed before the transmission module, the system consists of only passive fiber parts, and yet it preserves a satisfactory behavior for the payload signal, after a considerable propagation distance.

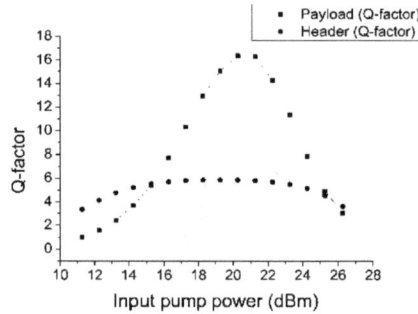

Fig. 4.13. Payload and header Q-factor vs. input pump power at the FWM module.

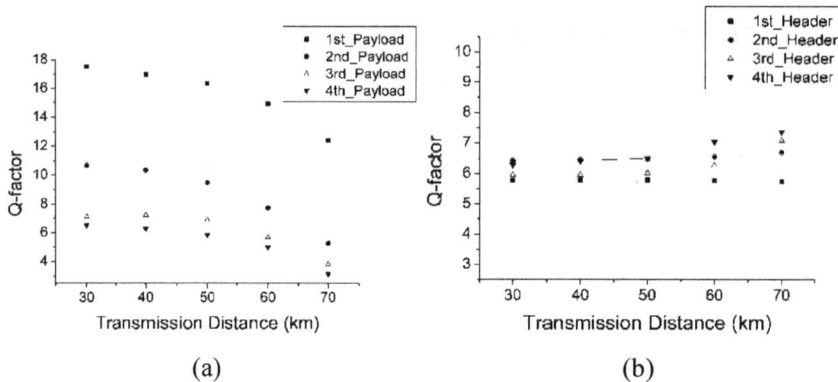

Fig. 4.14. Q-factor values vs. propagation distance a) payload, b) header.

Moreover, as shown in Fig. 4.15, the IM payload signal retains as well, a satisfactory performance, after having bypassed over four intermediate label swapping nodes, followed by their corresponding, 40, 50 and 60 km each, DCF fiber spans. Q factor performance of payload is extremely high at first spans and it gradually deteriorates moving forward to the edge node.

Actually it was expected, due to the lack of active regenerative modules, such as SOAs and due to the gradual decrease of the extinction ratio caused by the bidirectional fiber used as an XPM wavelength converter and label removal module as shown in Fig. 4.16.

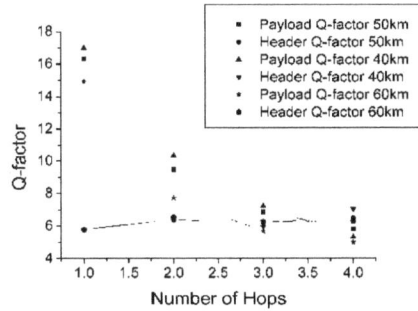

Fig. 4.15. Payload and header Q-factor evolution for 4 transmission hops and for various transmission distances.

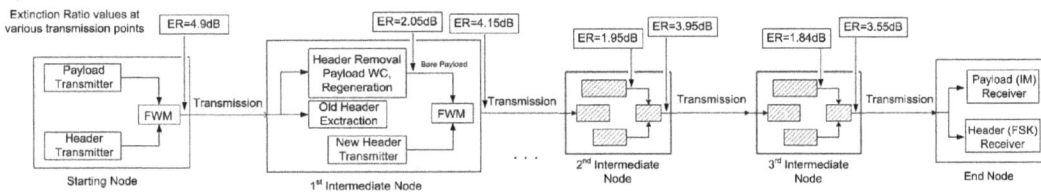

Fig. 4.16. Extinction ratio evolution at various transmission points.

Let us also, not forget that, apart from a few EDFAs, the system is characterized as an all fiber passive configuration, and the bidirectional HNL DSF based interferometric intermediate node does not regenerate the signal, but only converts its wavelength carrier and removes the old label. Moreover, the FSK modulated header performance is slightly decreasing, due to the IM signal's extinction ratio corresponding decrease, across the fiber spans, as shown in Fig. 4.16. Finally, Fig. 4.17 shows eye diagrams of the IM NRZ signal, before its entrance at the first label swapping node, and at the end of the fourth span, which is actually the end of the network route. The obvious eye closure at the end of the route imposes a limit for the system, as concerns the number of intermediate label swapping nodes and their successive propagation spans.

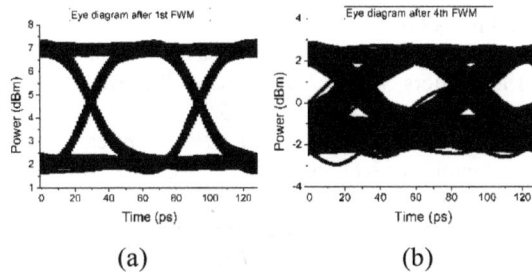

(a) (b)

Fig. 4.17. Eye diagram of the transmitted signal a) Before 1st transmission;
b) Before 4th transmission.

4.6. Conclusions

This chapter, attempts a short survey, consisting not only on conventional label swapping technique updates but also on references on new trends and solutions that overcome major label swapping limitations related to speed and capacity matters. Modern labeling techniques such as hybrid 2-D or 3-D labeling schemes, and SAC, with advanced coherent modulation techniques, seem to give adequately an effective solution to ultrahigh link capacities and packet switching speeds of scalable, big data interconnected DCs and HPC systems. Moreover, an all fiber orthogonal IM/FSK coding technique for AOLS applications is presented here. Apart from a few EDFAs, used for amplification of the signal prior to each propagation span all the system configuration is totally based on fiber. Both, the FWM module for the combination of IM modulated payload and FSK modulated header onto common optical carrier, as well as the interferometric wavelength converter module for the header removal and wavelength conversion are HNL DSF fiber based. For the first time, NRZ payload and header pulses are involved into the AOLS processing, at high 40 Gbs bit streams for the signal, due to the ultrafast fiber response time. Simulation results have proven the feasibility of the method and suggest a promising solution for high bitrate AOLS applications.

References

[1]. A. Viswanathan, N. Feldman, Z. Wang, R. Callon, Evolution of multiprotocol label switching, *IEEE Commun. Mag*, Vol. 36, Issue 5, May 1998, pp. 165-173.

[2]. A. Banerjee, et al., Generalized multiprotocol label switching: An overview of routing and management enhancements, *IEEE Commun. Mag*, Vol. 39, Issue 1, Jan 2001, pp. 144-150.

[3]. Z. Zhu, V. J. Hernandez, M. Y. Jeon, J. Cao, Z. Pan, S. J. Ben Yoo, RF photonics signal processing in subcarrier multiplexed optical label switching communication systems, *IEEE J. Lightwave Technol.* Vol. 21, Issue 12, December 2003, pp. 3155-3166.

[4]. D. J. Blumenthal, B. E. Olsson, G. Rossi, T. E. Dimmick, L. Rau, M. Masanovic, O. Lavrova, R. Doshi, O. Jerphagnon, J. E. Bowers, V. Kaman, L. A. Coldren, J. Barton, All-optical label swapping networks and technologies, *IEEE J. Lightwave Technol.*, Vol. 18, Issue 12, December 2000, pp. 2058-2075.

[5]. K. Vlachos, et al., An optical IM/FSK coding technique for the implementation of a label controlled arrayed waveguide packet router, *IEEE J. Lightwave Technol.*, Vol. 21, Issue 11, November 2003, pp. 2617-2628.

[6]. E. N. Lallas, N. Skarmoutsos, D. Syvridis, An optical FSK based label coding technique for the realization of the all optical label swapping, *IEEE Photon. Technol. Lett.*, Vol. 14, Issue 10, October 2002, pp. 1472-1474.

[7]. C. Joergensen, et al., All optical wavelength conversion at bit rates above 10Gb/s using semiconductor optical amplifiers, *IEEE J. Lightwave Technol.*, Vol. 3, Issue 5, Oct.ober1997, pp. 1168-1180.

[8]. H. Simos, A. Bogris, D. Syvridis, Investigation of a 2R all optical regenerator based on four wave mixing in a semiconductor optical amplifier, *IEEE J. Lightwave Technol.*, Vol. 22, Issue 2, January 2004, pp. 1-9.

[9]. A. Bogris, D. Syvridis, Regenerative properties of a pump modulated four wave mixing scheme in dispersion shifted fibers, *IEEE J. Lightwave Technol.*, Vol. 21, Issue 9, September 2003, pp. 1892-1902.

[10]. J. Yu, P. Jeppesen, 80 Gb/s wavelength conversion based on cross phase modulation in high nonlinearity dispersion shifted fiber and optical filtering, *IEEE Photon. Technol. Lett.*, Vol. 13, Issue 8, August 2001, pp. 833-835.

[11]. J. Yu, P. Jeppesen, Simultaneous all optical demultiplexing and regeneration based on self phase and cross phase modulation in a dispersion shifted fiber, *IEEE J. Lightwave Technol.*, Vol. 19, Issue 7, July 2001, pp. 941-949.

[12]. P. Ohlen, B. E. Olsson, D. J. Blumenthal, All optical header erasure and penalty free rewriting in a fiber based high speed wavelength converter, *IEEE Photon. Technol. Lett.*, Vol. 12, Issue 6, June 2000, pp. 663-665.

[13]. C. Kolleck, U. Hempelmann, All optical wavelength conversion of NRZ and RZ signals using a nonlinear optical loop mirror, *IEEE J. Lightwave Technol.*, Vol. 15, Issue 10, October 1997, pp. 1906-1913.

[14]. J. Yu, X. Zheng, C. Peucheret, A. Clausen, H. Poulsen, P. Jeppesen, All optical wavelength conversion of short pulses and NRZ signals based on a nonlinear optical loop mirror, *IEEE J. Lightwave Technol.*, Vol. 18, Issue 7, July 2000, pp. 1007-1017.

[15]. P. Boffi, L. Marazzi, M. Martinelli, A novel interferometric wavelength converter, *IEEE Photon. Technol. Lett.*, Vol. 11, Issue 11, November 1999, pp. 1393-1395.

[16]. C. Guillemot, et al., Transparent optical packet switching: The European ACTS KEOPS project approach, *IEEE J. Lightwave Technol.*, Vol. 16, Issue 12, December 1998, pp. 2117-2134.

[17]. A. Okada, All-optical packet routing in AWG-based wavelength routing networks using an out-of-band optical label, in *Proceedings of the Optical Fiber Communication Conference and Exhibition (OFC'02)*, Washington, DC, 2002, pp. 213-215.

[18]. C. Skoufis, S. Sygletos, N. Leligou, C. Matrakidis, I. Pountourakis, A. Stavdas, Data-centric networking using multiwavelength headers/labels in Packet-over-WDM networks: A comparative study, *IEEE J. Lightwave Technol.*, Vol. 21, Issue 10, October 2003, pp. 2110-2122.

[19]. Z. Zhu, V. J. Hernandez, M. Y. Jeon, J. Cao, Z. Pan, S. J. Ben Yoo, RF photonics signal processing in subcarrier multiplexed optical-label switching communication systems, *IEEE J. Lightwave Technol.*, Vol. 21, Issue 12, December 2003, pp. 3155-3166.

[20]. K. Fouli, M. Maier, OCDMA and optical coding: Principles, applications, and challenges, *IEEE Commun. Mag.*, Vol. 45, Issue 8, August 2007, pp. 27-34.

[21]. S. Huang, K. Baba, M. Murata, K. Kitayama, Variable bandwidth optical paths: Comparison between optical code labelled path and OCDM path, *IEEE J. Lightwave Technol.*, Vol. 24, Issue 10, October 2006, pp. 3563-3573.

[22]. D. Apostolopoulos, D. Petrantonakis, O. Zouraraki, E. Kehayas, N. Pleros, H. Avramopoulos, All-optical label/payload separation at 40 Gb/s, *IEEE Photon. Technol. Lett.*, Vol. 18, Issue 19, October 2006, pp. 2023-2025.

[23]. B. E. Olsson, P. Öhlén, L. Rau, G. Rossi, O. Jerphagnon, R. Doshi, D. S. Humphries, D. J. Blumenthal, V. Kaman, J. E. Bowers, Wavelength routing of 40 Gbit/s packets with 2. 5 Gbit/s header erasure/rewriting sing all-fiber wavelength converter, *Electron. Lett.*, Vol. 31, Issue 4, February 2000, pp. 345-347.

[24]. W. Hung, C. K. Chan, L. K. Chen, F. Tong, A bit-serial optical packet label-swapping scheme using DPSK encoded labels, *IEEE Photon. Technol. Lett.*, Vol. 15, Issue 11, November 2003, pp. 1630-1632.

[25]. D. Gurkan, S. Kumar, A. E. Willner, K. R. Parameswaran, M. M. Fejer, Simultaneous label swapping and wavelength conversion of multiple independent WDM channels in an all-optical MPLS network using PPLN waveguides as wavelength converters, *IEEE J. Lightwave Technol.*, Vol. 21, Issue 11, November 2003, pp. 2739-2745.

[26]. D. J. Blumenthal, A. Carena, L. Rau, V. Curri, S. Humphries, All-optical label swapping with wavelength conversion for WDM-IP networks with subcarrier multiplexed addressing, *IEEE Photon. Technol. Lett.*, Vol. 11, Issue 11, November 1999, pp. 1497-1499.

[27]. A. Carena, M. D. Vaughn, R. Gaudino, M. Shell, D. J. Blumenthal, OPERA: An optical packet experimental routing architecture with label swapping capability, *IEEE J. Lightwave Technol.*, Vol. 16, Issue 12, December 1998, pp. 2135-12145.

[28]. R. Gaudino, M. Shell, M. Len, G. Desa, D. J. Blumenthal, MOSAIC: A multiwavelength optical subcarrier multiplexed controlled network, *IEEE J. Select. Areas Commun.*, Vol. 16, Issue 7, September 1998, pp. 1270-1285.

[29]. B. Meagher, et al., Design and implementation of ultra-low latency optical label switching for packet-switched WDM networks, *IEEE J. Lightwave Technol.*, Vol. 18, Issue 12, December 2000, pp. 1978-1987.

[30]. H. J. Lee, S. J. B Yoo, V. K. Tsui, S. K. H. Fong, A simple all-optical label detection and swapping technique incorporating a Fiber Bragg Grating Filter, *IEEE Photon. Technol. Lett.*, Vol. 13, Issue 6, June 2001, pp. 635-637.

[31]. N. Wada, H. Harai, W. Chujo, F. Kubota, Multi-hop, 40 Gbit/s variable length photonic packet routing based on multiwavelength label switching, waveband routing, and label swapping, in *Proceedings of the Optical Fiber Communication Conference and Exhibition (OFC'02)*, Washington, DC, 2002, pp. 216-217.

[32]. E. Lallas, N. Skarmoutsos, D. Syvridis, Coherent encoding of optical FSK header for all optical label swapping systems, *IEEE J. Lightwave Technol.*, Vol. 23, Issue 3, March 2005, pp. 1199-1209.

[33]. C. W. Chow, C. S. Wong, H. K. Tsang, All-optical ASK/DPSK label-swapping and buffering using Fabry-Perot laser diodes, *IEEE Journal of Sel. Topics in Quant. Electronics*, Vol. 10, Issue 2, March/April 2004, pp. 363-370.

[34]. S. Aleksic, V. Krajinovic, Comparison of optical code correlators for all-optical MPLS networks, in *Proceedings of the European Conference on Optical Communication (ECOC'02)*, Copenhagen, Denmark, Sept. 2002.

[35]. H. Brahmi, G. Giannoulis, M. Menif, V. Katopodis, D. Kalavrouziotis, C. Kouloumentas, P. Groumas, G. Kanakis, C. Stamatiadis, H. Avramopoulos, D. Erasme, On the fly all-optical packet switching based on hybrid WDM/OCDMA labeling scheme, *Optics Communications*, Vol. 312, February 2014, pp. 175-184.

[36]. R. Matsumoto, T. Kodama, K. Morita, N. Wada, K. Kitayama, Scalable two- and three-dimensional optical labels generated by 128-port encoder/decoder for optical packet switching, *Optics Express*, Vol. 23, Issue 20, October 2015, pp. 25747-25761.

[37]. P. Seddighian, S. Ayotte, J. B. Rosas-Fernández, J. Penon, L. A. Rusch, S. LaRochelle, Label stacking in photonic packet-switched networks with spectral amplitude code labels, *IEEE J. Lightwave Technol.*, Vol. 25, Issue 2, February 2007, pp. 463-471.

[38]. K.-S. Chen, C. -C. Yang, J. -F. Huang, Orthogonal stacked spectral coding labels for fast packets routing over optical MPLS network, *Journal of El. Science and Technology*, Vol. 13, Issue 2, June 2015, pp. 130-134.

[39]. S.-H. Meng, K. -S. Chen, J. -F. Huang, C. -C. Yang, Orthogonal stacked composite M-sequence labels for quick packet routing over optical MPLS network, in *Proceedings of the IEEE 28th Canadian Conference on Electrical and Computer Engineering (CCECE'15)*, Halifax, Canada, May 2015, pp. 908-913.

[40]. I. A. Aboagye, F. Chen, Y. Cao, Performance analysis of 112 Gb/s×4-channel WDM PDM-DQPSK optical label switching system with spectral amplitude code labels, *Photonic Sensors*, Vol. 7, Issue 1, March 2017, pp. 88-96.

[41]. J. Zhang, N. Pablo, V. Nielsen, C. Peucheret, P. Jeppesen, An optical FSK transmitter based on an integrated DFB laser EA modulator and its application in optical labeling, *IEEE Photon. Technol. Lett.*, Vol. 15, Issue 7, July 2003, pp. 984-986.

[42]. T. Monroy, et al., Performance of a SOA-MZI wavelength converter for label swapping using combined FSK/IM modulation format, *Opt. Fiber Technol.*, Vol. 10, Issue 1, January 2004, pp. 31-49.

Chapter 5
The Fourier Transform Relation between Dirac Bras $< \vec{k}| = FT < \vec{r}|$ and Wave Optics

Do Tan Si

5.1. Introduction

Consider the diffraction by a 3D object of a plane wave with wave vector $\vec{k}_0$ propagating in a medium with refractive index n_1. The diffracted wave may be expanded into an infinity of plane waves each of which having a wave vector $\vec{k}$ and propagating in a medium with refractive index n_2. We are interested in the problem of calculating the amplitudes i.e. the coefficients of expansion of these $\vec{k}$-components of the diffracted wave.

For this purpose, one may utilize the Huygens-Fresnel principle where it is postulated that every point of a wave-front is considered as a centre of a secondary spherical wavelet, that the wavelets interfere mutually and the wave-front at any later instant may be constructed by drawing the envelope of all of these [1, 2]. Applications of this principle for obtaining the diffraction amplitudes corresponding to geometrically simple 2D objects such as pinholes, rectangular and circular apertures, systems of equidistant parallel planes, multiple-slit configurations, etc. and spheres, are well known in optics' textbooks, for example in [3, 4]. But we think that it is hard by this principle to obtain them for 3D objects such as cubes, tetrahedra, prisms, ellipsoids, cylinders and intersections, reunions, convolutions, transforms of them.

Besides, consider the diffractions by an aperture opened on an opaque screen Oxy of waves emitting from a point source P_0 far enough from the screen and beginning from a certain instant of time t_0. One may utilize the Kirchhoff's diffraction theory based on the time-independent wave equation, or Helmholtz equation, together with some boundary

Do Tan Si
Association of Physicists, Ho Chi Minh City, Vietnam

conditions at and outside of the aperture to express the amplitude of the diffracted wave at a point P far enough from the screen under the form of an integral over the aperture of some function. If in addition the distances from P to a point Q of the aperture may be approximated by quadratic expressions in terms of the coordinates of Q then one speaks of Fresnel diffraction. If these distances are linear instead of being quadratic, one speaks of Fraunhofer diffraction [1]. In the latter case, the amplitude of the $\vec{k}$ -component of the diffracted wave is the Fourier transform with respect to $\Delta\vec{k}$ of the pupil function which is equal to some constant at points in the aperture and zero at points outside it [1, 2]. This property of Fraunhofer diffraction by a plane aperture is largely exploited in Fourier optics, noticeably in the domain of image processing techniques [3]. But it is not yet applicable for 3D objects because that it was not known how to write down the functions representing them and how to calculate the Fourier transforms of these functions.

Furthermore, utilizing the Maxwell equations in the electromagnetic field theory [1-3], where light is assimilated to the electric field vector, together with boundary conditions concerning the electric and magnetic fields at the interface of two media, one may resolve rigorously the problem of scattering of light in relatively simple cases. For example, in the framework of this theory one may prove again the two groups of Fresnel formula [4] obtained in the past by Fresnel on the basis of his elastic theory of light.

In this chapter, we propose a method for obtaining the formula giving the amplitude of the $\vec{k}$ -component of the diffracted waves in a Fraunhofer diffraction of a plane wave $\vec{k}_0$ by a 3D object and how to apply this formula to concrete 3D objects thank to the calculations of the Fourier transforms of their object functions. This method is based on the reasoning about the diffraction of a plane wave by a material point then generalized for 3D objects. It may also be obtained by the hypothesis [5] that in the Hilbert space of quantum mechanics the bra $<\vec{k}|$ is the Fourier transform of the bra $<\vec{r}|$ so that

$$< \vec{k}|\hat{r}|\vec{k}_0 > = FT < \vec{r}|\hat{r}|\vec{k}_0 > = i\nabla_k < \vec{k}|\vec{k}_0 > .$$ The formula obtained by these methods gives

the amplitudes of a $\vec{k}$ -component diffracted wave as the Fourier transform calculated for $\Delta\vec{k} \equiv \vec{k} - \vec{k}_0$ of the object function describing the 3D diffracting objects. For applications, we propose also a new manner for representing a 3D object not by equations but by a function composed from the Dirac delta function $\delta(x)$ and the Heaviside function $H(x)$ as so as sums, products and convolution products of them. Moreover, we find that in an oblique system of coordinates $\vec{u}_1, \vec{u}_2, \vec{u}_3$ it is more convenient to represent an object by a function of the form $f(\vec{u}_1.\vec{r}, \vec{u}_2.\vec{r}, \vec{u}_3.\vec{r})$ because the Fourier transform of the latter is simply obtainable from the use of reciprocal vectors.

Briefly, this work allows us to understand from the principe $<\vec{k}| = FT <\vec{r}|$ the principle of holography, calculate the diffraction of objects delimited by planes and spheres such as tetrahedra, oblique pyramids, cones, cylinders, ellipsoids and reunions, intersections, convolutions between them, etc. Extension for diffractions by lattices of identical objects are possible and would give interesting interference fringes.

Finally as the function representing a sphere and its Fourier transform are known we will try to calculate the deflection of light by the form of the sun for comparison with the deflections by Newton force and by General Relativity of Einstein.

5.2. Physics for Wave Optics

5.2.1. Law of Diffractions of Plane Waves by an Object in Classical Optics

In a system of axis $Oxyz$ let

$$A(\vec{r},t) = A_0 e^{-i\vec{k}_0 \vec{r} - i\omega t} \tag{5.1}$$

be the function which represents a plane wave propagating in the direction of the wave vector $\vec{k}_0 = \hbar^{-1}\vec{p}_0$, having amplitude A_0, period T, wavelength λ, circular frequency $\omega = 2\pi T^{-1}$. For simplicity hereinafter we call it a $\vec{k}_0$ wave.

5.2.1.1. Diffraction by a Material Point

In the interaction of a plane wave $\vec{k}_0$ with a material point situated at the position $\vec{r}_0$ the output wave is compose of an incommensurable infinity of plane waves $\vec{k}$ centered at $\vec{r}_0$ (Fig. 5.1). Moreover because of invariance with respect to translation of the point $\vec{r}_0$ the amplitude of each component wave $\vec{k}$ depends only on the scalar product $\vec{k}\vec{r}_0$ so that it may be represented within a multiplicative constant by the function

$$e^{-i\vec{k}\vec{r}_0} e^{-i\vec{k}_0(\vec{r}-\vec{r}_0)-i\omega t} = e^{-i(\vec{k}-\vec{k}_0)\vec{r}_0} e^{-i\vec{k}_0\vec{r}-i\omega t} \tag{5.2}$$

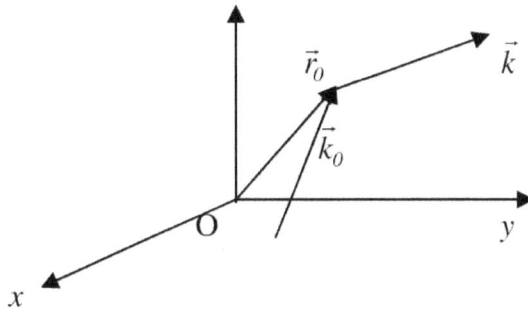

Fig. 5.1. Diffraction by a material point at $\vec{r}_0$.

Remarking that $e^{-i\vec{k}\vec{r}_0}$ is the Fourier transform [5] of the function $\delta(\vec{r} - \vec{r}_0)$ representing the point $\vec{r}_0$ we may conclude that: "In a diffraction of a plane wave $\vec{k}_0$ by a material point at $\vec{r}_0$ the diffracted wave is represented within a multiplicative constant by the function $\delta(\vec{r} - \vec{r}_0)e^{i\vec{k}_0(\vec{r}-\vec{r}_0)-i\omega t}$ and the amplitude of its $\vec{k}$ -component is the Fourier transform of $\delta(\vec{r} - \vec{r}_0)$ calculated for the deviation $\Delta\vec{k} = (\vec{k} - \vec{k}_0)$ of the wave vector".

5.2.1.2. Diffraction by Two Material Points

In an interaction of $\vec{k}_0$ with two material points represented by $\delta(\vec{r} - \vec{r}_1)$ and $\delta(\vec{r} - \vec{r}_2)$ the diffracted wave along $\vec{k}$ is represented by

$$(c(\vec{r}_1 - \vec{r}_0)e^{-i(\vec{k}-\vec{k}_0)\vec{r}_1} + c(\vec{r}_2 - \vec{r}_0)e^{-i(\vec{k}-\vec{k}_0)\vec{r}_2})e^{-i\vec{k}_0\vec{r}-i\omega t}, \qquad (5.3)$$

where $c(\vec{r}_i - \vec{r}_0)$ takes into account the effect of electromagnetic force depending on the distance between $\vec{r}_i$ and the point $\vec{r}$ where we observe $\vec{k}$.

In this chapter we propose to self limit on the study of the cases where the diffracting points and the observation point are far enough so that the $c(\vec{r}_i - \vec{r}_0)$ are quasi equal one another. With this condition which seemingly is equivalent to the Fraunhofer conditions, we see that the diffracted wave may be represented by the function

$$(\delta(\vec{r} - \vec{r}_1) + \delta(\vec{r} - \vec{r}_2))e^{-i\vec{k}_0(\vec{r}-\vec{r}_0)-i\omega t},$$

and the amplitude of $\vec{k}$ is

$$(e^{-i\Delta\vec{k}\vec{r}_1} + e^{-i\Delta\vec{k}\vec{r}_2})e^{-i\vec{k}_0\vec{r}-i\omega t} \qquad (5.4)$$

5.2.1.3. Diffraction by a 3D Object

Let D be a domain delimited by a 3D object composed of a number of points. We may represent it by the object function

$$f_D(\vec{r}) = 1 \ \text{for} \ \vec{r} \in D,$$

$$= 0 \ \text{for} \ \vec{r} \notin D \qquad (5.5)$$

Generalizing the case of two points to this case we see that the diffracted wave is

$$\sum_{\vec{r}_j \in D} \delta(\vec{r} - \vec{r}_j)\, e^{i\vec{k}_0(\vec{r}-\vec{r}_0)-i\omega t} = f_D(\vec{r})e^{i\vec{k}_0(\vec{r}-\vec{r}_0)-i\omega t},$$

and the amplitude of $\vec{k}$ is, denoting the Fourier transform of $f_D(\vec{r})$ by $F_D(\vec{k})$,

$$F_D(\Delta\vec{k})e^{i\vec{k}_0\vec{r}-i\omega t}, \tag{5.6}$$

so that after all we may propose The fundamental theorem on Fraunhofer diffractions: "In the diffraction of a plane wave $\vec{k}_0$ by a 3D object under Fraunhofer conditions the diffracted wave is composed of plane waves $\vec{k}$ each having amplitude equal to the Fourier transform of the object function $f_D(\vec{r})$ calculated for the deviation $\Delta\vec{k} = \vec{k} - \vec{k}_0$" (Fig. 5.2).

The above theorem may be proven in quantum mechanics' terminologies as discussed hereafter.

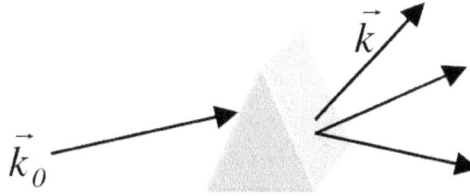

Fig. 5.2. Diffraction of a plane wave by a 3D object.

5.2.2. Quantum Mechanical Approach for Wave Optics

Let according to Dirac [6] $|x>$ and $|p>$ be the kets representing respectively a state with well-defined position x and momentum p of a particle; $\hat{X}$, $\hat{P}$ be the operators of measurements of x and p

$$\hat{X}|x> = x|x>, \tag{5.7}$$

$$\hat{P}|x> = p|p> \tag{5.8}$$

Let $<x|\alpha>$ be the position-representation, $<p|\alpha>$ the momentum-representation, $<E|\alpha>$ the energy- presentation, $<t|\alpha>$ the time-representation of a physical state $|\alpha>$. From the hypothesis that $<x|\alpha>, <E|\alpha>$ are respectively the Fourier transforms of $<p|\alpha>, <t|\alpha>$ we may deduced the principles in quantum mechanics including the Dirac fundamental commutation relation [8]. These results confirm that the hypothesis is right.

From now on we will utilize the vector $\vec{k}$ representing the wave vector related to the momentum vector by the relation

$$\vec{p} = \hbar\vec{k},\tag{5.9}$$

and readapt the Fourier transform so that $<\vec{k}|\alpha>$ is the transform of $<\vec{r}|\alpha>$

$$<\vec{k}|\alpha> = FT <\vec{r}|\alpha> = (2\pi)^{-\frac{3}{2}} \int_{R^3} e^{-i\vec{k}\vec{r}} <\vec{r}|\alpha> d\vec{r}\tag{5.10}$$

As $|\alpha>$ is arbitrary we may state

$$<\vec{k}| = FT <\vec{r}|\tag{5.11}$$

Jointed these relations with the following properties of the Fourier transform [5]

$$FT\, 1 = \sqrt{2\pi}\,\delta(k),\tag{5.12}$$

$$FTxf(x) = i\partial_k FTf(x) = i\partial_k F(k),\tag{5.13}$$

$$FTf(x) = FTf(x)1 = f(i\partial_k)FT\, 1 = \sqrt{2\pi}\, f(i\partial_k)\delta(k),\tag{5.14}$$

we get from the quantum mechanical relation

$$<x|\hat{X} = x<x|,\tag{5.15}$$

the formulae

$$<k|\hat{X}|k_0> = FT<x|\hat{X}|k_0> = FT\,x<x|k_0>,$$

$$= i\partial_k FT<x|k_0> = i\partial_k<k|k_0>,\tag{5.16}$$

$$<\vec{k}|\hat{r}|\vec{k}_0> = FT<\vec{r}|\hat{r}|\vec{k}_0> = i\nabla_k FT<\vec{r}|\vec{k}_0> = i\nabla_k<\vec{k}|\vec{k}_0>,\tag{5.17}$$

$$<\vec{k}|f(\hat{r})|\vec{k}_0> = f(i\nabla_k)\delta(\vec{k}-\vec{k}_0) = (2\pi)^{-\frac{3}{2}} F(\vec{k}-\vec{k}_0)\tag{5.18}$$

The above formula (5.18) reaffirms the fundamental theorem in wave optics (5.6) and shows that (5.11) may be considered as the origin of wave optics as we can see hereinafter.

5.2.3. Diffraction by an Aperture

In the particular case of an aperture situated in the $x, y - plane$ the object function $f_D(\vec{r})$ has the form $g(x, y)\delta(z)$ so that its Fourier transform has the form $G(k_x, k_y)u(k_z)$ where we have defined a new function $u(x) \equiv 1$. Thus the $\vec{k}$ diffracted wave doesn't depend on $(\varDelta \vec{k})_z$. We get then again the formula

$$F(\varDelta \vec{k}) = (2\pi)^{-\frac{3}{2}} u(\varDelta k_z) \int_{-\infty}^{+\infty}\int_{-\infty}^{+\infty} e^{-i(\varDelta k_x x + \varDelta k_y y)} g(x, y)dxdy, \qquad (5.19)$$

formerly proved in the framework of Kirchhoff's diffraction theory [1, 2].

Furthermore, if the incident wave $\vec{k}_0$ is perpendicular to the aperture we have

$$(\varDelta \vec{k})_x = k_x, (\varDelta \vec{k})_y = k_y, \qquad (5.20)$$

so that $F(\varDelta \vec{k})$ depends only on k_x, k_y. We get then the familiar theorem in Fourier optics [2, 3] saying that the amplitude of $\vec{k}$ (for $\vec{k}_0$ perpendicular to the aperture) is proportional to the Fourier transform of the aperture with respect to k_x, k_y. If $\vec{k}_0$ is not perpendicular to the aperture we must utilize the formula (5.6)

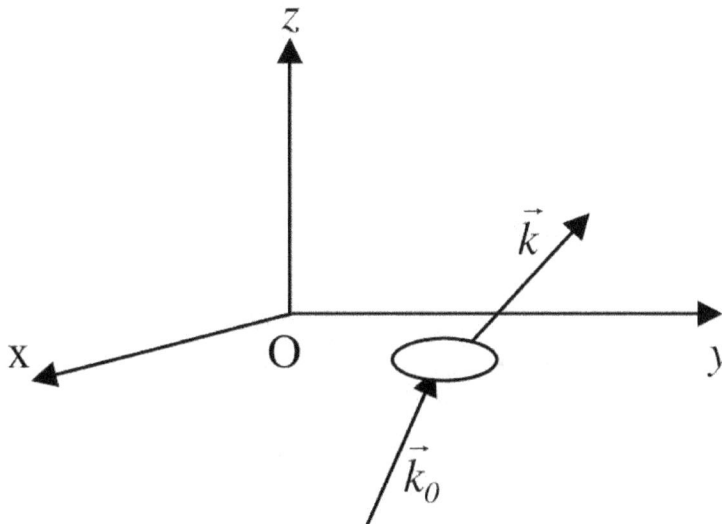

Fig. 5.3. Diffraction by an aperture.

5.3. Mathematics for Wave Optics

5.3.1. Useful Elementary Functions

5.3.1.1. The Unity Function and Object Function of Entire Space

The unity function $u(x)$ in this chapter is defined by

$$u(x) = \lim_{a \to 0} \frac{\sin ax}{ax} = 1 \qquad (5.19)$$

From this simple function we may represent the full space by the function $u(x)u(y)u(z)$ and because its Fourier transform is $(2\pi)^{3/2}\delta(x)\delta(y)\delta(z)$ we must have $\Delta\vec{k} = 0$ which confirms the principle that wave propagates straightly in a homogeneous space.

5.3.1.2. The Heaviside Function and Object Function of Subspaces Delimited by Planes

The Heaviside function [9] is defined by

$$H(x) = 1 \text{ for } x > 0, \ H(x) = 0 \text{ for } x < 0, \ H(x) = 1/2 \text{ for } x = 0 \qquad (5.20)$$

and has the properties

$$H(ax) = sign(a)H(x), \qquad (5.21)$$

$$H(a^2 - x^2) = H(x+a) - H(x-a) \qquad (5.22)$$

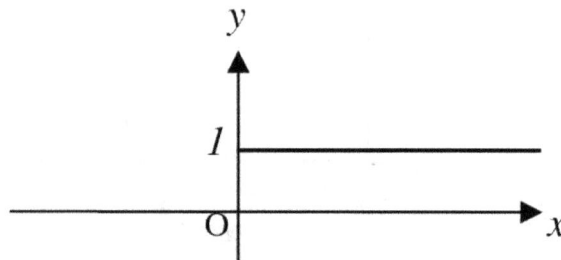

Fig. 5.4. The Heaviside function.

A simple example is that the region in space where $x > a, y > b, z > c$ is represented by

$$H(x-a)H(y-b)H(z-c),$$

and a circular cylinder along Oz is by $H(R^2 - x^2 - y^2)u(z)$.

5.3.1.3. The Rectangular or Step Function and Object Function of Tubes, Box

The step function is defined by

$$R_{2a}(x) \equiv \frac{H(x+a) - H(x-a)}{2a} = \frac{H(a^2 - x^2)}{2a} \tag{5.23}$$

Its graph is Fig. 5.5.

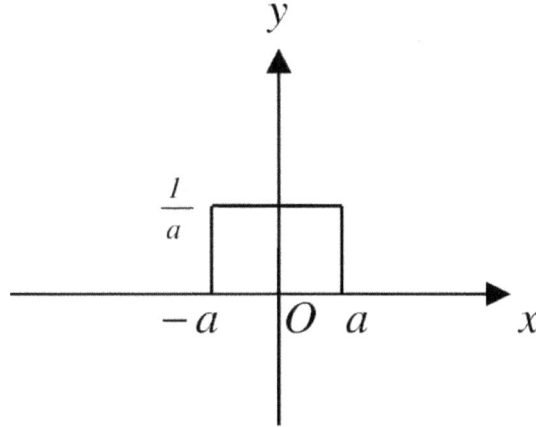

Fig. 5.5. The step function.

A box centered at the origin is represented by

$$R_{2a}(x)R_{2b}(y)R_{2c}(z) \tag{5.24}$$

A tube along Oz is by

$$R_{2a}(x)R_{2b}(y)u(z) \tag{5.25}$$

5.3.1.4. The Dirac Delta Function and Object Functions of Planes

The delta function $\delta(x)$ has been defined in physics by Dirac [6] and justified afterward by Schwartz [7]. Its particular properties are

$$\delta(x) \equiv \lim_{\varepsilon \to 0} R_\varepsilon(x) = H'(x), \tag{5.26}$$

$$x\delta(x) = 0, \tag{5.27}$$

$$\int_{-\infty}^{+\infty} f(x)\delta(x-a)dx = f(a), \tag{5.28}$$

$$\delta(ax) = a^{-1}H'(ax) = a^{-1}sign(a)H'(x) = |a|^{-1}\delta(x), \tag{5.29}$$

$$\int_a^b \delta(x)dx = H(b) - H(a), \tag{5.30}$$

$$\delta(f(x)) = \sum_i |f'(a_i)|^{-1}\delta(x-a_i), \text{ where } f(a_i) = 0, f'(a_i) \neq 0 \tag{5.31}$$

$$f(\vec{r}) = \int_{\Re^3} f(\vec{r}_0)\delta(\vec{r}-\vec{r}_0)d\vec{r}_0, \text{ for } f(\vec{r}) \text{ continue nearby } \vec{r}_0 \tag{5.32}$$

Let D be a diagonal matrix we have

$$\delta(D\vec{r}) = \delta(D_{11}x)\delta(D_{22}y)\delta(D_{33}z) = |D_{11}D_{22}D_{33}|^{-1}\delta(\vec{r}) = |det D|^{-1}\delta(\vec{r})$$

Let A be a non singular matrix then because $A\vec{r}$ is equal to zero if and only if $\vec{r}$ is equal to zero we may write

$$\delta(A\vec{r}) = \beta(A)\delta(\vec{r}),$$

$$\delta(AB\vec{r}) = \beta(A)\delta(B\vec{r}) = \beta(A)\beta(B)\delta(\vec{r})$$

$$\delta(AA^{-1}\vec{r}) = \beta(A)\beta(A^{-1})\delta(\vec{r}) = \delta(\vec{r})$$

Now let U the matrix which transforms A into a diagonal matrix D we have from the above formulae

$$\delta(A\vec{r}) = \delta(UDU^{-1}\vec{r}) = \delta(D\vec{r}) = |det D|^{-1}\delta(\vec{r}) = |det A|^{-1}\delta(\vec{r}) \tag{5.33}$$

By the same reason we may transform the infinitesimal displacement $d(A\vec{r})$ into

$$d(A\vec{r}) = d(UDU^{-1}\vec{r}) = d(D\vec{r}) = |det D|^{-1}d\vec{r} = |det A|^{-1}d\vec{r} \tag{5.34}$$

Thank to the delta function the plane Oxy may be represented by the function $u(x)u(y)\delta(z)$.

Now let $\vec{u}_1, \vec{u}_2, \vec{u}_0$ be three non coplanar vectors representing three points A, B, C in space and $\vec{v}_1, \vec{v}_2, \vec{v}_0$ their reciprocal vectors defined by

$$\vec{r}_i\vec{v}_j = \delta_{ij} \quad \forall i, j = 0,1,2 \tag{5.35}$$

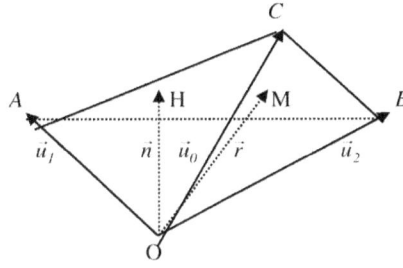

Fig. 5.6. Triangle in oblique axes system.

The vector $\vec{r}$ representing a point M of the plane ABC may be expanded by a useful formula

$$\vec{r} = \sum_i (\vec{u}_i.\vec{r})\,\vec{v}_i = \sum_i (\vec{v}_i.\vec{r})\,\vec{u}_i \, , \qquad (5.36)$$

saying, that "The coordinates of M in the oblique system of axes $(\vec{u}_1,\vec{u}_2,\vec{u}_\sigma)$ are $(\vec{v}_1\vec{r},\vec{v}_2\vec{r},\vec{v}_0\vec{r}\,)$ ".

From this lemma we obtain that like the plane Oxy is represented by the function $u(x)u(y)\delta(z)$ the plane OAB scanned by $\vec{u}_1,\vec{u}_2$ in the oblique axes system $(\vec{u}_1,\vec{u}_2,\vec{u}_0)$ may be represented by the function

$$f_{OAB}(\vec{r}) = u(\vec{v}_1\vec{r})u(\vec{v}_2\vec{r})\delta(\vec{v}_0\vec{r}) \qquad (5.37)$$

For obtaining the function of the plane ABC let $\vec{n}$ be the unity vector perpendicular to it we have

$$\vec{r}\vec{n} = \vec{u}_1\vec{n} = \vec{u}_2\vec{n} = \vec{u}_0\vec{n}, \qquad (5.38)$$

so that

$$\vec{r} = \alpha\vec{u}_1 + \beta\vec{u}_2 + \gamma\vec{u}_0 \; with \; (\alpha + \beta + \gamma) = 1 \qquad (5.39)$$

This gives

$$\vec{r} = (\vec{v}_1\vec{r})\vec{u}_1 + (\vec{v}_2\vec{r})\vec{u}_2 + (\vec{v}_0\vec{r})\,\vec{u}_0 \; with \; (\vec{v}_1 + \vec{v}_2 + \vec{v}_0)\vec{r} = 1 \quad , \qquad (5.40)$$

$$f_{ABC}(\vec{r}) = u(\vec{v}_1\vec{r})u(\vec{v}_2\vec{r})\delta((\vec{v}_1 + \vec{v}_2 + \vec{v}_0)\vec{r} - 1)\,, \forall \vec{v}_1\vec{r}, \vec{v}_2\vec{r} \qquad (5.41)$$

The formulae obtained from the latter by circular permutations of $(\vec{u}_1,\vec{u}_2,\vec{u}_0)$ represent also the object functions of the plane ABC. Its Fourier transform is discussed hereinafter.

Consider the plane ABC. We will convention that if a point M represented by $\vec{r}$ satisfies

$$\vec{r}\vec{n} > \vec{u}_1\vec{n} = \vec{u}_2\vec{n} = \vec{u}_0\vec{n}, \tag{5.42}$$

then we say that M situates upper the plane. The function of the half space under the plane *ABC* is then

$$S^-{}_{ABC}(\vec{r}) = u(\vec{v}_1\vec{r})u(\vec{v}_2\vec{r})H(1 - (\vec{v}_1 + \vec{v}_2 + \vec{v}_0)\vec{r}) \tag{5.43}$$

5.3.2. The Fourier Transform of $\delta(\vec{r})$ and $H(x)$

The Fourier transform of a function $f(x)$ and its properties are very well known for example in [5, 10, 11]. In this chapter we will cite only those which are useful. Let

$$F(k) = FTf(x) = (2\pi)^{-\frac{1}{2}}\int_{-\infty}^{\infty} e^{-ikx} f(x)dx, \tag{5.44}$$

we have

$$FT\delta(x) = (2\pi)^{-\frac{1}{2}}u(x), \quad FTu(x) = (2\pi)^{\frac{1}{2}}\delta(x),$$

$$FTxf(x) = (2\pi)^{-\frac{1}{2}}\int_{-\infty}^{\infty} xe^{-ikx} f(x)dx = i\partial_k F(k),$$

$$FTf(x) = FTf(x)u(x) = f(i\partial_k)FTu(x) = (2\pi)^{\frac{1}{2}} f(i\partial_k)\delta(x) \tag{5.45}$$

For $f(x)$ vanishing at infinity we have by integration by parts

$$FTf'(x) = (2\pi)^{-\frac{1}{2}}\int_R e^{-ikx} f'(x)dx = (ik)(2\pi)^{-\frac{1}{2}}\int_R e^{-ikx} f(x)dx = ik\,FTf(x) \tag{5.46}$$

For $g(\partial_x)$ well defined we have

$$FTg(\partial_x)f(x) = g(ik)FTf(x), \tag{5.47}$$

so that

$$FT\delta(x - x_0) = FTe^{-x_0\partial_x}\delta(x) = (2\pi)^{-\frac{1}{2}} e^{-ix_0 k} \tag{5.48}$$

For the Fourier transform of the Heaviside function let us first write down the formula

$$x = |x|e^{-i\pi H(-x)}, \tag{5.49}$$

which leads to

$$\ln x = \ln|x| - i\pi H(-x),$$

and by derivation to the strange but correct formula

$$\frac{1}{x} = \frac{1}{|x|} signx \ - i\pi H'(-x) = \frac{1}{x} + i\pi\delta(x) \tag{5.50}$$

Now, because the 3D objects we study are limited in dimension, the Heaviside function utilized in this chapter is understood to be equal to $H(x)$ at finite values of the variable and vanishes at infinity so that

$$FT\delta(x) = FTH'(x) = ikFTH(x),$$

$$FTH(x) = \frac{1}{ik}FT\delta(x) = (2\pi)^{-\frac{1}{2}}\frac{1}{ik} = (2\pi)^{-\frac{1}{2}}(\frac{1}{ik} + \pi\delta(k)), \tag{5.51}$$

which assures that the Fourier transform of the sum of $H(x)$ and $H(-x)$ is non vanishing.

5.3.3. Useful Properties of the Fourier Transform

5.3.3.1. The Fourier Transform of a Fourier Transform and Holography

From (5.45) we see that if $f(x)$ has Fourier transform then

$$FTFTf(x) = FTf(i\partial_k)FT\ 1 = f(iik)FTFT\ 1 = f(-k) \tag{5.52}$$

This property says that if $g(\vec{k}) = FTf(\vec{r})$ then the 3D object described by $g(\vec{r})$ would have as Fourier transform $FTg(\vec{r}) = f(-\vec{r})$. We think that this is the mathematical basis of holography. Moreover when we irradiate the hologram recorded on a planar film in order to see the original 3D object what we see is the convolution of it with the FT of the plate which is system of straight lines perpendicular to the plane. As result we will see the objet follows us.

5.3.3.2. The Fourier Transform of Object Functions in a Change of Arguments

Let U and V be the matrices constructed respectively from the non coplanar line vectors $\vec{u}_1, \vec{u}_2, \vec{u}_3$ and their reciprocal $\vec{v}_1, \vec{v}_2, \vec{v}_3$. By definition we have the following relation

$$U\tilde{V} \equiv I \tag{5.53}$$

Consider now the Fourier trans form

$$F(\vec{k}) = FTf(\vec{r}) = (2\pi)^{-3/2} \int_{R^3} e^{-i\vec{k}\vec{r}} f(\vec{r}) d\vec{r} \tag{5.54}$$

In a change of arguments where the column matrix representing $\vec{r}$ becomes $U\vec{r}$ and the line matrix $\vec{k}$ becomes $\vec{k}U^{-1} = \vec{k}\widetilde{V}$ so that the scalar product $\vec{k}\vec{r}$ would unchanged we have

$$FTf(U\vec{r}) = (2\pi)^{-3/2} \int_{R^3} e^{-i\vec{k}\vec{r}} f(U\vec{r}) dU\vec{r} = (2\pi)^{-3/2} \int_{R^3} e^{-i\vec{k}U^{-1}U\vec{r}} f(U\vec{r}) dU\vec{r} ,$$

or under the ordinary form utilizing (5.53)

$$FTf(U\vec{r}) = (2\pi)^{-3/2} |det\,V| \int_{R^3} e^{-i(V\vec{k})(U\vec{r})} f(U\vec{r}) d\vec{r} = |det\,V| F(V\vec{k}) \tag{5.55}$$

In particular, we see that

$$FTf(\vec{u}_1\vec{r})g(\vec{u}_2\vec{r})l(\vec{u}_0\vec{r}) = |detV|\, F(\vec{v}_1\vec{k})G(\vec{v}_2\vec{k})L(\vec{v}_0\vec{k}) \tag{5.56}$$

5.3.3.3. Fourier Transform of a Convolution Product

The convolution product of two functions is defined by the notation

$$f(\hat{r}) \otimes g(\vec{r}) = \int_{\Re^3} f(\vec{r}_0)g(\vec{r}-\vec{r}_0) d\vec{r}_0 \tag{5.57}$$

For example we have the formula

$$f(\hat{r}) \otimes \delta(\vec{r}-\vec{a}) = \int_{\Re^3} f(\vec{r}_0)\delta(\vec{r}-\vec{a}-\vec{r}_0) d\vec{r}_0 = f(\vec{r}-\vec{a}) , \tag{5.58}$$

which says that the convolution of an object with a point $\vec{a}$ is the same object centered at $\vec{a}$.

We have for $f(\vec{r})$ expandable into Taylor series

$$FTf(\vec{r})g(\vec{r}) = f(i\nabla_k)FTg(\vec{r}) = f(i\nabla_k)G(\vec{k})$$

$$= f(i\nabla_k)\int_{\Re^3} G(\vec{k}_0)\delta(\vec{k}-\vec{k}_0) d\vec{k}_0 = (2\pi)^{-3/2}\int_{\Re^3} G(\vec{k}_0)F(\vec{k}-\vec{k}_0) d\vec{k}_0$$

$$= (2\pi)^{-3/2} F(\vec{k}) \otimes G(\vec{k}) , \tag{5.59}$$

and see that "The Fourier transform of the intersection of two continuous objects is $(2\pi)^{-\frac{3}{2}}$ times the convolution product of their Fourier transforms".

Inversely the Fourier transform of a convolution product of two functions expandable into Taylor series is

$$FTf(\vec{r}) \otimes g(\vec{r}) = (2\pi)^{\frac{3}{2}} F(\vec{k})G(\vec{k})$$ (5.60)

5.4. Applications to Fraunhofer Diffractions

5.4.1. Diffraction by Two Points and Young's Experience

From (5.6) we see that the amplitude of diffraction by two points at $\vec{r}_0$ and $-\vec{r}_0$ verifies

$$e^{i\vec{r}_0(\vec{k}-\vec{k}_0)} + e^{-i\vec{r}_0(\vec{k}-\vec{k}_0)} = 2\cos\vec{r}_0(\vec{k}-\vec{k}_0),$$ (5.61)

and is maximum for

$$\vec{k}\vec{r}_0 = kr_0 \cos(Ox,\vec{k}) = \vec{k}_0\vec{r}_0 + m\pi$$ (5.62)

This indicates that the brightest diffracted waves $\vec{k}$ form the cones $(O,Ox,arccos\dfrac{\vec{k}_0\vec{r}_0 + m\pi}{kr_0})$ (Fig. 5.7).

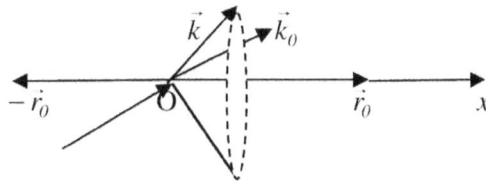

Fig. 5.7. Diffraction by two points.

The values of m vary from zero to the maximum integer m_{max} less than or equal to $(kr_0 - \vec{k}_0\vec{r}_0)/\pi$.

In the Young experience one makes $\vec{k}_0\vec{r}_0 = 0$ so that m_{max} is the biggest integer approaching kr_0/π.

If we add two other points at $\pm 2\vec{r}_0$ then we add a second system of cones $(O, Ox, arccos(\ n\pi\ /\ 2kr_0)\ )$. Let $m_{max} = 10$ then the cones corresponding to $m = 2n = 0,2,4,6,8,10$ are twice brighter than the others. By this way we think that it is possible to examine diffractions by systems of points distributed regularly on Ox, Oy, Oz.

5.4.2. Diffraction by Identical Objects, Bragg's Law

From the property (5.6) we see that N identical objects situated at $\vec{r}_i, i = 1,2,..., N$ has as Fourier transform

$$\sum_{j=1}^{N} e^{-i\vec{k}\vec{r}_i} FTf_D(\vec{r}) \tag{5.63}$$

In particular in a diffraction of $\vec{k}_0$ by $2N$ identical objects situated on Ox at $\pm \vec{r}_0, \pm 2\vec{r}_0, ..., \pm N\vec{r}_0$ because

$$\sum_{n=1}^{N} e^{-i\vec{k}n\vec{r}_0} = e^{-i\vec{k}\vec{r}_0} \frac{e^{-iN\vec{k}\vec{r}_0} - 1}{e^{-i\vec{k}\vec{r}_0} - 1} = e^{-i(N+1)\vec{k}\vec{r}_0)/2} \frac{\sin(N\vec{k}\vec{r}_0 / 2)}{\sin(\vec{k}\vec{r}_0 / 2)} , \tag{5.64}$$

the amplitude of $\vec{k}$ is

$$\sum_{n=1}^{N} (e^{-i\Delta\vec{k}n\vec{r}_0} + e^{i\Delta\vec{k}n\vec{r}_0}) = 2\cos\frac{(N+1)\Delta\vec{k}\vec{r}_0}{2} \frac{\sin(N\Delta\vec{k}\vec{r}_0 / 2)}{\sin(\Delta\vec{k}\vec{r}_0 / 2)} \tag{5.65}$$

It is maximum for $\Delta\vec{k}\vec{r}_0 = 2m\pi$. For example for $2N$ equidistant planes parallel to Oxy, for obtaining bright diffracted wave we must have $\Delta\vec{k} \mathbin{/\mkern-5mu/} Oz$ and d must be so that, with θ being the complementary of the incident angle

$$\Delta\vec{k}\vec{r}_0 = (\Delta\vec{k}_z)d = 2k_0d \sin\theta = 2m\pi \Rightarrow 2d \sin\theta = m\lambda, \quad m \in N \tag{5.66}$$

We find again then by another manner the well known Bragg's law in interference [3, 4].

For interferences by equidistant parallel fences, Dirac comb, aligned spheres, the quadruples points, systems of straight lines on Ox and Oy axis, de Moiré fringes and so all we think that one may utilized similar reasoning for obtaining corresponding interference patterns (Fig. 5.8).

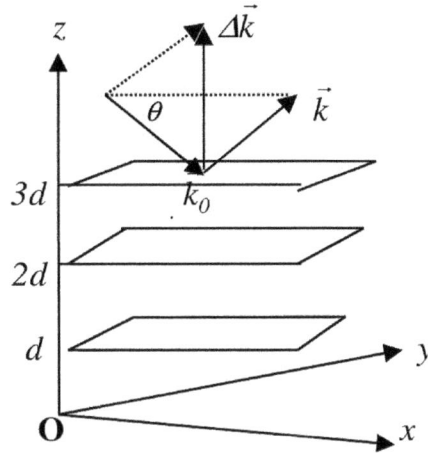

Fig. 5.8. Interference by equidistant parallel planes.

5.4.3. Diffraction by the Semi-space $z < 0$

5.4.3.1. Descartes' and Snell's Laws

Consider the diffraction of a plane wave $\vec{k}_1$ at the surface Oxy separating the region $z < 0$ with index of refraction n_1 and the region $z \geq 0$ with index n_2. The function representing the semi-space $z < 0$ is

$$f(\vec{r}) = u(x)u(y)H(-z) \tag{5.67}$$

Because

$$F(\Delta\vec{k}) = \sqrt{2\pi}\delta(\Delta\vec{k}_x)\delta(\Delta\vec{k}_y)(-i(\Delta\vec{k}_z)^{-1} + \pi\delta(\Delta\vec{k}_z)), \tag{5.68}$$

the diffracted waves must satisfied the conditions

$$\Delta\vec{k}_x = \Delta\vec{k}_y = 0, \tag{5.69}$$

i.e. all the vector $\Delta\vec{k}$ must be perpendicular to the plane Oxy.

Let $\vec{k}'$ be the refracted and $\vec{k}''$ the reflected waves as shown in Fig. 5.9. The line joining the extremities of $\vec{k}_1$ and $\vec{k}'$ supports $(\Delta\vec{k})' = (\vec{k}' - \vec{k}_1)$. That of $\vec{k}_1$ and $\vec{k}''$ supports $(\Delta\vec{k})'' = (\vec{k}'' - \vec{k}_1)$. These two supports are confounded into one line perpendicular to Oxy.

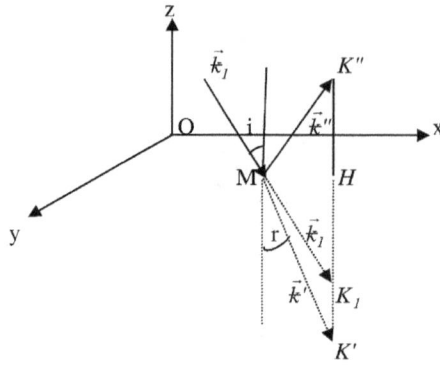

Fig. 5.9. Diffraction of wave by a half space.

The above considerations lead to the following results, let i, r be the angles that $\vec{k}_1$, $\vec{k}'$ made with Oz. Because $\vec{k}_1$, $\vec{k}'$, $\vec{k}''$ have equal projections on Oxy we get

$$k_1 \sin i = k' \sin r = k'' \sin(Oz, \vec{k}''),\qquad(5.70)$$

or because $k'' = k_1$ and $n_1 k' = n_2 k_1$

$$\sin i = \sin(Oz, \vec{k}_1) = \sin(Oz, \vec{k}''),$$

which leads to the Descartes' law for reflection and the Snell's law for refraction of light [3, 4]

$$n_1 \sin i = n_2 \sin r \qquad(5.71)$$

From the Snell law we have

$$i = \frac{\pi}{2} \Rightarrow r = arcsin \frac{n_1}{n_2},\qquad(5.72)$$

so that for $n_1 < n_2$ the angle of refraction is limited by $arcsin \frac{n_1}{n_2}$.

5.4.3.2. The Fresnel Equations and Polarization of Light

Now, let a, a', a" denoted the amplitudes of the incident, the refracted and the reflected waves. The amplitudes a', a" are proportional respectively to $aF(\overrightarrow{\Delta k}'), aF(\overrightarrow{\Delta k}'')$.

Considering the triangle $MK'K_1$ and the Snell's law we have

$$a' = \frac{va}{(\Delta \vec{k}')_z} = \frac{va}{K_1 K'} = \frac{va}{k' \cos r - k_1 \cos i} = \frac{va \sin r}{k_1 \sin(i_1 - i_2)}, \tag{5.73}$$

$$a'' = \mu a \frac{1}{(\Delta \vec{k}'')_z} = -\frac{\mu a}{2 k_1 \cos i} \tag{5.74}$$

In order to calculate the coefficients μ, v we will make use of the law of conservation of energies. The incoming density of energy at the interface Oxy is proportional to a^2, to the inclination $|\cos i_1|$ and the duration of time an incoming photon is in the vicinity of it, i.e. to v_1^{-1} or n_1. Similarly for the density of outgoing energies so that

$$n_1 a^2 \cos i = n_2 a'^2 \cos r + n_1 a''^2 \cos i \tag{5.75}$$

The above equations and the formula

$$4 \cos i \sin i \cos r \sin r = \sin^2(i+r) - \sin^2(i-r) \tag{5.76}$$

lead to the following, by (5.73) and (5.74),

$$4 k_1^2 n_1^2 \cos^2 i \sin^2(i-r) = \mu^2 \sin^2(i-r) + v^2 (\sin^2(i+r) - \sin^2(i-r)) \tag{5.77}$$

We have three interesting solutions of (5.77)

(i)
$$v = 0, \ \mu = 2 n_1 k_1 \cos i \tag{5.78}$$

In this case there is extinction of the diffracted wave because $a' = 0$

(ii)
$$\mu = v = 2 n_1 k_1 \cos i \sin(i-r) / \sin(i+r) \tag{5.79}$$

This solution allows us to find two of the Fresnel Equations obtainable from the electromagnetic theory of light [3, 4]

$$\frac{a'}{a} = \frac{2 n_1 \cos i}{n_1 \cos i + n_2 \cos r} = \frac{2 \cos i \sin r}{\sin(i+r)}, \tag{5.80}$$

$$\frac{a''}{a} = \frac{n_1 \cos i - n_2 \cos r}{n_1 \cos i + n_2 \cos r} = -\frac{\sin(i-r)}{\sin(i+r)}, \tag{5.81}$$

(iii)
$$\mu = -v \cos(i+r) \tag{5.82}$$

We get from (5.77)

$$v = 2 n_1 k_1 \cos i \tan(i-r) / \sin(i+r), \tag{5.83}$$

$$\mu = -2 n_1 k_1 \cos i \tan(i-r) / \tan(i+r) \tag{5.84}$$

135

This solution allows us to find again the other two Fresnel Equations [3, 4]

$$\frac{a'}{a} = \frac{2 n_1 \cos i}{n_1 \cos i + n_2 \cos r} = \frac{2 \cos i \sin r}{\sin(i + r)\cos(i - r)}, \tag{5.85}$$

$$\frac{a''}{a} = \frac{\tan(i - r)}{\tan(i + r)} \tag{5.86}$$

From the Fresnel Equations one may foresee the phenomenon of polarization of light [1-3] that there is extinction of the reflected wave when $i + r = \pi / 2$ or

$$\tan i = \frac{\sin i}{\cos i} = \frac{\sin i}{\sin r} = \frac{n_2}{n_1} \tag{5.87}$$

This precisely is the Brewster's angle for polarized light [3, 4].

5.4.4. Diffractions by Objects Delimited by Planes

5.4.4.1. Diffractions by Trihedra and Tetrahedra

Consider three non-collinear vectors having the same origin O and making angles not greater than 180° between them

$$\overrightarrow{OA_i} \equiv \vec{u}_i, \ i = 0,1,2, \tag{5.88}$$

and its reciprocal vectors $\vec{v}_0, \vec{v}_1, \vec{v}_2$.

The three planes defined by the straight lines supporting $\vec{u}_0, \vec{u}_1, \vec{u}_2$ as shown in Fig. 5.10. divide the space into eight trihedra. Consider the trihedron that has all the components of $\vec{r}$ positive on $\vec{u}_0, \vec{u}_1, \vec{u}_2$. It may simply be represented by the function

$$T^{(3)}{}_{\vec{u}_0,\vec{u}_1,\vec{u}_2}(\vec{r}) \equiv H(\vec{v}_1.\vec{r})H(\vec{v}_2.\vec{r})H(\vec{v}_0\vec{r}) \tag{5.89}$$

According to (5.51) we have

$$\tilde{H}(k) = (2\pi)^{-\frac{1}{2}}(\frac{1}{ik} + \pi\delta(k)), \tag{5.90}$$

$$\tilde{T}^{(3)}{}_{\vec{u}_0,\vec{u}_1,\vec{u}_2}(\vec{k}) = det(\vec{u}_0,\vec{u}_1,\vec{u}_2)\tilde{H}(\vec{u}_0.\vec{k})\tilde{H}(\vec{u}_1.\vec{k})\tilde{H}(\vec{u}_2.\vec{k}), \tag{5.91}$$

and may conclude that the amplitude of diffraction of a plane wave $\vec{k}_0$ into a plane wave $\vec{k}$ by a trihedron is maximum whenever $\Delta\vec{k}$ is perpendicular to two vectors of the set $\vec{u}_0, \vec{u}_1, \vec{u}_2$, i.e. to each one of its faces.

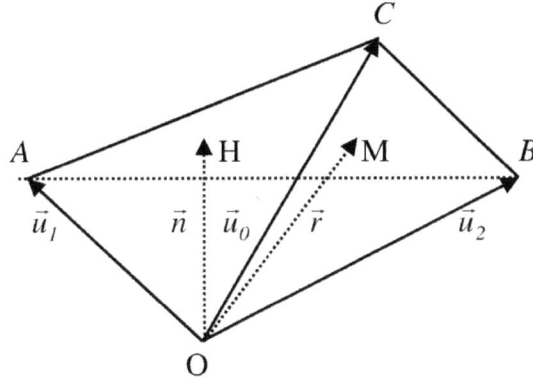

Fig. 5.10. Triangle in oblique axes system.

The tetrahedron $OABC$ being the part of the trihedron $OABC$ living under the plane ABC is represented by

$$T^{(4)}_{\vec{u}_1,\vec{u}_2,\vec{u}_0}(\vec{r}) \equiv T^{(4)}_{OABC}(\vec{r}) = T^{(3)}_{OABC}(\vec{r})S^{-}_{ABC}(\vec{r}) \qquad (5.92)$$

The Fourier transform of $T^{(4)}_{OABC}$ is the convolution product between $\tilde{T}^{(3)}_{OABC}(\vec{k})$ and $\tilde{S}^{-}_{ABC}(\vec{k})$.

In order to calculate the Fourier transform of

$$S^{-}_{ABC}(\vec{r}) \equiv u(\vec{v}_1.\vec{r})u(\vec{v}_2.\vec{r})H(-(\vec{v}_1 + \vec{v}_2 + \vec{v}_0).\vec{r} + 1), \qquad (5.93)$$

we see that it is more convenient to consider the axes system $(\vec{v}_1,\vec{v}_2,\vec{v}_1 + \vec{v}_2 + \vec{v}_0)$ and its reciprocal $(\vec{u}_1 - \vec{u}_0,\vec{u}_2 - \vec{u}_0,\vec{u}_0)$ in order to be able to utilize the formula (5.58) as followed

$$\tilde{S}^{-}_{ABC}(\vec{r}) = |\Delta_{012}|2\pi\delta((\vec{u}_1 - \vec{u}_0)\vec{k})\delta((\vec{u}_2 - \vec{u}_0)\vec{k})e^{-i\vec{k}\vec{u}_3}\tilde{H}(-\vec{u}_0\vec{r}), \qquad (5.94)$$

where

$$\Delta_{012} = det(\vec{u}_1 - \vec{u}_0,\vec{u}_2 - \vec{u}_0,\vec{u}_0) = det(\vec{u}_0,\vec{u}_1,\vec{u}_2) \qquad (5.95)$$

Utilizing the relation obtainable from (5.34)

$$d(U\vec{k}_0) = det(\vec{u}_0,\vec{u}_1,\vec{u}_2)d\vec{k}_0 = d(\vec{u}_0.\vec{k}_0)d(\vec{u}_1.\vec{k}_0)d(\vec{u}_2.\vec{k}_0), \qquad (5.96)$$

and the notations

$$z_i \equiv \vec{u}_i\vec{k}, \quad i = 0,1,2, \qquad (5.97)$$

$$z \equiv \vec{u}_2 \vec{k}_0, \tag{5.98}$$

we find that it is

$$\widetilde{T}^{(4)}{}_{OABC}(\vec{k}) = (2\pi)^{-3/2} 2\pi \Delta^2{}_{012} \times$$

$$\times \int_{R^3} dU \vec{k}_0 \widetilde{H}(\vec{u}_1(\vec{k}-\vec{k}_0)) \widetilde{H}(\vec{u}_2(\vec{k}-\vec{k}_0)) \widetilde{H}(\vec{u}_0(\vec{k}-\vec{k}_0)) e^{-i(\vec{k}_0)\vec{u}_0} \widetilde{H}(-\vec{u}_0 \vec{k}_0),$$

$$\dots\dots\dots\dots\delta((\vec{u}_1 - \vec{u}_0)\vec{k}_0)\delta((\vec{u}_2 - \vec{u}_0)\vec{k}_0$$

$$\widetilde{T}^{(4)}{}_{OABC}(\vec{k}) = (2\pi)^{-1/2} \Delta_{012} \times$$

$$\times \int_{-\infty}^{+\infty} dz e^{-iz} \widetilde{H}(z_0 - z) \widetilde{H}(z_1 - z) \widetilde{H}(z_2 - z) \widetilde{H}(-z) \tag{5.99}$$

The integral in (5.99) is a sum of four terms

$$I_0 \equiv \int_{-\infty}^{+\infty} dz e^{-iz}((z_0 - z)(z_1 - z)(z_2 - z)(-z))^{-1}, \tag{5.100}$$

$$I_2 \equiv -\pi^2 \sum_{j=0}^{2} (e^{-iz_j} \frac{\delta(z_{j-1} - z_j)}{(z_{j+1} - z_j)z_j} + \frac{\delta(z_j)}{(z_{j+1}z_{j-1})}) \; where \; z_{j\pm 3} \equiv z_j, \tag{5.101}$$

$$I_3 \equiv -i\pi^3 \sum_{j=0}^{2} \delta(z_{j-1})\delta(z_{j+1})(z_j)^{-1} \; where \; z_{j\pm 3} \equiv z_j \tag{5.102}$$

In order to calculate I_0 we have to decompose the product $((z_0 - z)(z_1 - z)(z_2 - z)(-z))^{-1}$ into a linear combination of terms $(z_i - z)^{-1}$ and eventually $z^{-2}, z^{-3}, (z_i - z)^{-2}, (z_i - z)^{-3}$ whenever one or two of the z_i are equal to zero or two or three of them are equal and non-vanishing, then make use of the formulae

$$\int_{-\infty}^{+\infty} \frac{e^{-ikz}}{z-a} dz = \sqrt{2\pi} FT \frac{1}{z-a} = \sqrt{2\pi} e^{-iak} FT \frac{1}{z} = \frac{\pi}{i} e^{-iak} signk, \tag{5.103}$$

$$\int_{-\infty}^{+\infty} \frac{e^{-ikz}}{z^2} dz = -\pi|k| + \int_{-\infty}^{+\infty} \frac{dz}{z^2} = -\pi|k| + 2\lim_{\varepsilon \to 0} \frac{1}{\varepsilon}, \tag{5.104}$$

$$\int_{-\infty}^{+\infty} \frac{e^{-ikz}}{z^3} dz = \frac{1}{2} i\pi k|k| - 2ik \lim_{\varepsilon \to 0} \varepsilon^{-1} \tag{5.105}$$

In (5.104) we may pledge that

$$J_\infty \equiv \int_{-\infty}^{+\infty} \frac{dz}{z^2} = 2 \lim_{z_j \to 0} \frac{1}{z_j} \qquad (5.106)$$

would be as large as $\delta(z_j)$ for any $z_j = 0$ as we can see hereinafter.

Explicitly we get

(i) for $\vec{k}$ not perpendicular to any edge, i.e. for $0 \neq z_j \neq z_{j+1} \neq 0$ where $j = 0,1,2$; $z_3 \equiv z_0$

$$\widetilde{T}^{(4)}{}_{OABC}(\vec{k}) = -i(2\pi)^{-\frac{3}{2}} \Delta_{012} \left(\sum_{j=0}^{2} \frac{e^{-iz_j}}{(z_{j-1} - z_j)(z_{j+1} - z_j)z_j} - \frac{1}{z_0 z_1 z_2} \right) \qquad (5.107)$$

(ii) for $\vec{k}$ perpendicular to the edge defined by $\vec{u}_p$ i.e. $z_p = 0$

$$\widetilde{T}^{(4)}{}_{\vec{u}_0,\vec{u}_1,\vec{u}_2,}(\vec{k}) = (2\pi)^{-\frac{5}{2}} \Delta_{012} \left(\frac{2\pi i e^{-iz_{p-1}}}{(z_{p+1} - z_{p-1})z^2{}_{p-1}} + \frac{2\pi i e^{-iz_{p+1}}}{(z_{p-1} - z_{p+1})z^2{}_{p+1}} - \right.$$

$$\left. - \frac{\pi i}{z^2{}_{p-1} z_{p+1}} - \frac{\pi i}{z_{p-1} z^2{}_{p+1}} + \frac{-\pi + J_\infty}{z_{p-1} z_{p+1}} - \frac{\pi^2 \delta(z_p)}{z_{p-1} z_{p+1}} \right) \qquad (5.108)$$

Similarly, for $\vec{k}$ perpendicular to the edge defined by $(\vec{u}_{p-1} - \vec{u}_{p+1})$ i.e. for $z_{p-1} = z_{p+1} \neq 0$

$$\widetilde{T}^{(4)}{}_{\vec{u}_0,\vec{u}_1,\vec{u}_2}(\vec{k}) \approx (2\pi)^{-\frac{5}{2}} \pi^2 \Delta_{0,1,2}((z_p - z_{p-1})z_{p-1})^{-1} \delta(z_{p+1} - z_{p-1})e^{-iz_{p-1}} \qquad (5.109)$$

(iii) for $\vec{k}$ perpendicular to the face defined by $\vec{u}_{p-1}$ and $\vec{u}_{p+1}$, i.e. $z_{p-1} = z_{p+1} = 0$

$$\widetilde{T}^{(4)}{}_{\vec{u}_0,\vec{u}_1,\vec{u}_2}(\vec{k}) \approx -(2\pi)^{-\frac{5}{2}} \pi^2 \Delta_{012} \delta(z_{p-1})(z_p z_{p+1})^{-1}) \qquad (5.110)$$

We may conclude then that the amplitude of diffraction of a plane wave $\vec{k}_0$ into a plane wave $\vec{k}$ by a tetrahedron is given by (5.107) where $\Delta\vec{k} \equiv \vec{k} - \vec{k}_0$ replaces $\vec{k}$ if $\Delta\vec{k}$ is not perpendicular to any edge; by (5.108) or (5.109) if $\Delta\vec{k}$ is perpendicular to one edge and by (5.110) if $\Delta\vec{k}$ is perpendicular to one face. We observe that the amplitude of diffraction is as infinitely large as $\lim_{\varepsilon \to 0} \delta(\varepsilon)$ if $\Delta \vec{k}$ becomes perpendicular to one edge and as $\lim_{\varepsilon \to 0} \dfrac{\delta(\varepsilon)}{\varepsilon}$ whenever $\Delta \vec{k}$ is perpendicular to one of its four faces.

5.4.4.2. Diffractions by Two Adjacent Tetrahedra

Consider two adjacent tetrahedra $T^{(4)}{}_{\vec{u}_0,\vec{u}_1,\vec{u}_2}(\vec{r})$, $T^{(4)}{}_{\vec{u}_0,\vec{u}_2,\vec{u}_3}(\vec{r})$

For calculating the Fourier transform of the sum of them, we first remark that there is a coplanar relation between $\vec{u}_1, \vec{u}_2, \vec{u}_3$ and by multiplying it with $\vec{u}_0 \times \vec{u}_1$, $\vec{u}_0 \times \vec{u}_2$ we get successively

$$\alpha \vec{u}_1 + \beta \vec{u}_2 + \gamma \vec{u}_3 = 0, \quad \Delta_{023} \vec{u}_1 + \Delta_{031} \vec{u}_2 + \Delta_{012} \vec{u}_3 = 0, \tag{5.111}$$

$$\frac{\Delta_{012}}{z_0 z_1 z_2} + \frac{\Delta_{023}}{z_0 z_2 z_3} = -\frac{\Delta_{031}}{z_0 z_1 z_3}, \tag{5.112}$$

$$\frac{\Delta_{012}}{z_1 - z_0} + \frac{\Delta_{023}}{z_3 - z_0} = \frac{\Delta_{013} z_2 - (\Delta_{012} + \Delta_{023}) z_0}{(z_1 - z_0)(z_3 - z_0)}, \tag{5.113}$$

so that in the case where $\vec{k}$ is not perpendicular to any edge

$$i(2\pi)^{\frac{3}{2}}(\widetilde{T}^{(4)}{}_{\vec{u}_0,\vec{u}_1,\vec{u}_2}(\vec{k}) + \widetilde{T}^{(4)}{}_{\vec{u}_0,\vec{u}_2,\vec{u}_3}(\vec{k})) = \frac{-\Delta_{013}}{z_0 z_1 z_3} +$$

$$+\frac{e^{-iz_0}}{(z_2 - z_0)z_0}\left(\frac{\Delta_{013} z_2}{(z_1 - z_0)(z_3 - z_0)} - \frac{(\Delta_{012} + \Delta_{023})z_0}{(z_3 - z_0)(z_1 - z_0)}\right) +$$

$$+\frac{e^{-iz_1}\Delta_{012}}{z_1(z_2 - z_1)(z_0 - z_1)} + \frac{e^{-iz_3}\Delta_{023}}{z_3(z_2 - z_3)(z_0 - z_3)} +$$

$$+\frac{e^{-iz_2}(\Delta_{013} - \Delta_{012} - \Delta_{023})}{(z_0 - z_2)(z_1 - z_2)(z_3 - z_2)} \tag{5.114}$$

In the case where ABD on Fig. 5.11 are aligned we have $(\Delta_{013} = \Delta_{012} + \Delta_{023})$ and if $z_2 = 0$ we get from (5.108)

$$\widetilde{T}^{(4)}{}_{\vec{u}_0,\vec{u}_1,\vec{u}_2}(\vec{k}) + \widetilde{T}^{(4)}{}_{\vec{u}_0,\vec{u}_2,\vec{u}_3}(\vec{k}) = \widetilde{T}^{(4)}{}_{\vec{u}_0,\vec{u}_1,\vec{u}_3}(\vec{k}) + (2\pi)^{-\frac{5}{2}}\Delta_{013}\frac{-\pi + J_\infty}{z_0 z_1 z_3} z_2, \tag{5.115}$$

which implies that $z_2 J_\infty = 0$ or J_∞ is equivalent to $\delta(z_2)$ when $z_2 = 0$ as pledged hereinabove.

The formula (5.115) shows also that for $\vec{u}_2 \Delta\vec{k} = z_2 = 0$ the amplitude of diffraction is not infinite which is correct because the interface is no longer a face of two adjacent tetrahedra.

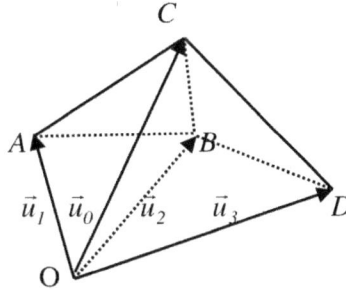

Fig. 5.11. Diffraction by adjacent tetrahedra.

5.4.4.3. Diffractions by Oblique Pyramids with Polygonal Bases

Consider a vector $\overrightarrow{OA_0} \equiv \vec{u}_0$ and a set of n vectors $\overrightarrow{OA_i} \equiv \vec{u}_i, i = 1,2,...,n$ situated in a plane which doesn't contain $\vec{u}_0$ and such that $\vec{u}_i, i = 1,2,..., n-1$ lies inside the angle $(\vec{u}_{i-1}, \vec{u}_{i+1})$, that $\vec{u}_1$ lies inside $(\vec{u}_n, \vec{u}_2)$ and $\vec{u}_n$ inside $(\vec{u}_{n-1}, \vec{u}_1)$.

Consider the pyramid $\Pi_{A_0;A_1A_2...A_n}$ with vertex A_0 and base the polygon $A_1A_2....A_n$. It is the reunion of n tetrahedral $OA_0A_iA_{i+1}$ with $\vec{u}_{i+n} \equiv \vec{u}_i$ by convention so that its Fourier transform is:

(i)   For $\vec{k}$ not perpendicular to any edge of the pyramid.

In this case we have $z_0 \neq z_i$ and $z_i \neq z_{i+1}, \forall i = 1,2,....,n$ where $z_{n+1} \equiv z_1$

$$2i(2\pi)^{\frac{3}{2}} \tilde{\Pi}_{A_0;A_1A_2...A_n}(\vec{k}) = \sum_{j=1}^{n} \Delta_{0,j,j+1} \left(\frac{e^{-iz_0}}{z_0(z_j - z_0)(z_{j+1} - z_0)} - \frac{1}{z_0 z_j z_{j+1}} \right) +$$

$$+ \sum_{j=1}^{n} \frac{(\Delta_{0,j-1,j} + \Delta_{0,j,j+1} - \Delta_{0,j-1,j+1})}{(z_j - z_{j-1+n})(z_j - z_{j+1})(z_j - z_0)} e^{-iz_j} \qquad (5.116)$$

But from (5.112)

$$\sum_{j=1}^{n} \frac{\Delta_{0,j,j+1}}{z_0 z_j z_{j+1}} = \frac{\Delta_{0,1,n}}{z_0 z_1 z_n} + \frac{\Delta_{0,n,n+1}}{z_0 z_n z_{n+1}} = \frac{\Delta_{0,1,1}}{z_0 z_1 z_1} = 0 \qquad (5.117)$$

so that we have the simplification

141

$$2i(2\pi)^{\frac{3}{2}}\widetilde{\Pi}_{A_0;A_1A_2...A_n}(\vec{k}) = \sum_{j=1}^{n}\frac{\Delta_{0,j,j+1}e^{-iz_0}}{z_0(z_j-z_0)(z_{j+1}-z_0)} +$$

$$+\sum_{j=1}^{n}\frac{(\Delta_{0,j-1,j}+\Delta_{0,j,j+1}-\Delta_{0,j-1,j+1})}{(z_j-z_{j-1+n})(z_j-z_{j+1})(z_j-z_0)}e^{-iz_j} \tag{5.118}$$

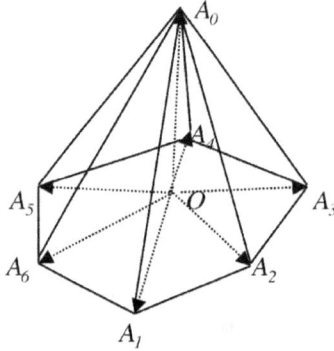

Fig. 5.12. Diffraction by an oblique pyramid.

(ii) For $\vec{k}$ perpendicular to an edge defined by $(\vec{u}_0-\vec{u}_{j'})$.

In this case we have $z_0 = z_{j'}$ and get from (5.116) that

$$\widetilde{\Pi}_{A_0;A_1A_2...A_n}(\vec{k}) \approx (2\pi)^{-\frac{5}{2}}\pi^2(\frac{\Delta_{0,j'-1,j'}}{z_{j'-1}-z_0}+\frac{\Delta_{0,j',j'+1}}{z_{j'+1}-z_0})\delta(z_{j'}-z_0)(z_0)^{-1}e^{-iz_0} \tag{5.119}$$

Similarly if $\vec{k}$ perpendicular to an edge defined by m $(\vec{u}_{j'}-\vec{u}_{j'+1})$ then $z_{j'}=z_{j'+1}$ and

$$\widetilde{\Pi}_{A_0;A_1A_2...A_n}(\vec{k}) \approx (2\pi)^{-\frac{5}{2}}\pi^2\Delta_{0,j',j'+1}((z_0-z_{j'})z_{j'})^{-1}\delta(z_{j'+1}-z_{j'})e^{-iz_{j'}} \tag{5.120}$$

(iii) For $\vec{k}$ perpendicular to a face defined by $(\vec{u}_{j'}-\vec{u}_0)$ and $(\vec{u}_{j'+1}-\vec{u}_0)$ that is the case where there exists two vectors $\vec{u}_{j'},\vec{u}_{j'+1}$ such that $z_{j'}=z_{j'+1}\neq 0$, we see that $\widetilde{\Pi}_{A_0;A_1A_2...A_n}(\vec{k})$ is infinitely large for $\vec{k}$ perpendicular to both $(\vec{u}_{j'}-\vec{u}_0)$ and $(\vec{u}_{j'}-\vec{u}_{j'+1})$ relatively to the case where it is perpendicular to $(\vec{u}_{j'}-\vec{u}_{j'+1})$ alone.

From the above results, we may conclude that the amplitude of diffraction of a plane wave $\vec{k}_0$ into a plane wave $\vec{k}$ by a pyramid $\Pi_{A_0;A_1A_2...A_n}$, given by (5.120) where $\Delta\vec{k}\equiv\vec{k}-\vec{k}_0$ replaces $\vec{k}$, is infinitely larger for $\Delta\vec{k}$ perpendicular to one of its faces relatively to the cases where $\Delta\vec{k}$ is perpendicular only to its interfaces or to its edges.

5.4.4.4. Diffractions by a Sphere

The equation of a sphere centered at O and having radius R as shown in Fig. 5.13 is

$$S(x, y, z) = H(R^2 - z^2)H(R^2 - y^2 - z^2)H(R^2 - x^2 - y^2 - z^2) \qquad (5.121)$$

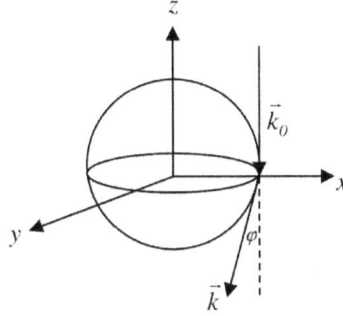

Fig. 5.13. Deflection of waves by a sphere.

It Fourier transform is invariant in a rotation around the origin so that

$$\tilde{S}(k_x, k_y, k_z) = \tilde{S}(0, 0, k) = 2(2\pi)^{-\frac{3}{2}} \int_{-R}^{R} e^{-ikz} dz \int_{-\sqrt{(R^2-z^2)}}^{\sqrt{(R^2-z^2)}} dy \sqrt{((R^2 - z^2) - y^2)} \qquad (5.122)$$

Because

$$\int_{-\sqrt{(R^2-z^2)}}^{\sqrt{(R^2-z^2)}} dy \sqrt{((R^2 - z^2) - y^2)} = (R^2 - z^2) \int_{-\pi/2}^{\pi/2} \sqrt{1 - \sin^2 \phi} d(\sin \phi) = \frac{1}{2}\pi(R^2 - z^2),$$

$$\int_{-R}^{R} dz e^{-ikz}(R^2 - z^2) = \frac{2}{-ik}(\frac{1}{-ik}(\mathrm{Re}^{-ikR} + \mathrm{Re}^{ikR}) - \frac{2}{-ik}\frac{1}{-ik}\frac{-1}{-ik}2i\sin kR,$$

we get

$$\tilde{S}(\vec{k}) = (2\pi)^{-\frac{1}{2}} \frac{2R}{k^2}(\frac{\sin Rk}{Rk} - \cos Rk) \qquad (5.123)$$

As conclusion we see that in a diffraction by a sphere the amplitude of diffraction is inversely proportional to $(\Delta k)^2$ with $\Delta k = \|\Delta \vec{k}\|$ and there is extinction if

$$\tan R\Delta k = R\Delta k \Rightarrow R\Delta k = 0.02 \qquad (5.124)$$

Let φ be the deviation angle in a diffraction, we have extinction for φ such that

$$\sin \frac{\varphi}{2} = \frac{\Delta k}{2k} = \frac{R\Delta k}{2Rk} = \frac{0.02}{2Rk} \qquad (5.125)$$

For example, for $\lambda = 10nm$ X-ray and $R = 2.5$ nm hemoglobin, there is extinction if

$$\sin\frac{\varphi}{2} = \frac{0.02}{\pi} = 0.0064$$

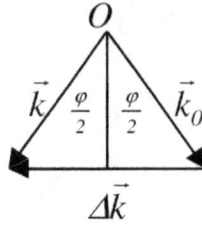

Fig. 5.14. Angle of deflection.

5.4.4.5. Deflection of Light by the Form of the Sun

In a diffraction by a sphere we have

$$\widetilde{S}'(0,0,k) = (2\pi)^{-\frac{1}{2}}\frac{2R}{k^3}((Rk - \frac{3}{Rk})\sin kR + 3\cos kR), \qquad (5.126)$$

so that there are maximum amplitude diffracted waves for

$$\tan R\Delta k_m = \frac{3R\Delta k_m}{3 - R^2(\Delta k_m)^2} \qquad (5.127)$$

The first root of this equation is $R\Delta k_m = 1.192$ which gives $\sin\frac{\phi}{2} = \frac{1.192}{2Rk}$.

For example, for 10 *nm* X-ray and 2.5 *nm* hemoglobin, there is bright diffracted wave corresponding to $\sin\frac{\phi}{2} = \frac{1.192}{2Rk} = \frac{1.192}{10\pi}10^{9-8} = 0.38$.

In particular in the diffraction of a planar wave $\vec{k}_0$ having $\lambda = 0.555\times10^{-6}\,m$ for instance by the sun we have

$$R = 6.96\times10^8\,m\,,$$

$$k_0 = \frac{2\pi}{\lambda} = \frac{2\pi}{0.555}10^6\,m^{-1}$$

Guessing that $R\Delta k_m \gg 1$ we have

$$\tan(R\Delta k_m) = \frac{3R\Delta k_m}{3 - (R\Delta k_m)^2} \approx \frac{-3}{R\Delta k_m} \Rightarrow (R\Delta k_m) = n\pi\,, \qquad (5.128)$$

$$\sin\frac{\varphi}{2} = \frac{\varDelta k_m}{2k} = \frac{R\varDelta k_m}{2Rk} = \frac{n\pi}{4\pi R}\lambda = \frac{0.555}{6.96}10^{-6-8},$$ (5.129)

$$\sin\frac{\phi}{2} = 0.020.10^{-14} \times n$$ (5.130)

Corresponding to $n = 10^9$ we have a deflection

$$\sin\frac{\phi}{2} = 0.020\times10^{-5}, \ \phi = 0.402\times10^{-5}\,rad = 8.29\,\text{arcsec}$$ (5.131)

In fact because of the multitude choice possible of n, instead of one deflected wave we have a beam of deflected waves. In this beam, there are waves with angles of deflection too small or intensity too feeble to be detected by the apparatus utilized by Eddington for detecting the deflection of light by the sun by General Relativity in May 1919 [13] which gives

$$\phi_{Einstein} = 0.848\times10^{-5}\,rad = 1.75\,\text{arcsec}$$ (5.132)

Nevertheless we think that there are detectable deflected rays of light causing by the form of the sun denoted by ϕ_S, such as the one corresponding to the above $\phi = 0.402.10^{-5}\,rad$, which may be add to the deflection of light caused by the Newtonian attraction of the sun [14]

$$\phi_{Newton} = 0.424\times10^{-5}\,rad = 0.875\,\text{arcsec},$$ (5.133)

in order to get the measured deflection of light by General Relativity, says

$$\phi_{Einstein} = \phi_{Newton} + \phi_S$$ (5.134)

5.5. Remarks and Conclusions

The first particularity of the present study on wave optics is that it essentially based on the study of diffraction by a material point then by generalization by a 3D object, described by a function called object function. Parallelely it may be done from the pioneered concepts of Dirac on kets, bras, measurement operators, etc. together with the hypothesis we introduce that $< \vec{k}|\alpha >$ where $\vec{k} = \hbar^{-1}\vec{p}$ is the Fourier transform of $< \vec{r}|\alpha >$. From these approaches we get the fundamental theorem in wave optics saying that the amplitude of a diffracted wave $\vec{k}$ in a Fraunhofer diffraction of a plane wave $\vec{k}_0$ is equal to the Fourier transform of the object function calculated for $\varDelta\vec{k} = \vec{k} - \vec{k}_0$.

The second particularity is that it mathematically leans only on the utilization of Heaviside and Dirac functions together with the notion of reciprocal vectors in order to describe the object functions and calculate their Fourier transforms.

The third particularity is that thank to this method we find again the laws of Descartes and Snell, the Fresnel Equations, the Brewster law on polarization, the Bragg's formula, the principal theorem in classical Fourier optics, the principle of holography. Moreover the amplitudes of Fraunhofer diffractions by oblique trihedra, tetrahedra, pyramids, spheres, sets of identical objects.

Finally deflection of light by the form of the sun is calculated and may be the answer of why there is difference between deflection by Newtonian force and General Relativity.

As conclusion we think that this chapter constitutes a contribution to the progress of wave optics in general and may have applications in acoustic, radiology, astronomy, etc. because it applies for any wavelength and any object's dimension.

Acknowledgments

The author respectfully dedicates this work to regretted Prof. Nguyen Chung Tu who had taught him optics at the Saigon Faculty of Sciences, now HoChiMinhcity Natural Sciences University. He dedicates it also to his adorable wife who daily encouraged him to be perseverant during the long months he perform this work.

References

[1]. M. Born, E. Wolf, Principles of Optics, *Pergamon*, 1970.
[2]. J. W. Goodman, Introduction to Fourier Optics, 3rd Ed., *Roberts & Co Publishers*, 2005.
[3]. E. Hecht, Optics, *Addison-Wesley*, 2002.
[4]. G. Bruhat, Optique, 6 Ed., *Masson*, 1965.
[5]. R. Bracewell, The Fourier Transform and Its Applications, 3rd Ed., *McGraw-Hill*, New York, 1999.
[6]. P. A. M. Dirac, Quantum Mechanics, 4th Ed., *Oxford Univ. Press*, 1960.
[7]. C. Eckart, Operator calculus and the solutions of the equations of quantum dynamics, *Phys. Rev.*, Vol. 28, 1926, pp. 711-726.
[8]. T. S. Do, The Fourier transform and principe of quantum mechanics, *Applied Mathematics*, Vol. 9, 2018, pp. 347-354.
[9]. F. D. Duff, D. Naylor, Heaviside unitfunction, in Differential Equations of Applied Mathematics, *John Wiley & Sons*, 1966, p. 42.
[10]. K. B. Wolf, Integral Transforms in Science and Engineering, *Plenum Press*, 1979.
[11]. L. Schwartz, Theorie des Distributions, Vol I-II, *Hermann et Cie*, 1951.
[13]. C. M. Will, The 1919 measurement of the deflection of light, *Classical and Quantum Gravity*, Vol. 32, 2015, 124001.
[14]. J. G. V. Soldner, On the deflection of a light ray from its rectilinear motion, by the attraction of a celestial body at which it nearly passes by, *Berliner Astronomisches Jahrbuch*, 1801, pp. 161-172.

Chapter 6
Wavefront Coding Technique for Imaging Systems

Vannhu Le, Cuifang Kuang, Zhigang Fan and Xu Liu

6.1. Introduction

The traditional imaging system with the clear aperture has always the finite depth of field (DOF) which corresponds to depth of focus in the image space. Sharp images of scene can be only captured within the DOF, and the images will degrade out of the DOF. As a result, the defocused images are lost information of object at high frequency region. A wide range of the DOF can generate the images more information of object space, and the recorded images show abundant features which are near to the real world. Some fields, such as microelectronics, machine vision, optical data storage, biotechnology and so on, require high DOF to capture three-dimensional features of scene [1-7]. In order to achieve the images more information of three-dimensional space, generally, optical imaging system should have high DOF or be less sensitive to defocus.

The pioneer method of DOF extension was presented by W. T. Welford in 1960 and he described the use of an annular aperture-stop to raise DOF for a simple lens so that more space details were acquired compared with a conventional camera [8]. He found that the annular stop improves the DOF compared to stopping down the aperture-stop. In 1963, Tsujiuchi introduced a method for extending DOF by modulating the incoming light in the aperture using a so-called amplitude mask [9]: a glass mask with the specific amplitude transmittance function. Then, a series of amplitude masks and their imaging performances to extend the DOF were reported [10-18]. All the imaging systems with these amplitude masks lead to an augmentation in exposure and decrease in image contrast and the modulation transfer function (MTF).

In 1972, one method for extending DOF was reported by Hausler involves by varying the focus continuously during the exposure so that MTF becomes a superposition with various

Vannhu Le
Le Quy Don Technical University, Cau Giay, Ha Noi, Viet Nam

foci [19]. A similar method was suggested in 2008, which instead of the specimen display, the detector was displayed in a camera during exposure [20]. However, this approach was still limited when employed a micro object manipulation imaging system because displaying parts are required in the vision system. The imaging system is not suitable for real-time observation.

In 1995, Dowski and Cathey proposed a method to increase the DOF, which a *cubic phase mask* (CPM) was placed in the pupil plane of an imaging system and the digital image post-processing was combined in the imaging procedure [21]. This technique was called as wavefront coding (WFC) technique and was associated with computational imaging. Authors have applied the stationary phase approximation method to analyze Ambiguity function (AF) and derived an imaging system with an asymmetric phase mask. This asymmetric phase mask was able to generate the invariant optical transfer function (OTF) or the invariant PSF to defocus so that the imaging property is less sensitive to defocus. However, the MTF of the CPM is much lower comparison to the diffraction-limit in-focus MTF of the traditional imaging system. So, the image of WFC system with a CPM is severely blurred. The phase mask parameters can be optimized such that it produces no zeroes in a desired MTF-frequency range. The digital processing with only one *point spread function* (PSF) can be used to obtain the high-quality image at all positions of defocus.

6.2. Theory of Wavefront Coding Technique

The generalized phase pupil function of an optical imaging system with a phase mask $f(x)$ and defocus parameter, ψ, can be described by

$$P(x) = \begin{cases} \dfrac{1}{\sqrt{2}} \exp\left\{ i\left[f(x) + \psi x^2 \right] \right\} & \text{if } |x| \leq 1 \\ 0 & \text{otherwise} \end{cases}, \tag{6.1}$$

where $i = \sqrt{-1}$ and ψ is the defocus parameter and can be presented by

$$\psi = \frac{\pi L^2}{4\lambda}\left(\frac{1}{f} - \frac{1}{d_0} - \frac{1}{d_i} \right) = \frac{2\pi}{\lambda} W_{20} = k W_{20}, \tag{6.2}$$

where $k = 2\pi\lambda$; λ is the wavelength of light; W_{20} is measured in unit of wave length of light; L is the one-dimensional length of the lens aperture; the distance d_0 is measured between the object and the first principal plane of the lens; the distance d_i is the distance between the second principal plane and the image plane.

Based on wavefront aberration theory, when a WFC imaging system suffered from defocus error ψ, its deviation of an effective path-length error $W(x)$ between the real wavefront and the ideal wavefront can be expressed as

$$W(x) = f(x) + \psi x^2 \tag{6.3}$$

The one-dimensional OTF of an optical imaging system with the phase pupil function can be presented as

$$H(u,\psi) = \int_{|u|-1}^{1-|u|} P(x+u)P^*(x-u)\,dx =$$

$$= \frac{1}{2}\int_{|u|-1}^{1-|u|} \exp\{i[f(x+u) - f(x-u)]\}\exp(i4\psi ux)\,dx, \tag{6.4}$$

where u is the spatial frequency and $|u| \le 1$.

The AF associated with the general phase masks can be employed as a polar display of the OTF for all values of defocus parameter. The AF associated with the OTF can be obtained by,

$$AF(u,t) = \frac{1}{2}\int \exp\{i[f(x+u) - f(x-u)]\}\exp(i2\pi t)\,dx \tag{6.5}$$

Based on Eqs. (6.4) and (6.5), the AF can be related with the OTF of an optical imaging system as

$$H(u,\psi) = AF(u, 2u\psi / \pi) \tag{6.6}$$

For a given defocus value, ψ, the corresponding OTF associated with the AF is the projection of the radial line with slope $2\psi/\pi$ onto the horizontal u axial. In this approach, the two-dimensional AF can be employed to determine the one-dimensional OTF for all values of defocus parameter.

In a WFC system, a phase pupil function with defocus error and the CPM can be given by

$$P(x,\psi) = \begin{cases} \dfrac{1}{\sqrt{2}}\exp\left[i\left(ax^3 + \psi x^2\right)\right] & \text{if } |x| \le 1 \\ 0 & \text{other} \end{cases} \tag{6.7}$$

The defocused PSF, h, is equal to the quartic modulus of the Fourier transform of the phase pupil function

$$h(x_0,\psi) = \left|FFT\left[P(x,\psi)\right]\right|^2, \tag{6.8}$$

where x_0 is the imaging plane coordinate.

The OTF, H, of WFC imaging system with the CPM can be presented in the following form

$$H(u,\psi) = \frac{1}{2} \int_{|u|-1}^{1-|u|} \exp\left\{ i \left[a(x+u)^3 - a(x-u)^3 + \psi(x+u)^2 - \psi(x-u)^2 \right] \right\} dx \qquad (6.9)$$

Through some changes, the OTF can be rewritten by

$$H(u,\psi) = \frac{1}{2} \exp\left(i2au^3 \right) \exp\left(-i\frac{2\psi^2 u}{3a} \right) \int_{|u|-1}^{1-|u|} \exp\left[i6au\left(x + \frac{\psi}{3a} \right)^2 \right] dx \qquad (6.10)$$

In order to assess the integral of Eq. (6.10), the stationary phase point approximation method can be used because it is always employed to judge the integral described by Eq. (6.11), which can be considered as the general from of Eq. (6.10) [22]

$$C = \int \left\{ \exp\left[iA(s) \right] B(s) \right\} ds, \qquad (6.11)$$

where A and B are the real-valued functions, s is the integral variable.

With the stationary phase point approximation method, the Eq. (6.11) can be assessed as the following:

$$C = \sqrt{\frac{2\pi}{\left| A''(s_s) \right|}} B(s_s) \exp\left\{ i \left[A(s_s) + \operatorname{sgn}\left[A''(s_s) \right] \frac{\pi}{4} \right] \right\}, \qquad (6.12)$$

where A'' is the second-order derivative of function A with respect to the integral variable s and s_s is the stationary point that satisfies the constraint equation as,

$$\left. \frac{\partial \left[A(s) \right]}{\partial s} \right|_{s=s_s} = 0 \qquad (6.13)$$

Thus, key to employ the stationary phase point approximation approach lies in the computation of stationary point, s_s, and the approximate analytical OTF can be acquired immediately, as long as the expression for s_s is available. By applying the stationary phase point approximation evaluation to Eq. (6.10), the stationary point x_i is given by

$$x_i = -\frac{\psi}{3a} \qquad (6.14)$$

By applying Eq. (6.10) with the station point x_i according to Eq. (6.12), the OTF of WFC imaging system with the CPM can be rewritten by

$$H(u,\psi) = \left(\frac{\pi}{24|au|} \right)^{1/2} \exp\left(i2au^3 \right) \exp\left(-i\frac{2\psi^2 u}{3a} \right) \qquad (6.15)$$

The equation (6.15) presents the one-dimensional OTF of WFC imaging system with the CPM. Note that the defocus term is excluded in Eq. (6.15), which means that the MTF (the real part of Eq. (6.15)) becomes invariant to defocus. However, the phase term

150

constraints two terms of the two three term in Eq. (6.15). One is independent with defocus, other depends on defocus. Specifically, the second of the phase terms, $\exp(-i2\,\psi^2 u/3a)$, as function of defocus ψ and it is a linear phase term to u variable. Such a term is the impact of merely shifting the location of the resulting PSF with big defocus. Luckily, this term can be controlled by employing the constant a. Large values of a ($a \geq 20$) minimize of the variation to displacement of the PSF with defocus [21]. In practice, this defocus-dependent term can be effectively controlled so as to be negligible. The final approximation for the OTF of WFC imaging system with the CPM is then

$$H(u,\psi) = \left(\frac{\pi}{24|au|}\right)^{1/2} \exp\left(i2au^3\right) \tag{6.16}$$

From Eq. (6.16), it can be seen that the OTF of WFC imaging system with the CPM does not depend on defocus. Based on the final expression of the OTF, the MTF of WFC imaging system with the CPM can be given by

$$\text{MTF}(u,\psi) = \left(\frac{\pi}{24|au|}\right)^{1/2} \tag{6.17}$$

From approximated expression of MTF shown in Eq. (6.17), it can be seen that the MTF does not depend on the defocus parameter. In other words, the MTF of WFC imaging system with the CPM is insensitive to defocus. Based on the use of the equation of the MTF shown in Eq. (6.17), we perform numerical simulation for some mask parameters of the CPM. Fig. 6.1 shows the MTF for different mask phase parameter values ($a = 60$, $a = 70$, $a = 80$, $a = 90$, and $a = 100$), where the MTF is in units of dB. As shown from Fig. 6.1, when the phase mask parameter is increasing, the MTF is lower. This means that the recorded image will be more blurred when the phase mask parameter is increasing.

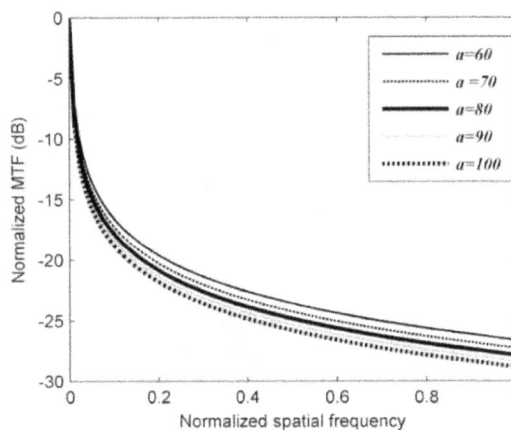

Fig. 6.1. Magnitude of the MTF for the phase mask parameters of the CPM, the phase mask parameters: $a = 60$, 70, 80, 90 and 100.

In the fact, the defocused MTFs of WFC imaging system with the CPM have a small oscillation to variation of defocus value. Each mask parameter of the CPM has a different defocus invariant characteristic. Based on using Eq. (6.10), the defocused MTF curves of the CPM for different five phase mask parameter values (a = 60, 70, 80, 90 and 100) and different defocus values are shown in Fig. 6.2, where the defocus value is set to 0, 15, and 30, respectively. For comparing, the corresponding defocused MTFs of a traditional imaging system with a clear aperture are also shown in Fig. 6.2(f). As Fig. 6.2 indicates, the defocused MTFs of WFC imaging system with the CPM are less sensitive to defocus in comparison with the defocused MTFs of a traditional imaging system. Although the defocused MTFs of WFC imaging system with the CPM have invariant property to defocus, but the magnitude of the defocused MTFs of WFC imaging system with the CPM is lower compared with the magnitude of the diffraction-limit in-focus MTF of a traditional imaging system. However, the defocused MTFs at all spatial frequencies are no zeros, and therefore the information of object is not lost. We can use the simple deconvolution filter to restore sharper images at in-focus and out-focus positions. As Figs. 6.2(a)-(e) show, when phase mask parameter is smaller, the magnitude of the defocused MTFs is bigger, but the defocused MTFs have more oscillations. This can be obviously seen in high frequency part.

The intensity image of traditional imaging system or WFC imaging system with the CPM can be given by

$$g = o \otimes h + n,\tag{6.18}$$

where o is the object; h is the PSF; n is the noise; the symbol $\otimes$ is the convolution operation.

From Eq. (6.18), it can be seen that the image captured by the detector is equal to convolution of the object with the PSF. In addition, the noise is added in the image-recorded process. Thus the difference between WFC imaging system and a traditional imaging system with a clear aperture lies in the PSF. To show difference between the optics part of WFC imaging system with the CPM and a traditional imaging system with a clear aperture, we perform numerical simulations of the defocused PSFs of the two imaging systems. Fig. 6.3 shows the defocused PSFs of WFC imaging system with the CPM and a traditional imaging system with a clear aperture for different defocus values of ψ = 0, 10, 20 and 30. For a traditional imaging system, the defocused PSFs are sensitive to defocus, the size and the shape of the defocused PSFs spread very fast with increase of defocus value. At the same time, for WFC imaging system with the CPM, the shape and size of the defocused PSFs are invariant to defocus, but the defocused PSFs have big size. Therefore, the intermediate images of the CPM are blurred. Additionally, the defocused PSFs of a traditional imaging system with a clear aperture are symmetric, while the defocused PSFs of WFC imaging system with the CPM are asymmetric.

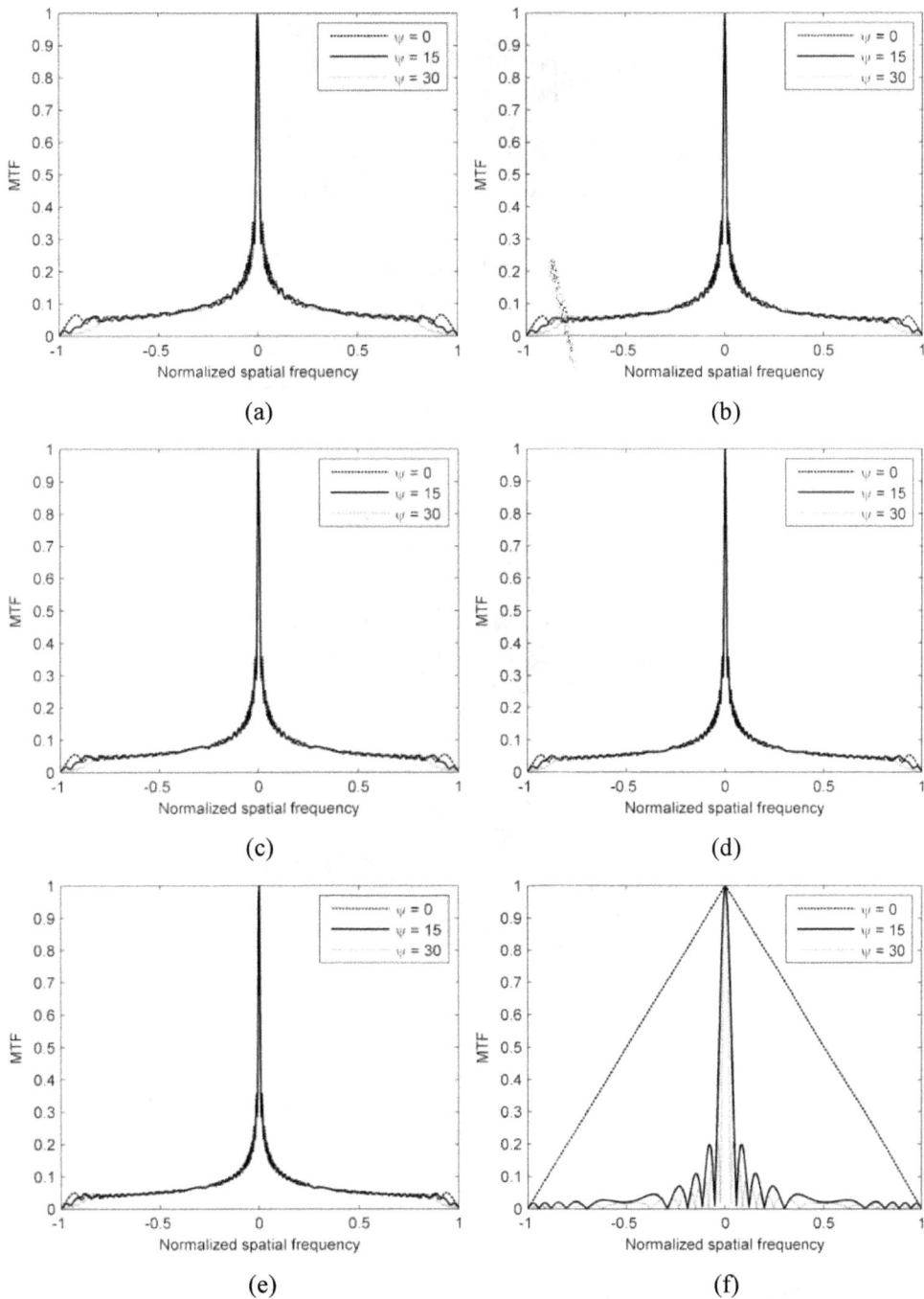

Fig. 6.2. Defocused MTF curves of a traditional imaging system with a clear aperture without any phase mask and WFC imaging system with the CPM. The phase mask parameters: (a) $a = 60$; (b) $a = 70$; (c) $a = 80$; (d) $a = 90$; (e) $a = 100$.

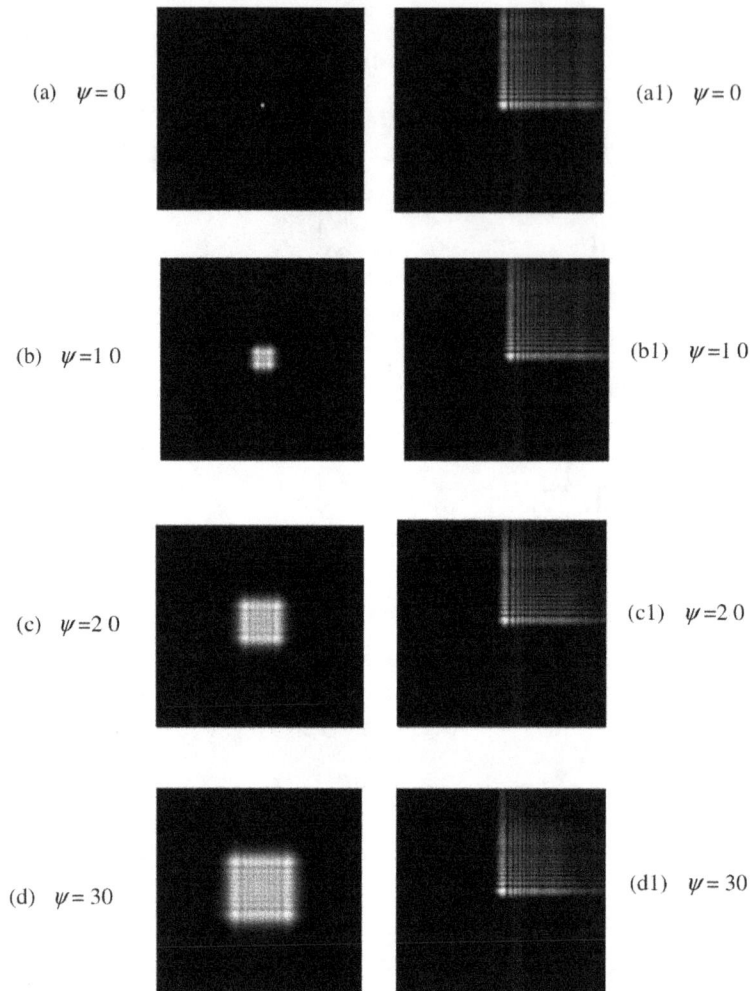

Fig. 6.3. Two-dimensional PSFs of: left column: a traditional imaging system with a clear aperture, right column: the CPM with the phase mask parameter of $a = 60$.

Employing the convolution, the Fourier transform of the image is the product of the Fourier transform of the object, O, the Fourier transform of the PSF, H, and the Fourier transform of the noise, N

$$G = O \times H + N, \tag{6.19}$$

where G is the Fourier transform of image, g.

To perform analysis of decoded characteristics in the digital processing, here the noise is ignored. The digital processing with the inverse filter driven from the in-focus OTF of WFC imaging system with the CPM is applied to restore sharp image at all positions of defocus. The inverse filter can be presented by

154

$$F = \frac{H_{ideal}}{H_{in-focus}},$$ (6.20)

where $H_{in\text{-}focus}$ is the in-focus OTF of WFC imaging system; H_{ideal} is the diffraction-limit in-focus OTF of traditional imaging system.

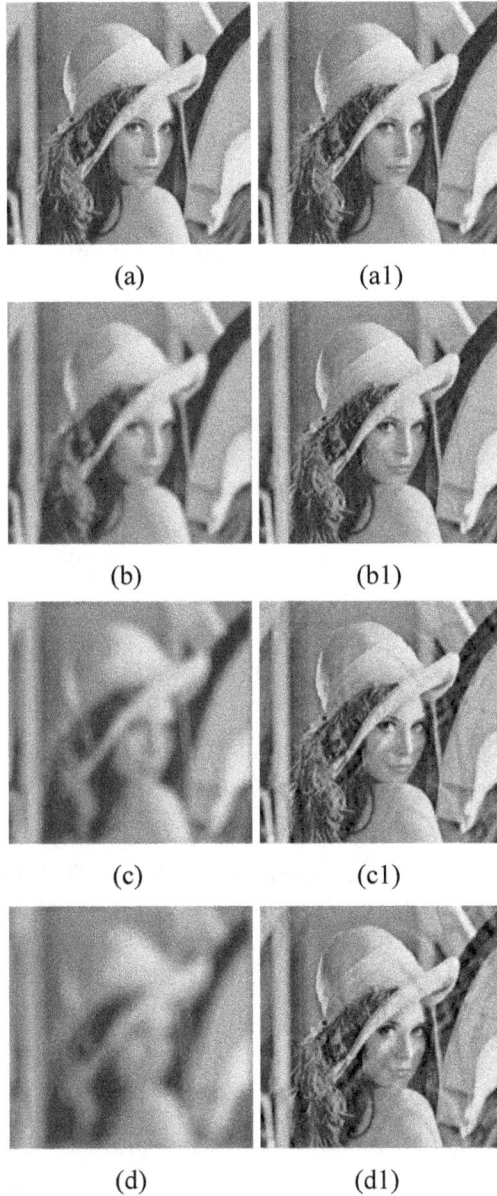

Fig. 6.4. Simulation images: left a traditional imaging system, right WFC imaging system with the CPM for phase mask parameter of $a = 60$: (a), (a1) defocus of $\psi = 0$; (b), (b1) defocus of $\psi = 10$; (c), (c1) defocus of $\psi = 20$; (d), (d1) defocus of $\psi = 30$.

Finally, in order to show evidently imaging performance of WFC imaging system with the CPM over a big range of defocus, the restored images for phase mask parameter of $a = 60$ and some different defocus values of $\psi = 0, 10, 20$ and 30 are shown in Fig. 6.4. For comparison, the corresponding recorded images of a traditional imaging system are also indicated in Fig. 6.4. For the recoded images of a traditional imaging system, when defocus value is increasing, the quality of the images degrades very fast. This means that a traditional imaging system is sensitive to defocus. While, for WFC imaging system with the CPM, the restored images over the big range of defocus are sharp and clear. This indicates that WFC imaging system with the CPM can be used to obtain high-quality image in extending the DOF.

6.3. Development of Wavefront Coding Technique

For WFC technique, researches are mainly concentrated into finding new phase masks, optimization of phase mask and reducing image artifacts due to digital processing. In the following, we will introduce research status about developments of phase masks, optimization methods of phase mask, and the digital processing of WFC imaging systems.

6.3.1. Development of Phase Mask Profiles

In WFC systems, a phase mask is placed in the pupil plane to modulate the wavefront so that the PSF or OTF is not sensitive to defocus and after applied the digital processing to restore sharp image. The most important part to WFC imaging lies in the design of suitable phase masks to obtain defocus invariant imaging characteristics. These phase masks are separated into two kinds: symmetrical phase mask and asymmetrical phase mask.

Sochacki *et al.* suggested the use of a phase mask to code wavefront for the DOF extension. They introduced a radially symmetric pupil function in the pupil plane of imaging system [23]. The phase function of this phase mask was called the radially symmetric logarithmic phase mask. In 1998, Zalvidea and Sicre presented another radially symmetric phase mask, which was called the quartic phase mask (QPM) to produce a similar axial irradiance [24, 25]. The difference with the previously radially symmetric phase logarithmic mask is that a phase shift is increased. In 2009, Feng Zhou et al. reported a radially symmetric phase mask which is called the rational phase mask [26]. A rational phase mask with 12 terms is optimized. This phase mask produces insensitive intensity distribution in comparison with the QPM. In 2001, Chi and George introduced another radially symmetric phase mask which is a lens with separated annular rings of different focus [27]. The proposed phase mask was called logarithmic asphere. In 2009, Mouroulis and Robinson et al. proposed a spherical phase mask to achieve the increase of the DOF [28]. This mask has the shape of spherical aberration, and ability of DOF improvement was demonstrated. Actually the spherical mask is a simplified QPM. In 2012, Xutao Mo suggested a radially symmetric phase mask composed of several annular zones with equal area based on the incoherent imaging theory from Fourier optics, and this phase mask is called APM [29]. Another radially symmetric mask similar to APM is also introduced. These two phase masks generate the MTF insensitive to defocus. All phase masks produce the final images with reducing contrast but images remain sharpness over enhanced the

DOF. The image contrast can be restored by employing image processing, but image sharpness is retained even in the unrestored image.

The CPM was first proposed in 1995 by Dowski and Cathey [21]. The CPM has been derived by an analytic method based on the stationary point approximation method to realize the imaging property for extending the DOF. Until now, many researchers all over the world focus their research to seek for new phase masks to enhance imaging performance of WFC imaging systems in extending the DOF. In 2003, pupil-phase-engineering associated with WFC technique was proposed [30, 31]. Another CPM was introduced, which could be described by a more general third order polynomial. This phase mask was called the generalized CPM. In 2001, the Dowski group developed a new kind of asymmetrical phase mask that owned the form of the logarithmic function by using the spatial domain computation routine [32]. The mathematical derivation was rigorous, and the depth extension effect was demonstrated by the corresponding simulation results. In addition, three years later, in 2004, other paper of this group was published to investigate the characteristics of the logarithmic phase mask again [33]. In this chapter, more analysis tools were employed and more aspects were considered. In 2008, Hui Zhao et al. introduced improved logarithmic phase mask by making a very simple modification to the conventional logarithmic phase mask [34]. The modified phase mask still had the logarithmic form, but it was superior to the conventional logarithmic phase mask. Two years later, in 2010, expanded research was carried out in [35], and better characteristic of the modified logarithmic phase mask was further confirmed. However, Hui Zhao and Yingcai Li shown that the defocused MTF of the previously logarithmic phase masks still exhibited remarkable oscillations [36]. Therefore, in 2010, the phase profiles of two phase masks were further modified and a more stable defocused MTF was obtained. In 2004, Angel Sauceda and Jorge Ojeda-Castañeda explored the use of phase profiles with fractional power for tailoring MTFs with high focal depth at pupil aperture [37]. Several fractional-power phase profiles are presented and numerical evaluations of their MTFs to focus errors are achieved. In 2007 Qingguo Yang et al. suggested the use of an exponential phase pupil mask to increase the DOF [38]. This phase mask has two variable parameters to control the shape of the mask so as to modulate the wave-front more flexible. In 2007, Nicolas Caron and Yunlong Sheng introduced a polynomial phase mask, which is still separated in the Cartesian coordinate system [39]. This phase mask is similar with the CPM. The phase function contains 16 terms of odd powers and was optimized by the simulated annealing. The phase mask leads invariant defocused MTF within a large range of DOF. In 2008, in order to acquire a further enhancement of the DOF in WFC imaging by decreasing the impact of focus error in the OTF, Yasushisa Takahashi and Shinichi Komatsu in Waseda University proposed the use of a free-form phase mask [40]. The specific shape of a free-form phase mask with 24 terms is optimized by using the simulated annealing algorithm. The optimized free-form phase mask produces a much larger focal tolerance and better final images than the CPM in absence of noise. In 2009, Feng Zhou et al. proposed a phase mask to increase the DOF in optical-digital hybrid imaging systems [41]. This phase mask was called the rational phase mask. Using the stability of defocused MTF as merit function, an optimized rational phase mask with 12 terms is obtained. The simulated results are shown that the restored images of rational phase mask have high contrast and high resolution performance. In 2010, Hui Zhao and Yingcai Li proposed a sinusoidal phase mask to raise the DOF [42]. This phase mask provides a smaller shifting effect of the PSF and small size of the PSF. After two years, an improved sinusoidal phase mask was proposed by Jiagang Wang et al. by adding a new parameter into the conventional sinusoidal phase mask [43]. The improved sinusoidal phase mask can further extend the DOF in hybrid imaging systems. Additionally, the thickness of the improved sinusoidal phase mask is lower than the conventional sinusoidal phase mask and hence it has an advantage in fabrication. Recently, a lot of phase masks to extend depth of field have been introduced to obtain the more invariant defocusMTFs, such as tangent phase mask, square root phase mask, acrsin phase mask, the imaging performance is improved further [44-46]. Some function types of asymmetrical phase masks are shown in Table 6.1.

Table 6.1. Phase types of asymmetrical phase masks.

Name	Phase function
Cubic	$f(x,y) = ax^3 + by^3$
Generalized cubic	$f(x,y) = a(x^3 + y^3) + b(x^2 y + xy^2)$
Symmetrical Quadratic	$f(x,y) = \text{sgn}(x)ax^2 + \text{sgn}(y)ay^2$
Logarithmic	$f(x,y) = \text{sgn}(x)ax^2(\log|x| + b) + \text{sgn}(y)ay^2(\log|y| + b)$
Improved-1 logarithmic	$f(x,y) = \text{sgn}(x)ax^2(\log\||x| + b\|) + \text{sgn}(y)ay^2(\log\||y| + b\|)$
Improved-2 logarithmic	$f(x,y) = \text{sgn}(x)ax^4(\log|x| + b) + \text{sgn}(y)ay^4(\log|y| + b)$
Improved logarithmic	$f(x,y) = \text{sgn}(x)ax^4(\log\||x| + b\|) + \text{sgn}(y)ay^4(\log\||y| + b\|)$
Exponential	$f(x,y) = ax\exp(bx^2) + ay\exp(y^2)$
Sinusoidal	$f(x,y) = ax^4 \sin(bx) + ay^4 \sin(by)$
Improved sinusoidal	$f(x,y) = a|x|^c \sin(bx) + a|y|^c \sin(by)$
Fraction	$f(x,y) = \text{sgn}(x)a|x|^b + \text{sgn}(y)a|y|^b$
Free-form	$f(x,y) = \sum_{n=0}^{K}\left(\sum_{m=0}^{n} C_{nm} x^n y^{m-n}\right)$
Rational	$f(x,y) = \text{sgn}(x)\dfrac{\sum_{n=0}^{N} a_n|x|^n}{\sum_{m=0}^{M} b_m|x|^m} + \text{sgn}(y)\dfrac{\sum_{n=0}^{N} a_n|y|^n}{\sum_{m=0}^{M} b_m|y|^m}$
Polynomial	$f(x,y) = \text{sgn}(x)\sum_{n=0}^{N} a_n|x|^n + \text{sgn}(y)\sum_{n=0}^{N} a_n|y|^n$
High-order	$f(x,y) = \text{sgn}(x)a|x|^n + \text{sgn}(y)b|y|^n \quad n \geq 3$
Tangent	$f(x,y) = ax^2 \tan bx + ay^2 \tan by$
Square-root	$f(x,y) = ax\sqrt{b - x^2} + ay\sqrt{b - y^2}$
Arsin	$f(x,y) = f(x,y) = ax^2 \arcsin bx + ay^2 \arcsin by$

6.3.2. Optimization Phase Mask

Generally, the phase mask parameters should be optimized before the use of phase mask because the imaging performance of the phase mask can be reduced by chosen phase mask parameters randomly. The evaluated methods of imaging quality of WFC systems are the similar as the conventional imaging systems. The mainly evaluated methods include the use of the PSF, the OTF, the MTF and the phase transfer function (PTF). These evaluated methods are used to construct the merit functions for assessing the similarity and the decoded ability of WFC imaging systems.

The PSF is a powerful tool which is used to evaluate imaging performance of imaging system. In wavefront coding technique, the optimization functions based on the use the PSF are built. In 2007, Wenzi Zhang et al. suggested a method based on the standard deviation to judge the stability level of the PSF over a wide range of defocus [47]. This method is to optimize the phase profile of phase mask and the sharper image is calculated for comparison. The optimized method provided better and more invariant imaging property than the original system without changing the position of the image plane. Another method based on the use of the simple Strehl ratio for optimization of pupil phase mask in WFC imaging system was introduced [48-52]. Based on the evaluation the of stability level of the Strehl ratio over a wide range of defocus, the phase mask parameters are optimized. Another method, *Fisher information* (FI) of the PSF has been described [53, 54]. The FI metric uses the full PSF, not just its on-axis values and measures the sensitivity of the PSF to defocus. Based on this optimization method, the optimal mask parameters of some phase masks are obtained. In 2010, wavefront encoding with several CPM designs was researched in three-dimensional PSF engineering to decrease their sensitivity to depth-induced spherical aberration which affects computational complexity in three-dimensional microscopy imaging [55]. The variation of wavefront encoding PSFs to defocus and to spherical aberration was evaluated as a function of phase mask parameters by employing mean-square-error (MSE) metric of the PSF. This method facilitates the selection of mask profiles for extended DOF microscopy and computational optical sectioning microscopy. Further investigates on pupil phase distribution and simulated wavefront encoding microscope images judged the engineered PSFs and demonstrated spherical aberration invariance over sample depths of 30 μm. In 2011, Mads Demenkov introduced optimization method of WFC imaging systems based on maximization of kurtosis of the restored PSF [56]. Employing this measure, author optimized several phase masks for extended DOF in hybrid imaging systems and the acquired results are nearly identical to optimization results based on full-reference image measures of restored images. The kurtosis measure is fast to calculate and requires no noise distributions, images, or alignment of the restored images, but only the signal-to-noise-ration (SNR).

Evaluation functions based on the MTF have been employed to optimize phase masks. In 2007, Hui Zhao et al. introduced the mean-square-derivation of the MTF to optimization of phase mask parameters [57]. Based on the optimization way, the phase masks are used to perform optimization. In 2009, Yan Feng and Zhang Xue-Jun introduced the MTF similarity which is defined to exhibit the relationship among MTF at different position or different field [58]. The invariant MTF of optimized imaging system drops to 0.0119 while the MTF of original imaging system is larger than 0.018. When the all the fields of view is taken into consideration, the MTF invariant of optimized imaging system and the MTF invariant of original imaging system are less than 0.015 and larger than 0.02, respectively. In 2010, Guillem et al. suggested an optimization process to seek for the most suitable phase mask strength by minimizing a metric function based on the restored and aberration-degraded MTF [59]. This process accounts for the trade-off between invariant acquired and imaging fidelity and permits selection of both the phase mask strength and the parameter of the restoration filter of a hybrid imaging system.

Evaluation functions related to the OTF have been proposed to assess extending DOF performance. A convenient mathematical tool to measure the sensitivity level of the OTF to focus error is the FI of the OTF [60]. FI is commonly employed to exhibit the information content of a given signal pertaining to a certain parameter. The smaller the value of FI related to the OTF in certain defocus image plane, the lower change of the OTF to the focus errors near this image plane. For an ideal defocused invariant optical imaging system, FI would be zero at all defocus state. Thus the integral of FI can be employed as a metric to measure the sensitivity level of an optical imaging system to defocus. The method is popularly used to optimize of phase mask parameters. In 2008, Shane Barwich introduced an optimization function based on the second derivative of the OTF with respect to defocus at the origin as a mean of decreasing computational overhead [61, 62]. The above derivation indicates that this metric can be calculated with four Fourier transfers per phase mask. The metrics defined over a big range of defocus value generally require at least one or two Fourier transfers per defocus value. Thus, the second-derivative metric provides calculation relief. A design is performed to demonstrate that this metric can be employed to predict invariant of the WFC imaging system to large values of defocus. In 2017, V. Le *et al.* proposed an optimization method based on the use of overall OTFs with multitarget optimization, the method optimize together all defocus positions to obtain optimal phase mask parameters [63].

Recently, several image evaluation functions based on the quantification of restored images over a given defocus range have been used to optimize phase masks. In 2009, Frederic Diaz *et al.* suggested an optimization method taking into account digital processing [64, 65]. Authors use an image quality criterion that takes into account the variable of the system's PSF along the expected defocus range and the noise improvement induced by deconvolution. By considering the CPM, this criterion may produce filter parameters that are remarkably different from previously proposed methods to ensure the stringent invariance of the PSF. In 2010, Tom Vettenburg et al. introduced an optimization method to seek for the phase-profile and modulation depth that maximizes the image fidelity for a given aberration tolerance [66]. Authors defined the image fidelity as the inverse of the expected MSE between the original image and the restored images or the mean structural similarity metric (MSSIM) between the original image and the restored images. By using these optimization techniques, commonly studied asymmetrical and radially-symmetric phase masks have been compared and analyzed. Mads Demenikov introduced another optimization technique based on the earth mover's distance, which has been extensively employed in image retrieval [67]. The earth mover's distance is a real metric and it is related to human visual perception. Furthermore, the earth mover's distance is not a pixel-to-pixel comparison, and it is hence invariant to both the translation and the rotation, which is why it has been many in image retrieval.

In 2013, Xutao Mo and Jinjiang Wang shown that if the PTF is independent with the defocus as much as possible, the effect of image artifacts can be decreased [68]. From that, an optimization model based on the defocus invariance of the PTF of WFC imaging system was introduced. The defocus invariance of the PTF can be expressed by the MSE of all the defocus. In 2016, V. Le, *et al.*, introduced an optimization based on the only use of the PTF [69]. The cubic phase mask and asymmetrical tangent phase mask was optimized. The tangent phase mask has less artifact images than the cubic phase mask.

6.3.3. The Digital Processing

In wavefront coding technique, the digital processing is usually used to restore sharp image, but this generates image artifacts on the final images. There are two major types of image artifacts caused by different restoration techniques including ring artifacts due to regularization and boundary artifacts near edges of the image due to deconvolution filter based on the use of the Discrete Fourier transform. To decrease ring artifacts, some iterative restoration techniques have been introduced [70]. The main problem of the iterative restoration techniques is their computational speed. Recently, another fast non-iterative technique, which employs Fourier transform based deconvolution followed by wavelet de-noising, has been presented [71]. Iterative restoration techniques for ringing artifacts decrease have been introduced for WFC imaging system with phase masks [72]. To decrease boundary artifacts, several image restoration techniques employ deconvolution filter with different types of boundary conditions [73]. Image restoration technique with reflexive and anti-reflexive boundary conditions to boundary reflections has been employed for a WFC imaging system with the CPM [74, 75]. In 2013, Peng Liu et al. suggested *bi*-conjugate gradient stabilized method in image deconvolution of a WFC imaging system [76]. The method can obtain smooth deconvoluted image without ringing effect on the border and vibration on the edge of the recovered image and smoother than the Wiener filter, additionally preserving more details than the Truncated Singular Value Decomposition method. WFC imaging systems with CPM have obvious image artifacts. In 2010, Mads Demennikov and Andrew R. Harvey presented the analytical solution of image artifacts for WFC system with the CPM [76]. Authors have indicated that image artifacts are the form of image replications and are caused by a net phase modulation of the OTF. Therefore, artifact-free image restoration can be achieved by displaying phase modulation of system-transfer function. In 2010, authors introduced a parametric blind-deconvolution algorithm based on minimization of the high-frequency content of the restored image that enables recovery of artifact-free images for a big range of defocus [77]. In 2013, Xutao Mo and Jinjiang Wang suggested the use of PTF to alleviate image artifacts on the restored for WFC imaging system [78]. In order to alleviate the image artifacts in WFC imaging systems with phase mask, an optimization model based on the PTF was proposed to make the PTF invariance to defocus. Thereafter, an image restoration filter based on the average PTF in the designed DOF was suggested a long with the PTF-based optimization. The combination of the optimization with the image restoration can alleviate the image artifacts on the restored images. However, this method still has image artifacts on the restored images. In 2013, to overcome noise and ring artifacts in linearly filtered microscope images, a nonlinear filtering scheme minimized the background noise and image artifacts in the post-processed image was suggested [79]. The nonlinear filter is generated by using a training algorithm and iterative optimizer. Biological microscope images processed with the nonlinear filter give a remarkable improvement in image quality and SNR over the conventional linear filter. Recently, in 2014, a method was called Complementary Kernel Matching, which exploits the range-dependent translation of two images recorded with dissimilar imaging PSFs obtained from two CPMs, followed by inference of defocus and recovery of an image with correct kernel filter and hence without image-replication artifacts or translation [80]. By employing Complementary Kernel Matching, authors demonstrated that the method can be used to obtain high-quality macroscopic and microscopic imaging of scenes presenting an extended defocus of up to two waves and the generation of defocus maps with an uncertainty of 0.036 waves.

6.4. Applications

WFC system has been used in many practical applications. WFC system is used to extend depth of filed in many optical systems. For infrared optical systems, the index refraction has sensitive to temperature change. When temperature has variation, the focal plane has shift. This produces the reduction of image quality of optical system. WFC system can be applied in infrared optical systems in order to obtain image invariant over a wide range of temperature, so the effect of temperature change can be corrected. Many experiences have shown the effectiveness of WFC system for correcting the effect of temperature change in infrared optical system [81-85]. In order to obtain superresolution and large depth of field, the WFC technique is combined with superresolution techniques as shown in Refs. [86-88]. For light sheet fluorescence microscopy, there are trade-off between thickness and field of view of light sheet. Order to overcome the limit, the wavefront coding technique is applied in light sheet fluorescence microscopy. CPM is used to extend field of view in light sheet microscopy for 3D imaging [89-91]. A thickness sample can be imaged by adding spherical and cubic phase mask in illumination path of light sheet microscopy. In 3D imaging, the imaging speed is an important parameter. In order to improve the imaging speed in light sheet fluorescence microscopy, the WFC technique is used. The cubic phase mask and spherical mask are used to obtain fast speed in light sheet microscopy [92, 93]. By using the CPM or spherical mask in detector path, the imaging speed in light sheet fluorescence microscopy is fast in comparison with conventional light sheet flurescence microscopy. Other application of WFC technique is to be used for correcting aberrations including Petzval curvature, astigmatism, assembly related misfocus, spherical and chromatic aberration [94-97]. When wavefront coding system is used, optical components in optical system can be reduced, while the image quality is also guaranteed. Therefore, the cost reduces. Some experience results are shown. Other application of wavefront coding system is in zoom optical system [98-100]. In zoom optical system with a phase mask, only an optical component requires shift, so the optical system is simple and perform easy. However, the image quality is also ensured.

6.5. Conclusion

In this chapter, we have provided an overview about WFC technique for imaging systems. The imaging theory of WFC technique is presented and discussed. The development of WFC technique including phase masks, optimization of the phase masks and the digital processing is expressed. Many practical applications of WFC technique in imaging systems are performed. The effectiveness of WFC technique in imaging systems is demonstrated by experimental results. WFC system has a wide range for practical applications and can offer great potential development of a huge range of new imaging systems in the future.

Acknowledgment

This work is supported by the National Key Research and Development Program of China (2016YFF0101400); National Basic Research Program of China (973 Program) (2015CB352003); National Natural Science Foundation of China (NSFC) (61335003, 61427818, 61851110762); Natural Science Foundation of Zhejiang province (LR16F050001); the Fundamental Research Funds for the Central Universities (2017FZA5004); and Vietnam National Foundation for Science and Technology Development (NAFOSTED) under Grant number (103.03-2018.08).

References

[1]. R. Hild, M. J. Yzuel, J. C. Escalera, et al., Influence of nonuniform pupils in imaging periodical structures by photolithographic systems, *Optical Engineering*, Vol. 37, Issue 4, 1998, pp. 1353-1363.

[2]. J. Campos, J. C. Escalera, R. Hild, et al., The assessment of uniform pupils in photolithography, *Microelectronic Engineering*, Vol. 30, Issue 14, 1996, pp. 103-106.

[3]. R. Hild, M. J. Yzuel, J. C. Escalera, High focal depth imaging of small structures, *Microelectronic Engineering*, Vol. 34, Issue 2, 1997, pp. 195-214.

[4]. R. Narayanswamy, G. E. Johnson, P. E. X. Silveira, et al., Extending the imaging volume for biometric iris recognition, *Applied Optics*, Vol. 44, Issue 5, 2005, pp. 701-712.

[5]. R. Hild, J. C. Escalera, M. J. Yzuel, et al., Optical imaging of small structures in optical lithography, high focal depth with optical filter, in *Proceedings of the 17th Congress of the International Commission for Optics: Optics for Science and New Technology*, 1996, pp. 79-80.

[6]. R. Hild, J. C. Escalera, M. J. Yzuel, et al., High focal depth imaging by beam shaping optical elements, *Microelectronic Engineering*, Vol. 35, Issue 1-4, 1997, pp. 205-208.

[7]. G. Saavedra, I. Escobar, R. Martinez-Cuenca, et al., Reduction of spherical-aberration impact in microscopy by wavefront coding, *Optics Express*, Vol. 17, Issue 16, 2009, pp. 13810-13818.

[8]. W. T. Welford, Use of annular apertures to increase focal depth, *Journal of the Optical Society of America A*, Vol. 5, Issue 8, 1960, pp. 749-753.

[9]. J. Tsujiuchi, Correction of optical images by compensation of aberrations and by spatial filter, in Progress in Optics (E. Wolf, Ed.), *North-Holland*, Amsterdam, The Netherlands, 1963.

[10]. M. Mino, Y. Okano, Improvement in the OTF of a defocused optical system through the use of shaded apertures, *Applied Optics*, Vol. 10, Issue 10, 1997, pp. 2219-2225.

[11]. J. Ojeda-Castañeda, L. R. Berriel-Valdos, E. L. Montes, Line-spread function relatively insensitive to defocus, *Optics Letters*, Vol. 8, Issue 8, 1983, pp. 458-460.

[12]. H. O. Bartelt, J. Ojeda-Castañeda, E. E. Sicre, Misfocus tolerance seen by simple inspection of the ambiguity function, *Applied Optics*, Vol. 23, Issue 16, 1984, pp. 2693-1696.

[13]. J. Ojeda-Castañeda, L. R. Berriel-Valdos, E. L. Montes, Spatial filter for increasing the depth of focus, *Optics Letters*, Vol. 10, Isue 11, 1985, pp. 520-522.

[14]. J. Ojeda-Castañeda, P. Andres, A. Diaz, Annular apodizerss for low sensitivity to defocus and to spherical aberration, *Optics Letters*, Vol. 11, Issue 8, 1985, pp. 487-489.

[15]. J. Ojeda-Castañeda, P. Andres, E. Montes, Phase-space representation of the Strehl ratio: ambiguity function, *Optical Society of America Am. A*, Vol. 4, Issue 2, 1987, pp. 313-317.

[16]. J. Ojeda-Castañeda, L. R. Berriel-Valdos, E. Montes, Ambiguity function as a design tool for high focal depth, *Applied Optics*, Vol. 27, Issue 4, 1988, pp. 790-795.

[17]. J. Ojeda-Castañeda, E. Tepichin, A. Pons, Apodization of annular apertures: Strehl ratio, *Applied Optics*, Vol. 27, Issue 24, 1988, pp. 5140-5145.

[18]. J. Ojeda-Castañeda, R. Ramos, A. Noyola-Isgleas, High focal depth by apodization and digital restoration, *Applied Optics*, Vol. 27, Issue 12, 1988, pp. 2583-2586.

[19]. G. Hausler, A method to increase the depth of focus by two step image processing, *Optics Communications*, Vol. 6, 1972, pp. 34-42.

[20]. H. Nagahara, S. Kuthirummal, C. Zhou, et al., Flexible depth of field photography, in *Proceedings of the European Conference on Computer Vision (ECCV'08)*, 2008, pp. 60-73.

[21]. E. R. Dowski, W. T. Cathey, Extended depth of field through wave-front coding, *Applied Optics*, Vo. 34, Issue 11, 1995, pp. 1859-1866.

[22]. H. Zhao, Y. Li, Approximate analytical optical transfer function for wavefront-coded imaging system with two-dimensional rectangularly separable phase masks, *Optical Engineering*, Vol. 49, Issue 9, 2010, 093201.

[23]. J. Sochacki, S. Bara, Z. Jaroszewicz, et al., Phase retardation of the uniform-intensity axilens, *Optics Letters*, Vol. 17, Issue 1, 1992, pp. 7-9.

[24]. D. Zalvidea, E. E. Sicre, Phase pupil functions for focal-depth enhancement derived from a Wigner distribution function, *Applied Optics*, Vol. 37, Issue 17, 1998, pp. 3623-3627.

[25]. D. Zalvidea, C. Colautti, E. E. Sicre, Quality parameters analysis of optical imaging systems with enhanced focal depth using the winner distribution function, *Journal of the Optical Society of America A*, Vol. 17, Issue 5, 2000, pp. 867-873.

[26]. F. Zhou, R. Ye, G. Li, et al., Optimized circularly symmetric phase mask to extend the depth of focus, *Journal of the Optical Society of America A*, Vol. 26, 2009, pp. 1889-1895.

[27]. W. Chi, N. George, Electronic imaging using a logarithmic asphere, *Optics Letters*, Vol. 26, Issue 12, 2000, pp. 875-877.

[28]. P. Mouroulis, Depth of field extension with spherical optics, *Optics Express*, Vol. 16, Issue 17, 2008, pp. 12995-13004.

[29]. X. Mo, Optimized annular phase masks to extend depth of field, *Optics Letters*, Vol. 37, Issue 11, 2012, pp. 1808-1810.

[30]. S. Prasad, T. C. Torgersen, V. P. Pauca, et al., Engineering the pupil phase to improve image quality, *Proceedings of SPIE*, Vol. 5108, 2003, pp. 1-12.

[31]. M. Demenikov, A. R. Harvey, A technique to remove image artifacts in optical systems with wavefront coding, *Proceedings of SPIE*, Vol. 7429, 2009, 74290N.

[32]. S. S. Sherif, E. R. Dowski, W. T. Cathey, A logarithmic phase filter to extend the depth of field of incoherent hybrid imaging systems, *Proceedings of SPIE*, Vol. 4471, 2001, pp. 272-279.

[33]. S. S. Sherif, W. T. Cathey, E. R. Dowski, Phase plate to extend the depth of field of incoherent hybrid imaging systems, *Applied Optics*, Vol. 43, Issue 13, 2004, pp. 2709-2721.

[34]. H. Zhao, Q. Li, H. J. Feng, Improved logarithmic phase mask to extend the depth of field of an incoherent imaging system, *Optics Letters*, Vol. 33, Issue 11, 2008, pp. 1171-1173.

[35]. H. Zhao, Y. C. Li, Performance of an improved logarithmic phase mask with optimized parameters in a wavefront-coding system, *Applied Optics*, Vol. 49, Issue 2, 2010, pp. 229-238.

[36]. H. Zhao, Y. C. Li, Optimized logarithmic phase masks used to generate defocus invariant modulation transfer function for wavefront coding system, *Optics Letters*, Vol. 35, Issue 15, 2010, pp. 2630-2632.

[37]. A. Sauceda, J. Ojeda-Castañeda, High focal depth with fractional-power wave fronts, *Optics Letters*, Vol. 29, 2004, pp. 560-562.

[38]. Q. Yang, L. Liu, J. Sun, Optimized phase pupil masks for extended depth of field, *Optics Communications*, Vol. 272, 2007, pp. 56-66.

[39]. N. Caron, Y. Sheng, Polynomial phase mask for extending depth-of-field optimized by simulated annealing, *Proceedings of SPIE*, Vol. 6832, 2007, 68321G.

[40]. Y. Takahashi, S. Komatsu, Optimized free-form phase mask for extension of depth of field in wavefront-coded imaging, *Optics Letters*, Vol. 33, Issue 13, 2008, pp. 1515-1517.

[41]. F. Zhou, G. W. Li, H. T. Zhang, et al., Rational phase mask to extend the depth of field in optical-digital hybrid imaging systems, *Optics Letters*, Vol. 34, Issue 3, 2009, pp. 380-382.

[42]. H. Zhao, Y. C. Li, Optimized sinusoidal phase mask to extend the depth of field of an incoherent imaging system, *Optics Letters*, Vol. 35, Issue 2, 2010, pp. 267-269.

[43]. J. Wang, J. Bu, M. Wang, et al., Improved sinusoidal phase plate to extend depth of field in incoherent hybrid imaging systems, *Optics Letters*, Vol. 37, 2012, pp. 4534-4536.

[44]. V. Le, S. Chen, Z. Fan, Optimized asymmetrical tangent phase mask to obtain defocus invariant modulation transfer function in incoherent imaging systems, *Optics Letters*, Vol. 39, Issue 7, 2014, pp. 2171-2173.

[45]. V. Le, Z. Fan, N. P. Minh, S. Chen, Optimized square root phase mask used to generate defocus invariant modulation transfer function in hybrid imaging systems, *Optical Engineering*, Vol. 54, Issue 3, 2015, 035103.

[46]. L. Wang, Q. Ye, J. Nie, X, Sun, Optimized asymmetrical arcsine phase mask for extending the depth of field, *IEEE Photonics Technology Letters*, Vol. 30, Issue 14, 2018, pp. 1309-1312.

[47]. W. Z. Zhang, Y. P. Chen, T. Y. Zhao, et al., Simple PSF based method for pupil phase masks optimization in wavefront coding system, *Journal of Zhejiang University Science A*, Vol. 8, Issue 2, 2006, pp. 180-185.

[48]. S. Mezouari, A. R. Harvey, Phase pupil function for reduction of defocus and spherical aberrations, *Optics Letters*, Vol. 28, Issue 10, 2003, pp. 771-773.

[49]. S. Mezouari, G. Muyo, A. R. Harvey, Amplitude and phase filters for mitigation of defocus and third-order aberrations, *Proceedings of SPIE*, Vol. 5249, 2004, pp. 238-248.

[50]. S. Mezouari, G. Muyo, A. R. Harvey, Circular symmetric phase filters for control of primary third-order aberrations: coma and astigmatism, *Journal of the Optical Society of America A*, Vol. 23, Issue 5, 2006, pp. 1058-1062.

[51]. W. Zhang, Y. Chen Y, T. Zhao, et al., Simple Strehl ratio based method for pupil phase mask's optimization in wavefront coding system, *Chinese Optics Letters*, Vol. 4, Issue 9, 2006, pp. 515-517.

[52]. S. Chen, Z. Fan, H. Chang, et al., Nonaxial Strehl ratio of wavefront coding systems with a cubic phase mask, *Applied Optics*, Vol 50, Issue 19, 2011, pp. 3337-3345.

[53]. V. Pauca, R. J. Plemmons, S. Prasad, et al., An integrated optical-digital approach for improved image restoration, in *Proceedings of the AMOS Technical Conference*, Maui, HI, September 2003, pp. 105-170.

[54]. V. P. Pauca, R. J. Plemmons, S. Prasad S, et al., Integrated optical-digital approaches for enhancing image restoration and focus invariant, *Proceedings of SPIE*, Vol. 5205, 2003, pp. 348-357.

[55]. S. Yuan, C. Preza, Point-spread function engineering to reduce the impact of spherical aberration on 3D computational fluorescence microscopy imaging, *Optics Express*, Vol. 19, Issue 23, 2011, pp. 23298-23314.

[56]. M. Demenikov, Optimization of hybrid imaging systems based on maximization of kurtosis of the restored point spread function, *Optics Letters*, Vol. 36, Issue 24, 2011, pp. 740-742.

[57]. H. Zhao, H. Feng H, Q. Li, Research on design of optimum phase mask for wave-front coded imaging system, *Proceedings of SPIE*, Vol. 6834, 2007, 68342P.

[58]. F. Yan, X. Zhang, Optimization of an off-axis three-mirror anastigmatic system with wavefront coding technology based on MTF invariant, *Optics Express*, Vol. 17, Issue 19, 2009, pp. 16819-16819.

[59]. G. Carles, A. Carnicer, S. Bosch, Phase mask selection in wavefront coding systems: A design approach, *Optical and Laser in Engineering*, Vol. 48, 2010, pp. 779-785.

[60]. H. Zhao, Y. Li, H. Feng, et al., Cubic sinusoidal phase mask: Another choice to extend the depth of field of incoherent imaging system, *Optical and Laser Technology*, Vol. 42, Issue 4, 2010, pp. 561-569.

[61]. D. S. Bariwick, Efficient metric for pupil-phase engineering, *Applied Optics*, Vol. 46, Issue 29, 2007, pp. 7258-7261.

[62]. D. S. Barwick, Defocus sensitivity optimization using the defocus Taylor expansion of the optical transfer function, *Applied Optics*, Vol. 47, Issue 31, 2008, pp. 5893-5902.

[63]. V. Le, C. Kuang, L. Xu, Optimization of wavefront coding systems based on the use of multitarget optimization, *Optical Engineering*, Vol. 56, Issue 9, 2017, 093102.

[64]. F. Diaz, F. Goudail, B. Loiseaux, et al., Increase is depth of field taking into account deconvolution by optimization of pupil mask, *Optics Letters*, Vol. 34, 2009, pp. 2970-2972.

[65]. F. Diaz, F. Goudail, B. Loiseaux, et al., Comparison of depth-of focus-enhancing pupil masks based on a signal-to-noise-ratio criterion after deconvolution, *Journal of the Optical Society of America A*, Vol. 27, Issue 10, 2009, pp. 2123-2131.

[66]. T. Vettenburg, N. Bustin, A. R. Harvey, Fidelity optimization for aberration-tolerant hybrid imaging system, *Optics Express*, Vol. 18, 2010, pp. 9220-9228.

[67]. M. Demenikov, Development of compact optical zoom lenses with extended depth-of-field, *Proceedings of SPIE*, Vol. 8488, 2012, 84880C.

[68]. X. Mo, J. Wang, Phase transfer function based method to alleviate image artifacts in wavefront coding imaging system, *Proceedings of SPIE*, Vol. 8907, 2013, 89074H.

[69]. V. Le, Z. Fan, S. Chen, et al., Optimization of wavefront coding imaging system based on the phase transfer function, *Optik*, Vol. 127, Issue 3, 2016, pp. 1148-1152.

[70]. R. L. Lagendijk, J. Biemond, D. E. Boekee, Regularized iterative image restoration with ring reduction, *IEEE Transactions on Acoustics, Speech and Signal Processing*, Vol. 36, Issue 12, 1988, pp. 1874-1888.

[71]. R. Neelamani, H. Choi, R. Baraniuk, Fourier-wavelet regularized deconvolution for ill-conditioned systems, *IEEE Transaction on Signal Processing*, Vol. 52, 2004, pp. 418-433.

[72]. J. V. D. Gracht, J. G. Nagy, V. Pauca, et al., Iterative restoration of wavefront coded imagery for focus invariance, in Integrated Computational Imaging Systems, *Optical Society of America*, 2001.

[73]. M. Donatelli, S. Serra-Capizzano, Anti-reflective boundary conditions and re-blurring, *Inverse Problems*, Vol. 21, Issue 1, 2005, pp. 169-182.

[74]. J. G. Nagy, M. K. Nguyen, L. Perrone, Kronecker product approximations for image restoration with reflexive boundary conditions, *SIAM J. Matrix Anal. Appl.*, Vol. 25, Issue 3, 2003, pp. 829-841.

[75]. Q. X. Liu, T. Y. Zhao, W. Zhang, et al., Image restoration based on generalized minimal residual methods with anireflective boundary conditions in a waveform coding system, *Optical Engineering*, Vol. 47, Issue 12, 2008, 127005.

[76]. P. Liu, Q. Liu, T. Zhao, et al., Bi-conjugate gradient stabilized method in image deconvolution of a wavefront coding system, *Optics and Lasers Technology*, Vol. 47, 2013, pp. 329-335.

[77]. M. Demennikov, A. R. Harvey, Image artifacts in hybrid imaging systems with a cubic phase mask, *Optics Express*, Vol. 18, 2010, pp. 8207-8212.

[78]. M. Demenikov, A. R. Harvey, Parametric blind-deconvolution algorithm to remove image artifacts in hybrid imaging systems, *Optics Express*, Vol. 18, 2010, pp. 18035-18040.

[79]. R. N. Zahreddine, R. H. Cormack, C. J. Cogswell, Noise removal in extended depth of field microscope images through nonlinear signal processing, *Applied Optics*, Vol. 52, Issue 10, 2013, pp. D1-D11.

[80]. P. Zammit, A. R. Harvey, G. Carles, Extended depth-of-field imaging and ranging in a snapshot, *Optica*, Vol. 1, 2014, pp. 209-216.

[81]. K. Kubala, E. Dowski, W. Cathey, Reducing complexity in computational imaging systems, *Optics Express*, Vol. 11, Issue 18, 2003, pp. 2102-2108.

[82]. S. Chen, Z. Fan, Z. Xu, et al., Wavefront coding technique for controlling thermal defocus aberration in an infrared imaging system, *Optics Letters*, Vol. 36, Issue 16, 2011, pp. 3021-3023.

[83]. S. Bradburn, W. T. Cathey, E. R. Dowski, Realizations of focus invariance in optical-digital systems with wave-front coding, *Applied Optics*, Vol. 36, Issue 35, 1997, pp. 9157-9166.

[84]. B. Feng, Z. Shi, B. Xu, et al., ZnSe-material phase mask applied to athermalization of infrared imaging systems, *Applied Optics*, Vol. 55, Issue 21, 2016, pp. 5715-5720.

[85]. B. Feng, Z. Shi, Y. Zhao, et al., A wide-FoV athermalized infrared imaging system with a two-element lens, *Infrared Physics & Technology*, Vol. 87, 2017, pp. 11-21.

[86]. Y. Zhou, P. Zammit, G. Carles, et al., Computational localization microscopy with extended axial range, *Optics Express*, Vol. 26, Issue 6, 2018, pp. 7563-7577.

[87]. S. Jia, J. C. Vaughan, X. Zhuang, Isotropic three-dimensional super-resolution imaging with a self-bending point spread function, *Nature Photonics*, Vol. 8, 2014, pp. 302-306.

[88]. B. Schroeder, S. Jia, Frequency analysis of a self-bending point spread function for 3D localization-based optical microscopy, *Optics Letters*, Vol. 40, Issue 13, 2015, pp. 3189-3192.

[89]. T. Vettenburg, H. I. C. Dalgarno, J. Nylk, et al., Light-sheet microscopy using an Airy beam, *Nature Method*, Vol. 11, 2014, pp. 541-544.

[90]. S. Quirin, N. Vladimirov, C. T. Yang, et al., Calcium imaging of neural circuits with extended depth-of-field light-sheet microscopy, *Optica*, Vol. 41, Issue 5, 2016, pp. 855-858.

[91]. O. E. Olarte, J. Andilla, D. Artigas, et al., Decoupled illumination detection in light sheet microscopy for fast volumetric imaging, *Optica*, Vol. 2, Issue 8, 2015, pp. 702-705.

[92]. R. Tomer, M. Lovett-Barron, I. Kauvar, et al., SPED light sheet microscopy: Fast mapping of biological system structure and function, *Cell*, Vol. 163, 2015, pp. 1796-1806.

[93]. F. Diaz, M. S. L. Lee, X. Rejeaunier, G. Lehoucq, et al., Real-time increase in depth of field of an uncooled thermal camera using several phase-mask technologies, *Optics Letters*, Vol. 36, Issue 3, 2011, pp. 418-420.

[94]. H. B. Wach, E. R. Dowski, W. T. Cathey, Control of chromatic focal shift through wave-front coding, *Applied Optics*, Vol. 37, Issue 23, 1998, pp. 5359-5367.

[95]. E. Dowski, K. Kubala, Reducing size, weight, and cost in a LWIR imaging system with wavefront coding, *Proceedings of SPIE*, Vol. 5407, 2004, pp. 66-73.

[96]. G. Saavedra, I. Escobar, R. Martinez-Cuenca, et al., Reduction of spherical-aberration impact in microscopy by wavefront coding, *Optics Express*, Vol. 17, Issue 16, 2009, pp. 13810-13818.

[97]. J. Arines, R. O. Hernandez, S. Sinzinger, Wavefront-coding technique for inexpensive and robust retinal imaging, *Optics Letters*, Vol. 39, Issue 13, 2014, pp. 3986-3988.

[98]. M. Demenikov, E. Findlay, A. R. Harvay, Miniaturization of zoom lenses with a single moving element, *Optics Express*, Vol. 17, Issue 8, 2009, pp. 6118-6127.

[99]. M. Demenikov, E. Findlay, A. R. Harvay, Experimental demonstration of hybrid imaging for miniaturization of an optical zoom lens with a single moving element, *Optics Express*, Vol. 36, Issue 6, 2011, pp. 969-971.

[100]. S. Colburn, A. Zhan, A. Majumdar, Varifocal zoom imaging with large area focal length adjustable metalenses, *Optica*, Vol. 5, Issue 7, 2018, pp. 825-831.

Chapter 7
Laser Modification of Multilayer Thin Films

Suzana Petrović and Davor Peruško

7.1. Introduction

Numerous of the important chemical reactions in nature and technology occur at surfaces and interfaces. The surface is defined as a boundary between a material and its environment, where is found the last atom layer before the adjacent phase (vacuum, vapour, liquids or another solid), enabling interaction of the material with that surrounding [1, 2]. The interfaces and grain boundaries are considered as an inner surface between two different materials often the same aggregate states or the same material with different structures, which are often critical to the behaviour of the material. The surface of solids plays a crucial role in many technologically important materials, processes and phenomena, for solids surface reaction with all types of gases, liquids and other solids. The most economically important applications of the solid surface are reflected in the follow processes such as catalysis, corrosion, passivation, adhesion, tribology and friction [3, 4].

The behaviour of the materials together with the nature and mechanism of processes at their surfaces are defined in large extent by surface composition (types of atoms, concentrations), surface chemistry (chemical states, types of chemical bonds) and surface structure (arrangement of surface atoms). In any interaction between the surface of solids and another phase, the surface atoms are the first to be encountered [5]. Therefore, it is important to know the properties of atoms at the surface, which are usually different from the same atoms in the bulk materials. The surface atoms have a different chemical environment, simply they have less number of nearest neighbours from those in the bulk, and their atomic and electronic distributions are changed and often exhibit high chemical reactivity. Consequently, the surfaces and interfaces make a preferred medium for chemical and biological processes in nature and technology [6]. In the real systems, polycrystalline or glassy solids composed from alloys or chemical compounds, their surface characteristics are very complex with quite different surface composition and structure in terms of main components and impurities. In the most cases, the solid surfaces

Suzana Petrović
Institute of Nuclear Science Vinča, University of Belgrade, Belgrade, Serbia

are not smooth, the roughness in nano and micro scale enabling that many surface atoms are located in some proturberances (surface defects) denoting positions with increased reactivity [7, 8]. Surface and sub-surface regions can be considered as a separate quasi-two-dimensional phase, which are covering normal bulk material. Due to the limitation of many physico-chemical processes on the surface of solids and their surface properties, thin films and coatings are considered as useful alternative in many fields of technology and industry.

Materials in the form of thin films are very interesting due to their properties, such as high hardness or unusual phase composition, which cannot be obtained in uniform bulk materials [9, 10]. Generally, metallic thin films possess specific physico-chemical and mechanical properties as high corrosion resistance, good radiation stability, hardness and porosity [11, 12]. Thin films are suitable material for a wide range of application, as protective coatings, catalytic components, optical devices, photovoltaic gas sensors, dye sensitized solar cells, in biomedicine as implants and tools and especially for environmental purposes [13-16]. Studies of the specific and complex material such as multilayer thin films deposited on a substrate are highly desirable today primarily due to the existence of new, improved properties [17-19]. Nanoscale multilayer thin films are usually produced by physical vapor deposition or magnetron sputtering in range from tens to hundreds alternating layers, with layer thicknesses ranging from a few to 100 nm [19-21]. Multilayer thin film is inherently metastable state and susceptible to various modification, especially thermal degradation. Many processes such as interdiffusion, intermixing, electron density gradation, chemical phase formation, etc. are activated in the multilayer structure due to thermal treatment [22].

Nowadays research and industry require more and more sophisticated and versatile methods/tools for surface engineering and synthesis of nanoscale objects. Despite the fact that traditional technologies such as ion implantation, plasma treatment, molecular beam epitaxy, have attained an advanced stage of development, novel methods like material modification using pulsed lasers have been studied [23-25]. Laser surface modification techniques are one of the important groups of surface engineering tools due to their characteristics such as rapid heating and melting, which facilitates the possibility of extended solid solution, fine microstructure, composition homogenization, appearance of excellent interfaces, etc. Laser processing of thin films based on metals allows machining on the micro- and nanoscale by direct pattern processing in a very fast and cost effective manner [26, 27].

7.2. Laser-material Interaction

The specific changes accompanied with the interaction between laser beam and the crystal lattice of a solid depends on the nature and characteristics of the incident laser beam, as well as the properties of the solid material. The most important laser parameters that determine the induced changes in the solid material are: the energy of the incident laser pulse, the incident angle between the laser beam and the normal at target, and the bombardment/irradiation time. What changes will be caused in a solid material depends on: atomic weight of target component, the binding energy in the crystal lattice, the target

170

temperature and the specific conditions on the surface of the target [26]. During the interaction of laser radiation with a solid target, the following changes can occur: radiation damage in the crystal lattice, which involves the creation of vacancies, interstices and their accumulations, structural changes inducing amorphization and recrystallization in the collision zone, changes in the chemical composition of the target due to the different sputtering rate of the target components, changes in topography on the surface of the target caused by erosion and redeposition [27].

To define thermal (photothermal) and non-thermal (photochemical) laser processes, a systematic knowledge of the fundamental interactions between laser radiation and matter is needed, as well as the knowledge of the various relaxation times of the processes is involved [26, 28]. This information is only available for several specific systems. Laser-induced processes are designated as thermal activated if the thermalization of the excited energy is faster than the excitation speed and the initial stage of the process. The photochemical term is used when laser-induced processes are not thermal. However, if both thermal and non-thermal mechanisms are existed, processes are designated as photophysical [29].

In the case of laser processing of metals with short laser pulses, the incident photons heat and evaporate the target only superficially, via the thermal process, thus the final consequence is removal/ablation of the material for high enough values of fluences [30, 31]. The laser–material treatment depends on several factors such as laser parameters (fluence, wavelength, pulse duration, etc.) and material characteristics (its physico-chemical properties, surface state, absorptivity, thermal diffusivity, etc., including even whether the sample is in bulk or layered form) [26, 32]. Also, the interaction is a function of environment conditions, e.g. gas (air, nitrogen, vacuum, etc.) or liquid (water, a solution, etc.). Commonly, in dependence on the applied laser intensity two regimes can be distinguished: (i) a regime of low laser intensity, which can be characterized with appearance of surface periodic or columnar structures, and (ii) a regime of high laser intensity, to which can be attributed the significant ablation leading to the formation of craters/cones at the target [32,33]. Laser-induced modification of solid materials, particularly limited on the sub-surface region, is used for material processing in many ways: micro- and nano-scale machining in microelectronics, formation of stable metallic and oxide nanoparticles, creation of desired surface patterns on nano- and micrometer level modifying the composition and properties of the irradiated surface.

Processes which can be activated in the materials depend on the pulse duration; and this is the first guidelines in achieving the desired modifications in the material (Fig. 7.1):

- Femtosecond pulses involve generation of free electrons in dielectric and semiconductor materials, absorption of laser radiation by free electrons in metals, causing electron photo-emission, possibilities of Coulomb explosion in dielectrics and ultrafast melting of semiconductors;

- After irradiation with picosecond pulses was occurred electron-lattice temperature relaxation and normal melting. As a result, melted material can reach a metastable state and material can be ablated via phase explosion;

- Plastic deformations can be caused by the thermal stresses in both subsurface region and bulk materials due to high temperature gradients during the irradiation with nanosecond pulses.

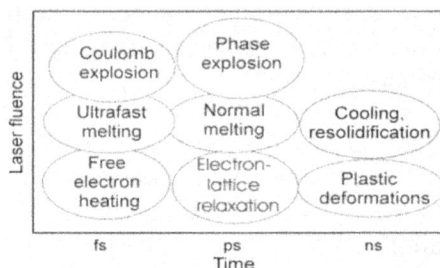

Fig. 7.1. Schematic view of possible occurred processes induced by laser radiation for the different fluences and pulse duration.

7.3. Laser-induced Changes in Composition

Laser processing of the material may modify the composition, chemical state, and structure of the irradiated surface [34-36]. Laser-assisted treatment, present particular interest as it allow the formation of non-equilibrium compounds on well localized areas in short time [37]. Laser radiation has a very narrow spectral band, it is possible to cause excitation of particular molecules, to active the oscillation of their atoms and thus to provoke or accelerate some photochemical reaction [38].

7.3.1. Laser-induced Surface Oxidation

By laser irradiation of metals and alloys in air or in controlled oxygen-rich atmosphere leads to surface oxidation, with formation of passive and protective ultra-thin oxide layer. The laser-induced process of surface oxidation depends on the initial material composition and microstructure, laser parameters and ambient conditions [39]. Recently, much attention is paid to produce novel oxide ultra-thin layer from metallic materials by direct laser-induced oxidation processes. The formed oxide layer is changed the surface composition, microstructure and properties of the treated material. Beside mechanical and corrosion properties, which are mainly changed, the formation of the oxide layers affects some optical properties as well surface absorption of the materials [40, 41]. On the other hand, laser surface oxidation involves rapid melting, intermixing and solidification of the pre-deposited elements and co-deposited oxygen to form a reaction zone confined to only the near-surface region within a very short interaction time. The process is characterized by extremely high heating/cooling rate (10^4-10^{10} Ks^{-1}), thermal gradient (10^5-10^8 Km^{-1}), and solidification velocity (1-30 ms^{-1}). As a consequence, laser surface oxidation may extend the solid solubility limit, and result is appearance of metastable and novel phases [26]. The thermal-activated diffusion of the reacting species at the metallic surface is relatively high and resulting in the formation of adhesive and thermal stable ultra-thin oxide layer. Tailoring the microstructure of such oxide layers can be a powerful tool in the functionalization of ultra-thin oxide layer as applied in nanotechnologies like tunnel junctions, gas sensors, model catalysts, and thin diffusion barriers for corrosion resistance [42-44].

One of the important examples for the formation of surface oxide is specific and complex target such as nickel-titanium (NiTi) alloy, which are high required today primarily due to possesses extra-ordinary physico-chemical and mechanical characteristics like highly corrosion resistance, electrical conductivity, super-elasticity, high strength and ductility, etc. Nickel-titanium alloy is placed in specific/unique functional group of materials as shape memory or smart materials. Moderately high biocompatibility as well as adequate corrosion resistance is main characteristics which make that NiTi alloy is acceptable for medicine purposes. However, the high Ni content can caused some allergenic reaction and/or release of toxic ions for the human body. Laser surface treatment provides the deposition of absorbed laser energy only in a thin surface layer, creating protective oxide layer, thus practically improving the corrosion resistance and decreasing Ni-ion release [45]. The inert protective oxide layer, TiO_2, was generated on the surface, because the NiTi show improved biocompatibility which is good or better than stainless steel and titanium implants [46].

Comparing the concentration depth profiles (AES spectra) of intact and laser modified Ni/Ti multilayer thin films (Fig. 7.2), it can be recognized different distribution of components at the surface and inside of thin film after laser modification. After the successive laser treatment of sample with defocused 100 picosecond pulses at the laser pulse fluence of 1 mJ cm^{-2}, initially well separated Ni and Ti layers (thickness of individual layers about 18 nm) were intermixed with Ti-oxide layer of 20 nm was formed on the surface. This is an interesting result for potential biocompatibility applications, especially since it also prevents release of toxic Ni ions. Beneath the Ti-oxide layers, homogenized and Ni-enriched thin film was generated. As a result, the oxygen reacted with components in multilayer Ni/Ti system during the laser irradiation in air, and triggered the titanium surface segregation. The effect of surface segregation of titanium is very pronounced, especially if one takes into account that the initial system had on the surface Ni as a top layer, due to faster diffusion of Ni component into the thin film. Formation of only the surface Ti-oxide layer could be connected with a faster diffusion of oxygen than diffusion of the thin film components, as well as high oxygen solubility in nickel and titanium [47]. The values of thermal diffusivity for Ni (0.24 cm^{2}s^{-1}) and Ti (0.094 cm^{2}s^{-1}) are of similar order of magnitude, so that inter-diffusion between the components could be expected. Also, the components of thin film possess a high affinity to the oxygen, especially titanium, which suggests the formation of oxides [48, 49].

The chemical states of titanium at the surface were determined by analysis of the chemical shifts of the binding energy in the corresponding XPS spectra, after irradiation of the (Ni/Ti)/Si system in air. The XPS Ti 2p spectra, of (Ni/Ti)/Si sample after laser action with 10, 50 and 100 accumulated pulses, are presented in Fig. 7.3. At the modified surface, titanium is present mainly as TiO_2, at the binding energy of E_b = 459.2 eV, due to its high affinity to the oxygen and relatively faster diffusion of Ni atoms into the thin film. The peak intensity of the corresponding oxide phase is increased with number of applied pulses, indicating greater participation of TiO_2 at the modified surface [48].

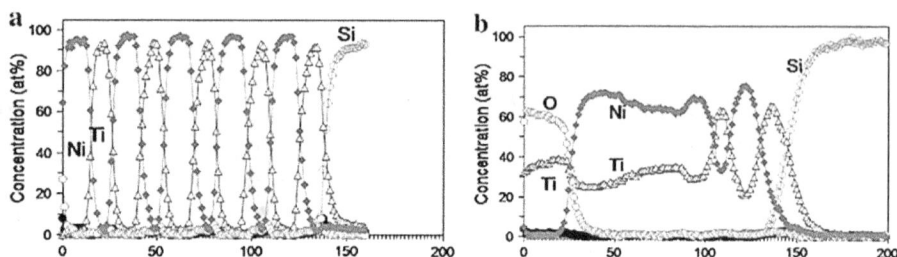

Fig. 7.2. AES depth profiles of the 5×(Ni/Ti)/Si system:
(a) As-deposited, and (b) After action of 100 laser pulses [49].

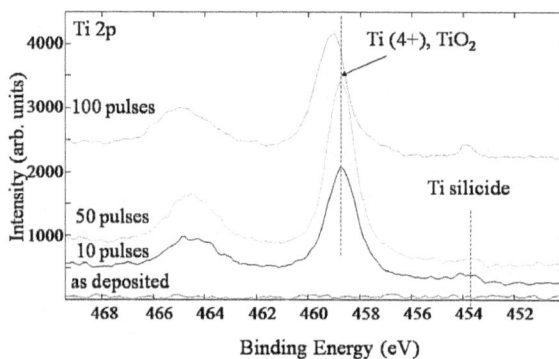

Fig. 7.3. XPS spectra of 2p Ti regions obtained on the surface before and after laser treatment with 10, 50 and 100 pulses [48].

With increased number of accumulated pulses the oxygen atoms were deeply penetrated in Ni/Ti system. The oxygen penetration depth is estimated via the thickness of the sample from the surface to the point where the oxygen concentration decreases almost to zero, using the concentration depth profiles. However, it must be pointed out that the thickness of the formed oxide layer on the surface cannot be determined based on this penetration depth. Disappearance of the multilayer structure with uniformly distribution of the components and penetration depth of oxygen after laser irradiation coincide with the estimated heat-affected-zone (HAZ). The calculated HAZ for this system was about 88 nm, for a single laser pulse, so it can be expected that the thickness of formed oxide layer would be close to HAZ depth [26, 48].

Another important system, in which it is essential to existence an oxide layer on its contact surface, is combination of aluminium and iron. Moreover, mixed aluminium and iron oxides are used as catalysts for the oxidation of various organic pollutants from wastewaters [50, 51]. A new anion-selective catalyst develops for future applications, principally using metal-oxides or surfactants, combining a good catalytic activity with a high stability against leaching. More economical process, metal-oxide Al_2O_3–Fe_2O_3-based catalyst was recently developed for the elimination of volatile organic compounds at lower temperatures (200-600 °C). Mixtures of iron and aluminium oxides are the most widespread and excellent candidates as heterogeneous catalysts for the removal of organic water pollutants such as phenol and phenol derivatives, homolytic decomposition of peroxide, and arsenic adsorption in aqueous solution [52]. The physico-chemical properties of bimetal Al–Fe-oxide differ from those of their single components and become a promising material for removal of phosphate by adsorption reaction from contaminated water in engineering systems.

In this part will be shown how pronounced changes in spatial and chemical distribution of components through the (Al/Fe)/Si multilayer structure after the laser treatment strongly depend on the fluences of the applied femtosecond laser radiation. The distribution of components at the subsurface region indicates effect of surface oxidation, which results primarily in reduction of Al concentration as top layer on the surface (Fig. 7.4). The surface of the laser-treated sample is covered by an oxide layer (12 nm) composed of Al- and Fe-oxide phases. The highest concentration of oxygen is reached at a depth of 10 nm, and then the concentration of oxygen decreases gradually further in the interior of the intermixed Al/Fe multilayer structure. The same concentrations of Al and Fe were noted at a depth of about 20 nm, which could be assumed that approximately equal ratios of the Al- and Fe-oxide phases are achieved at this depth. Laser processing of the $3\times$(Al/Fe)/Si system has reduced Al concentration and increased Fe concentration in the subsurface region, in respect to laser-untreated sample. Therefore, oxygen atoms reacted with the material during the laser irradiation in air and triggered the iron surface segregation and deep penetration of aluminium. An additional reason for reducing the surface concentration of Al could be the segregation of Al on the surface accompanied by the preferential ablation of the Al, which has a lower atomic mass, as well as lower melting point and the evaporation point than the iron [53].

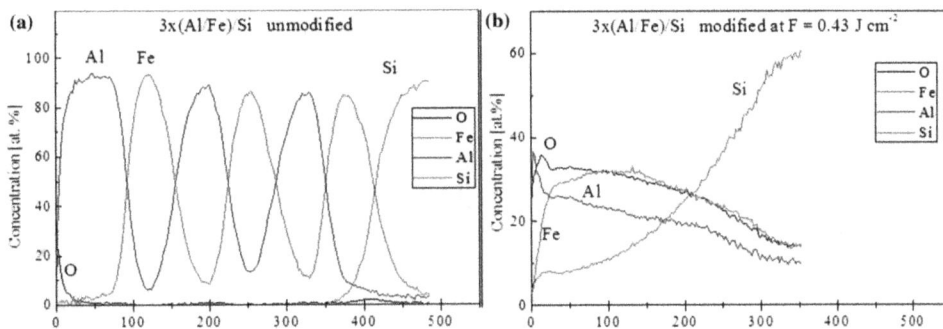

Fig. 7.4. AES depth profile of $3\times$(Al/Fe)/Si system: (a) As-deposited sample; (b) After femtosecond laser processing with fluence of $F = 0.43$ J cm^{-2} [53].

The chemical states of constitutive elements (Al and Fe) based on the chemical shifts of the binding energy pointed on laser-induced oxidation in sub-surface region after laser treatment of the $3\times$(Al/Fe)/Si system in air. In the corresponding XPS Al 2p spectra at the surface and at the depth of 3 nm, for unmodified sample and at three different values of laser pulse fluences, were detected changes in binding energy of Al atom (Fig. 7.5). At the surface of unmodified $3\times$(Al/Fe)/Si system, native Al-oxide phase was formed, while at depth of 3 nm Al exists in metallic state. After applying any of these values of laser pulse fluences, Al atoms are in oxide state as Al_2O_3 with corresponding binding energy of 75.8 eV [54]. With applied higher laser pulse fluences, the concentration of Al atoms in the metallic state is slowly reduced, and Al appearing only in Al-oxide phase.

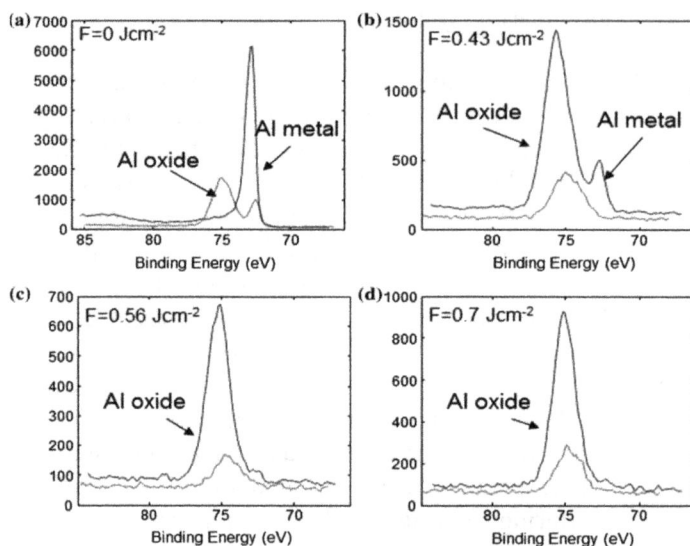

Fig. 7.5. XPS spectra of 2p Al regions obtained at the sample surface (red line) and at depth of 3 nm (blue line) before and after laser treatment at 0.43, 0.56, and 0.7 J cm^{-2} fluences [53].

For the reason that Fe is not the top surface layer of the as-deposited sample, in the XPS Fe 2p spectra, the low concentration of Fe component is expected at the surface and in the sub-surface region after laser processing (Fig. 7.6). At a low laser pulse fluence (0.43 J cm^{-2}), the Fe species are not registered at the surface, but metallic Fe was detected under the surface in a depth of 3 nm. During the laser processing at higher fluence (0.56 J cm^{-2}), the oxidation of Fe is enhanced toward formation of the Fe_2O_3 phase, but without its existence at the surface. Iron oxide is appeared on the surface of a laser modified $3\times$(Al/Fe)/Si system only for fluence of (0.7 J cm^{-2}). Increasing the laser fluence has caused a greater mobility (diffusion) of components and made the conditions suitable for the formation of oxide phases on the surface and into thin film. Generation of metal-oxides is further confirmed from the fact that oxygen only occurs with the binding energy of 532.5 eV, which is attributed its stable oxidation state.

A histogram (Fig. 7.7) of surface concentration of components before and after treatment at different applied fluences is shown that the concentration of oxygen gradually decreases with the application of increasing fluences. The surface concentration of Al is decreased with increasing laser fluence, while at the same time a surface concentration of Fe is increased. Almost the same trend of change in component concentration is observed in sub-surface region (shown in the histogram), but with somewhat higher relative values. In sub-surface (3 nm depth) area, the concentration of oxygen is very low (3 %) for as-deposited $3\times$(Al/Fe)/Si system in comparison with oxygen concentrations after fs-laser modification, which is the additional confirmation of laser-assisted surface oxidation. The difference in composition in the sub-surface area caused by various fluences is reflected with different ratios of formed metal-oxides. The concentration of Al-oxide is reduced due to increasing Fe-oxide concentration, after modification with higher fluence. During the irradiation by the high laser fluence, a higher concentration of Fe_2O_3 in the subsurface region could be associated with the fact that the diffusion and thus the intermixing of the components had intensified, due to increasing energy delivered to the $3\times(Al/Fe)/Si$ system [53].

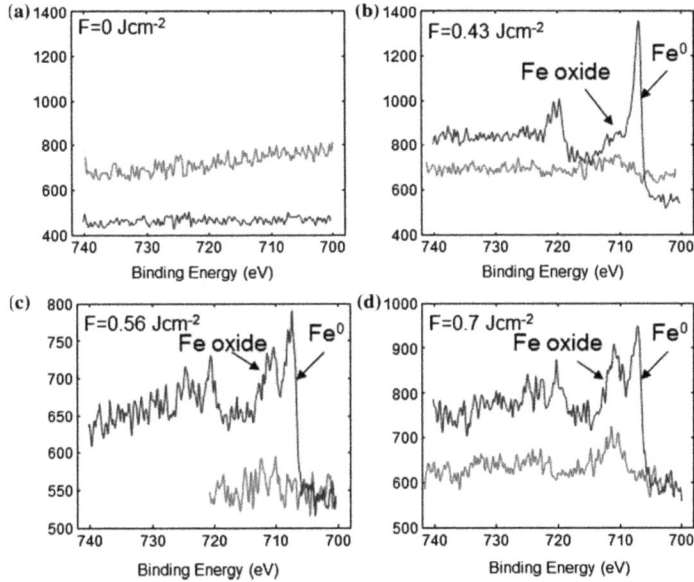

Fig. 7.6. XPS spectra of 2p Fe regions obtained at the sample surface (red line) and at depth of 3 nm (blue line) before and after laser treatment at 0.43, 0.56, and 0.7 J cm^{-2} fluences [53].

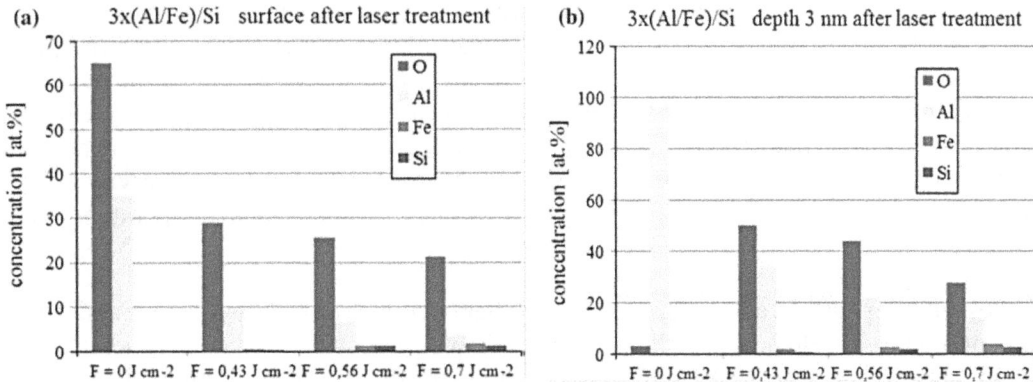

Fig. 7.7. Histograms of the component concentration for 39 (Al/Fe)/Si system before and after laser processing with three different values of fluences: (a) At the surface, and (b) In the depth of 3 nm [53].

177

The most likely order of processes during the interaction of low intensity ultra-short laser pulses (pico- and femtosecond) with metal is absorption of the laser energy by free electrons, thermalization within the electron subsystem, energy transfer to the lattice, and energy losses due to the electron heat transport into the target [55]. However, in the case of multi-pulse laser modification, the local temperature rise caused by the first pulse could enhance energy absorption of the following ones and reduce the ablation threshold by inducing the melting and material removal [56]. The interaction between metal surface and air during fs-laser processing occurred in a period of time between successive fs pulses. The assumed mechanism of the laser-assisted oxidation could include previously adsorbed oxygen molecules on the top sites at surface, which subsequently began to deform and dissociate, thereby allowing creation of a strong metal–oxygen bond. The gradient of oxygen concentration assists in its diffusion into the metal lattice at high temperatures, whereby the dissociated oxygen had diffused through longer distances in the metal. Diffusion of oxygen into the metal lattice structure might help to break the metal–metal bond and form a new metal–oxygen bond; as a result, the surface oxidation is initiated [57]. Furthermore, during the interaction between metal and adsorbed oxygen, the deeply diffused oxygen atoms can get electrons from the metallic atoms and becoming negatively charged. This electron transfer process chemically activates the metallic structure as a positively charged species. As a result, the dissociated oxygen can form a bond with the chemically active metal atoms [58]. These bonds reduce the oxygen mobility within the subsurface region and initiate formation of thin oxide film, which is created on the metal surface.

7.3.2. Laser-induced Surface Alloying

Among the various surface protection techniques, laser surface alloying is a relatively new technique that has been employed to improve hardness, wear, as well as corrosion resistance, and some other surface characteristics of both metallic and composite materials [59-61]. The usual procedure for the production of high-entropy alloys by laser surface alloying is irradiation of the mixture with pure elemental powders or powder with all alloying agents. Laser surface alloying is a method for processing of materials based on changing chemical composition of metallic samples by melting a thin superficial layer with laser beam. The refinement of composition and microstructure for metallic materials are inherently determined by the cooling rates which can be controlled by precise selection of laser processing parameters. The resultant cooling rates mostly depend on thermal gradient and solidification speed, but also depend on the thermal conductivity of the substrate, physical properties of substrate, laser process parameters and temperature of the substrate.

Successful laser surface alloying requires sufficient depth of laser-alloyed zone and solute content of alloying elements, with providing suitable processing parameters such as laser power, beam diameter, scanning speed, and degree of overlapping. Under the laser-induced surface alloying implies significant mixing between the coating material and the some kind of sub-layer or substrate with aim to form an alloyed surface layer with new phases and microstructures [62]. The materials between which the alloying is achieved should have similar thermo-physical properties, including diffusion coefficient, melting point, thermal expansion coefficient, thermal conductivity and modulus of elasticity. Too large difference in melting point or modulus of elasticity between the alloying components makes it difficult to form a good metallurgical bonding due to cracks and pores. Multilayer systems composed on different metallic components with thicknesses of individual layers of about 10-20 nm are a good starting point for synthesis of alloys and intermetallic compounds by laser processing. In general, reactive multilayers could provide an approach to self-propagating high temperature synthesis (SHS) of new alloys and show potential for use in alternative joining processes [63]. Multilayer systems must be composed of components with a large negative mixing enthalpy to achieve a satisfactory intercourse of the reactive species. Local ignition from an external energy source like electric spark or laser pulse results in a rapid exothermic reaction in Al/Pt, Ni/Al and Ti/Al multilayer foils [63-66]. The advantage of these processes is laid in fact that the mixing components are not exposed to high temperatures. The generated heat is limited to the inside bonding interface between reactive components in a time period of a few milliseconds. Therefore, temperature sensitive components and materials with different coefficients of thermal expansion can potentially be fused without thermal or thermo-mechanical damage. On the other hand, an undoubted advantage of applying laser surface alloying comparing with other techniques of modifying surface layer like plasma nitriding, paste boriding, ion implantation, is precision of operations and significant reduction of the time when an alloyed layer with determined thickness and microstructure is created without changes of the substrate.

Titanium-based alloys possess an attractive combination of low density, moderately high strength and good aqueous corrosion resistance for industrial and technological applications. Nickel-titanium (NiTi) alloys are one of the most important groups of structural materials in various modern industries. In fact, NiTi alloy and its intermetallic compound are competitive candidates for optical components in the field of soft X-ray and neutron optics, such as super mirror, polarizer and monochromator [67]. These alloys have found application as powerful actuators in micromechanical systems (MEMS) such as micro-valves, micro-fluid pumps and micro-grippers [68, 69]. The nearly equal-atomic NiTi alloy as a relatively new biomaterial has attracted immense research interest because of its unique properties such as super-elasticity, clamping capacity and shape memory effect [70]. However, metallic biomaterials have a tendency to corrode in a physiological environment, thereby accelerating the release of Ni ions which may cause toxic reactions in the body [71]. Quite an effective way to impede the leaching of nickel ions is to produce a barrier layer on the NiTi alloy by laser surface alloying [72].

The possibilities of forming a NiTi alloy and an intermetallic compound are illustrated in the case of various multilayer structures consisted of one, five, and ten (Ni/Ti) bilayers, irradiated by a non-focused picosecond laser pulses. The laser pulse fluence of 0.9 J cm^{-2} was sufficient to induce a noticeable damage after 50 pulses. On that occasion, the high reflectivity was lost with accumulating the pulses, so the damage on the Ni/Ti multilayer thin film, after laser irradiation, was recognizable as a broad blurred area, visible to the naked eye. The results of RBS analyses of different (Ni/Ti)/Si multilayer structures, before and after laser treatment with 50 successive pulses with fluence of 0.9 J cm^{-2}, are shown in Fig. 7.8. The signals arising from Ni and Ti layers in as-deposited (Ni/Ti)/Si bilayer sample were well separated. Significant changes in the RBS spectrum were recorded after irradiation with 50 laser pulses, where signals for Ti and Ni disappeared and a new broad peak with a visible plateau was appeared. The broad peak indicates that intensive mixing occurred between the components of the individual Ni and Ti layers. The appearance of the plateau can be contributed to the formation of the NiTi alloy or intermetallic compound. The sample of as-deposited multilayered 5×(Ni/Ti)/Si has shown clearly visible individual layers of Ni and Ti, although three layers of Ni and Ti overlap in the RBS spectrum (Fig. 7.8), which are manifested by increased intensity of the three middle peaks. The RBS spectrum of the laser treated sample indicates that 4 individual layers closer to the Si substrate retained the layered structure, while mixing of the components occurred between the Ni and Ti layers close to the surface [69].

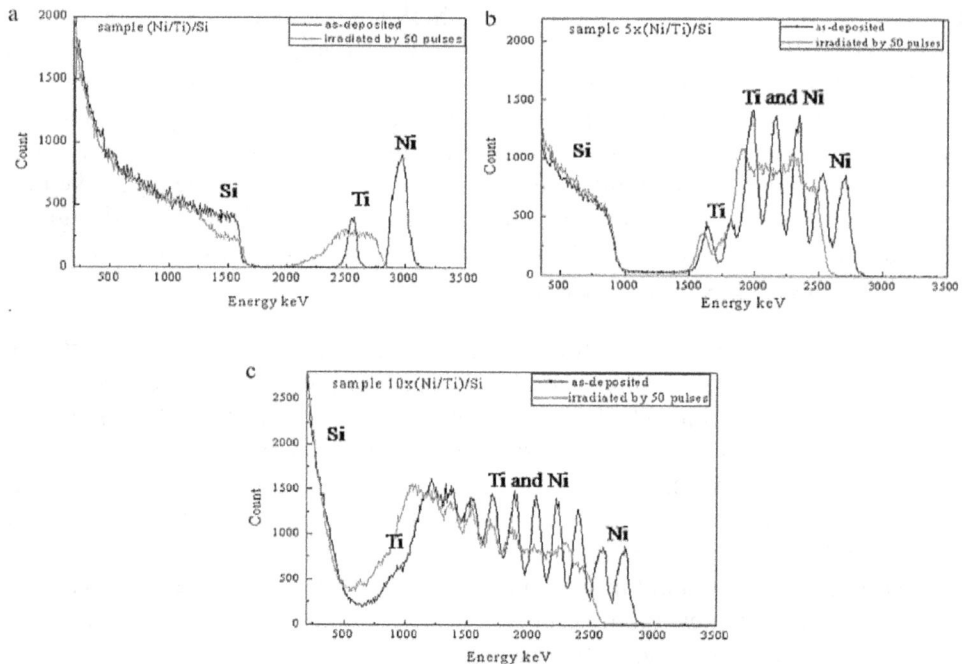

Fig. 7.8. RBS analyses before and after laser treatment with 50 successive pulses of fluence 0.9 J cm^{-2} for the following multilayer structures: (a) 1×(Ni/Ti)/Si, (b) 5×(Ni/Ti)/Si and (c) 10×(Ni/Ti)/Si [69].

By observing the high-energy edge that corresponds to the surface of the sample, it can be concluded that the Ni atoms were pulled back from the surface. The signal of the Ti component is slightly shifted towards higher energies, which can be attributed to diffusion of Ti atoms toward the surface. These facts directly indicate that Ti is the dominant component on the surface after laser treatment. Intensive mixing of components with formation of alloy between Ni and Ti occurred in the layers near the surface, while the layers closer to the Si substrate remained intact, meaning that these Ti and Ni layers did not undergo any changes. The third sample composed of 10 bilayers was irradiated with the same laser parameters and the composition changes before and after the treatment can be compared via RBS spectra. Loss of layered structure is observable in the layers near the surface (about 8 layers in total), with the appearance of diffusion of titanium towards the surface. In this case also, the Ni component was at the surface in relatively low concentration, even though the Ni layer was the top layer before laser modification. It can be said that the significant changes of the composition in the 5×(Ni/Ti)/Si and 10×(Ni/Ti)Si systems occurred to a depth of 110 nm, which includes a total of 7 layers. Disappearance of the multi-layered structure in the first seven layers after laser irradiation coincides with the estimated heat-affected-zone (HAZ) value. This value was calculated to be about 92 nm, for a single laser pulse, according to the method proposed by other authors [26], and taking into account Ni and Ti thermal diffusivities and the laser pulse duration of 150 ps. Multi-pulse actions on the surface causes reflectivity changes, and the fraction of the laser energy absorbed increases.

The 5-bilayered sample was selected for additional XRD analysis, because almost the whole of the thin film was included in the composition change. RBS spectra for laser-induced modification of 5×(Ni/Ti)/Si sample after 50 subsequent laser pulses with fluences of 0.9 and 1.2 J cm^{-2} have not indicated great differences in the number of intermixing layers. In fact, there were no significant differences in the depth of the modification after irradiation by two values of the fluences. As expected, the slightly higher laser fluence of 1.2 J cm^{-2} induced more intensive formation of the NiTi intermetallic compound, but not additional intermixing of the layers. The comparative view of the XRD analysis are given in Fig. 7.9, for the following samples: (a) as-deposited 5×(Ni/Ti)/Si, (b) irradiated with a fluence of 0.9 J cm^{-2}, and (c) irradiated with a fluence of 1.2 J cm^{-2}. The XRD pattern (Fig. 7.9) corresponding to the as-deposited 5×(Ni/Ti)/Si sample shows three well-defined diffraction peaks, which can be assigned to a face-centred cubic (fcc) structure of the Ni (111) plane and a hexagonal close packed (hcp) structure of the Ti (002), and Ti (100). After irradiation of the 5×(Ni/Ti)/Si sample with a fluence of 0.9 J cm^{-2}, the XRD pattern (Fig. 7.9) showed a few changes, primarily the appearance of a new peak which was superimposed on the existing Ni peak.

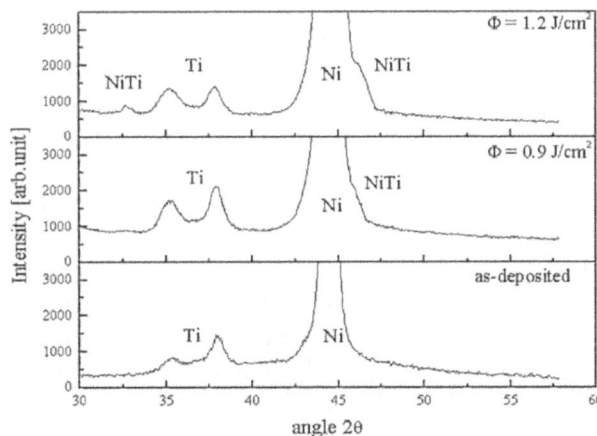

Fig. 7.9. XRD spectra of the 5×(Ni/Ti)/Si multilayer structure: (a) As-deposited; (b) Irradiated with a fluence of 0.9 J cm^{-2}, and (c) A fluence of 1.2 J cm^{-2} [69].

This peak can be attributed to an intermetallic compound NiTi. The crystal structure of the newly formed intermetallic compound is monoclinic and the observed peak originates from the reflection of the NiTi (111) plane. The observed peak is weakly expressed, indicating that the formed intermetallic compound consists of small grains. At the higher fluence, the XRD pattern (Fig. 7.9) shows the appearance of another peak, which can be attributed to a reflection of the NiTi (110) alloy phase. This result suggests that the higher fluence induced formation of alloys in the irradiated zone.

Alloys and intermetallic compounds combined from titanium and aluminium are very promising materials for high temperature applications owing to their low specific weight and high specific strength. Titanium aluminides are existed in three different phases, Ti_3Al, $TiAl$ and $TiAl_3$ with great engineering significance. Titanium aluminides can be synthesized from Al/Ti multilayers at temperatures up to 600 °C, when species undergo intermixing as a result of thermally induced atomic diffusion [73]. Once the reaction has been initiated with a laser pulse, a large amount of energy can be released in exothermal reaction and reaction can become self-propagating [74-77].

On the example of the formation of intermetallic compounds between Al and Ti, an effect of pulse number and applied pulse energy will be shown the ration of the formed intermetallic phases. The XRD pattern (Fig. 7.10) for as-deposited 5×(Al/Ti)/Si multilayer system contained only fairly wide diffraction peaks that correspond to metallic titanium and aluminium, indicating their small grains. Formation of intermetallic compound $AlTi_3$ was achieved at 10 laser pulse with energy of 85 mJ, detecting corresponding diffraction lines $AlTi_3$ (200), $AlTi_3$ (201) and $AlTi_3$ (102) of the titanium rich phase. Treating of 5×(Al/Ti)/Si multilayer with a larger number of pulses (50 and 100 pulses) of the same energy promotes the formation of intermetallic phases between Al and Ti species. A new quite board diffraction lines were appeared in XRD pattern, which are attributed to the following $AlTi$ (111) and $AlTi_3$ (002) planes [78]. Coexistence of the equilibrium among $AlTi$ and $AlTi_3$ phases was reported for heat treated Al/Ti multilayers [79] and high temperature ion beam synthesis [77], therefore, it can be safely expected to strike a balance between these intermetallic phases in laser processed 5×(Al/Ti)/Si multilayer system. After the laser irradiation of 5×(Al/Ti)/Si multilayer system at lower value of pulse energy (65 mJ), the changes in composition were almost the same. The appearance of the same diffraction lines can be noticed, but their intensities are smaller, indicating a less pronounced effect of intermetallic formation at this lower pulse energy.

Very well defined multilayer structure of 5×(Al/Ti)/Si system was illustrated and confirmed by cross-sectional transmission electron microscopy (TEM) analysis (Fig. 7.11). From TEM image can be concluded that Al/Ti multilayer thin film has polycrystalline structure with small crystalline grains oriented in different directions. The interface between the first Ti layer and the silicon substrate is sharp. Laser irradiation of the 5×(Al/Ti)/Si system with 10 pulses at both values of energies (65 mJ and 85 mJ), leads to complete disappearance of the layered Al/Ti structure except the Ti layer close to the Si substrate which was remain unaffected. During the irradiation with lower pulse energy (65 mJ), in the laser-alloyed zone is formed a quasi- amorphous structure composed of Ti and Al species. For the same number of pulses but at higher energy (85 mJ), the large crystal grains with dimensions up to 50 nm are formed, which are probably correspond to

$AlTi_3$ and $AlTi$ intermetallic compounds according to XRD analysis. The crystal grains extended through the whole laser-alloyed zone after accumulation of 100 pulses with energy of 85 mJ. In all observed laser modifications, the induced structural and composition changes were happened in a thin film, as indicated by the sharp interface without of diffusion between the Si substrate and the first Ti layer.

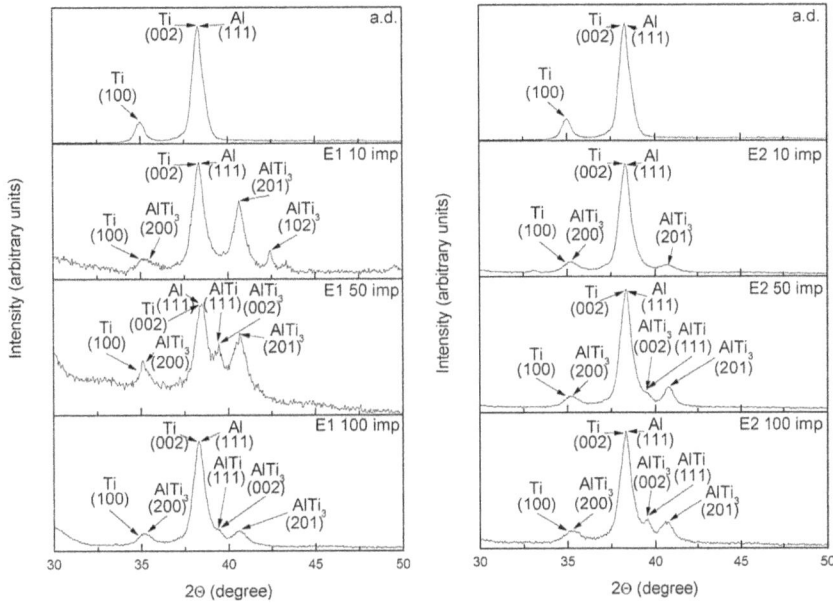

Fig. 7.10. XRD spectra of $5\times(Al/Ti)/Si$ samples irradiated with different numbers of laser pulses at energies (a) E_1 D 85 mJ, and (b) E_2 D 65 mJ [76].

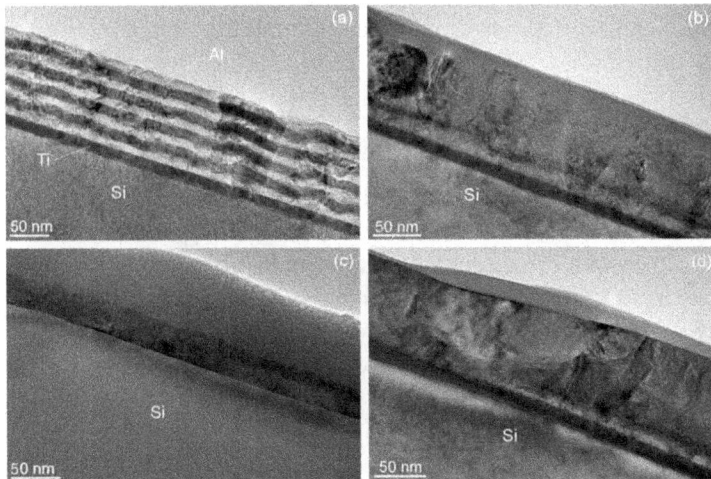

Fig. 7.11. TEM bright field cross-sectional images of (a) $5\times(Al/Ti)/Si$ samples: As-deposited, (b) After irradiation with 10 laser pulses at energy E_1, (c) After irradiation with 10 laser pulses at energy E_2 and (d) after irradiation with 100 laser pulses at energy E_1 [76].

183

Fluences in the performed laser modification were 0.35 and 0.45 J cm^{-2} for pulse energies of 65 and 85 mJ, respectively. In the both case, the fluences were below the value of the laser ablation threshold for aluminium (0.538 J cm^{-2}), as the first layer on which the laser beam encounters during modification. This choice of the fluence values contributed to the preservation of the thin film thickness after the laser modification. On the other hand, using the average value of thermal diffusivity for Al and Ti and laser pulse duration, the calculated heat-affected zone of 129.8 nm is coincided with thickness of the deposited multilayer structure (130 nm). This calculation and TEM analysis confirm the assumption that in multilayer 5×(Al/Ti)/Si systems only layers within the heat-affected zone participate in intermixing and intermetallic formation. The experimental parameters give the value of energy transferred to the material for the first laser pulse (initial reflectivity 95.54 %) about 3.4 mJ. The quantity of material included in the reaction is 8.9 10^{-8} mol (five bilayers within the cylinder exposed to the laser beam diameter of 3 mm), so the total amount of transferred energy was approximately 40 kJ mol^{-1} which is below the ignition threshold for a self-propagating reaction. After 100 laser pulses the reflectivity decreases to 73.92 %, and the transferred energy was 247 kJ mol^{-1}. This value is very close to that referred to in the literature [80] (268.6 kJ mol^{-1} for an individual film thickness of 10 nm), which is necessary for a self-propagating reaction in the Al/Ti multilayer system. However, a self-propagating reaction was not observed, probably reaction was started in some sample areas as large crystal grains. Likely absence of self-propagating reaction through the multilayer structure originates from the relatively high thermal conductivity of the silicon substrate.

7.4. Laser Surface Texturing

Laser surface texturing by formation of organized surface structures in the nanosecond to femtosecond regimes is dedicated to improve surface properties of different kinds of materials, including extraordinary surface wettability, reduction of friction and wear, improve corrosion resistance, colorization of metallic surface, improve solar cell performance and activation of biomaterials. Laser surface texturing is an excellent tool for modifying surface roughness on nearly all types of materials [81, 82]. Using laser radiation for surface texturing of materials is a low waste, single-step procedure with potentially high processing rate and the ability to control roughness or wettability of the processed surface, directly on the original materials. [83]. Surface topography with complex patterns or motives can be created by upgrading or removing materials with various physical chemical processes including cold ablation or thermal reaction, i.e. melting and vaporization depending on laser pulse duration. Some of the most common patterns that are generated on the surface are laser-induced periodic surface structure, arranged hole/crater grids, line arrays, grooves and spikes [84].

7.4.1. Laser-induced Periodic Surface Structure

In recent years, research on the generation of laser-induced periodic surface structures (LIPSS) indicated that their formation is a common phenomenon, appearing for a wide range of wavelengths and pulse durations as well as to a large variety of materials, including semiconductors, metals, and dielectrics [85-87]. Among the laser processing parameters used for LIPSS formation include wavelength, number of pulses, laser fluence, pulse duration, incident angle, and polarization. It is shown that laser processing of solid materials with multiple, linearly polarized mostly ultrafast laser pulses can lead to the formation of two distinct types of LIPSS, exhibiting low spatial frequency LIPSS (LSFL) and high spatial frequency LIPSS (HSFL) spatial frequency, respectively [88]. The LSFLs, characterized by a spatial period close to the irradiation wavelength, can be attributed to the optical interference between the incident laser beam and a surface-electromagnetic wave created during the irradiation. On the other hand, the HSFL have markedly lower periodicity compared with the irradiation wavelength [89]. Several physical mechanisms have been proposed to contribute to the HSFL formation using ultrafast laser pulses, including self-organization, second harmonic generation (SHG), excitation of surface plasmon polaritons, melt hydrodynamics, Coulomb explosion, and thermoplastic deformation [90-92]. Periodic surface structures have been found application for anti-counterfeiting, decorating, sensing, in catalysis, optical data storage, and optical gratings. LSFL for colour marking or colour display has been successfully demonstrated because the LSFL periods can effectively diffract white light to exhibit intensely bright colour [93].

Creation of surface periodic structure creation is illustrated on the case of multilayer $15\times(Al/T)/Si$ and $16\times(Ni/Ti)/Si$ thin films after irradiation with femtosecond laser pulses (Fig. 7.12) [94]. For the both multilayer systems irradiated in static regime (made spots), three zones with different morphological characteristics are identified: (i) the central part indicates the area of strong ablation and LIPSS formation; (ii) the transition zone exhibiting a sharp boundary between the ablated area and area where both types of LIPSS, the LSFL and the HSFL, are formed and (iii) the periphery, with a weak surface modification. The origin of LSFL is explained as optical interference between the incident laser radiation and a surface electromagnetic wave generated by surface defects or/and by surface irregularities. The periodic structure as low spatial frequency LIPSS created in the central zone were perpendicular to the laser polarization, and covered with nanoparticles in accordance to re-deposited material [89, 95]. The density of accumulated nanoparticles is significantly higher at the surface of the $16\times(Ni/Ti)/Si$ system. The higher yield of nanoparticles during the laser modification of $16\times(Ni/Ti)/Si$ system is conditioned to the difference in the absorption coefficient values between Ni (6.7×10^5 cm^{-2}) and Al (1×10^6 cm^{-2}). The spatial period of LSFL ripples formed at the surface of $15\times(Al/T)/Si$ multilayer sample has a value about 730 nm, while for the $16\times(Ni/Ti)$ system, the less regular LSFL ripples exhibit a lower periodicity of 590 nm. The different LSFL spatial periods obtained at the same experimental conditions could be explained by the different composition of the respective multilayers. In the transition zones of the irradiated spots, periodic structures with the characteristics of HSFL are formed. These ripples with periodicity of 155 nm are parallel to the laser polarization. On the peripheries, related to the wings of the laser beam Gaussian distribution, mild modifications were observed. Partial ablation/exfoliation of one or more layers of multilayer $15\times(Al/Ti)/Si$ system is seen at this area. On the contrary, no exfoliation of a few ultra-thin layer was observed in the case of $16\times(Ni/Ti)/Si$ system [94].

In the dynamic regime of laser processing of multilayer 15×(Al/Ti)/Si and 16×(Ni/Ti)/Si systems was achieved the formation of LIPSS (Fig. 7.13). In the scanned areas irradiated with a laser fluence of 0.56 J cm^{-2} and a certain degree of spots overlapping, LIPSS with 500 nm periodicities were created perpendicular to the laser polarization for both multilayer systems. The creation of low spatial frequency LIPSS was accompanied by surface melting, followed by a rapid resolidification but also to a certain degree by material ablation [95]. However, the formation of HSFL in a spatial vacation between two LSFLs was observed, indicating that the formation of HSFL preceded the formation of LSFL, especially expressed in the Al/Ti multilayer thin film.

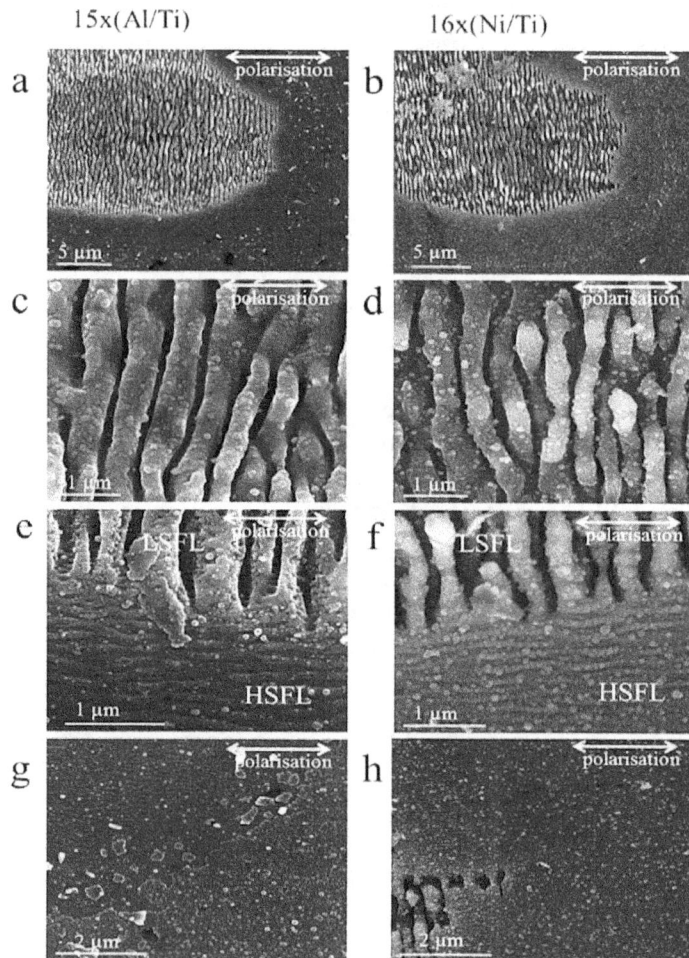

Fig. 7.12. FE-SEM analysis of single spots at the 15×(Al/Ti)/Si and 16×(Ni/Ti)/Si multilayers, after irradiation with 1000 fs laser pulses at a fluence of 0.56 J cm^{-2}. (a) and (b) Low magnification view; (c) and (d) spot centers; (e) and (f) transition between LSFL and HSFL zones; and (g), and (h) the spot peripheries. The direction of laser polarization is indicated by a double arrow [94].

On the contrary, for both multilayer systems irradiated with the lower fluence of 0.24 J cm^{-2} resulted in the formation of high spatial frequency LIPSS, oriented parallel to the laser polarization (Fig. 7.13). The HSFL ripples are better defined on the surface of 15×(Al/Ti)/Si multilayer, since a large density of nanoparticles were observed onto the 16×(Ni/Ti)/Si film surface. This observation suggests that the incident laser fluence and, in turn, the density of excited carriers are important parameters affecting the LIPSS properties. Lower laser-induced electron density would favor HSFL formation, oriented parallel to the laser polarization with periods close to λ/n (n-refractive index) [85, 89, 96]. On the other hand, LSFL ripples showing a period closer to laser wavelength and oriented perpendicular its polarization, can be the result of a high laser-induced electron density.

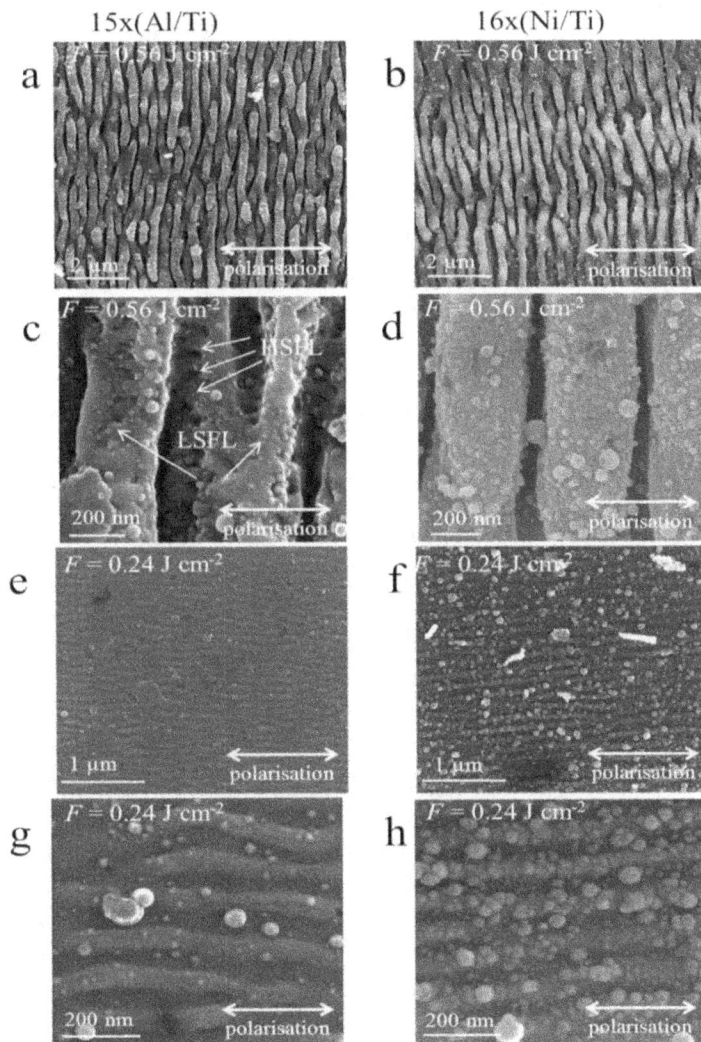

Fig. 7.13. FE-SEM images of the scanned areas of the 15×(Al/Ti)/Si and 16×(Ni/Ti)/Si films, after irradiation with the 40 fs laser pulses at the fluences of 0.56 J cm^{-2} (a)–(d) and 0.24 J cm^{-2} (e)–(h). The direction of laser polarization is indicated by a double arrow [94].

By analysing the EDS spectra taken from three different positions, the unmodified, the HSFL, and the LSFL areas, it was noticed that the laser processing did not lead to material ablation during the HSFL formation. Nonetheless, elemental composition at the LSFL regions revealed that material had been ablated from these areas. Initial concentrations of Al and Ti were decreased approximately 50 % in the Al/Ti system, and Ni and Ti concentrations in the Ni/Ti multilayer system were reduced almost 80 %. The differences in composition after LSFL formation for these systems most probably come from a higher absorption coefficient for Ni, 38 which caused that higher part of the laser energy is delivered to Ni/Ti system, and therefore, modification/ablation were intensive [94].

A large area with a uniform LIPSS, in principle can be obtained by lateral displacement of the sample in relation to the laser beam or vice versa during laser processing. Large surfaces of a few millimetres, homogenously covered with LIPSS, have been obtained by certain combinations of different irradiation parameters (repetition rate, fluence, focused beam radius, and scanning velocity), satisfying accumulated fluence boundary conditions which play a decisive role in the spatial emergence of LIPSS [97]. In addition, surface nanopatterning of metals and semiconductors with ultrafast laser pulses near ablation threshold creates a novel inter-perpendicular co-existence periodic structure with different spatial periods (LSFL and HSFL) at the same time [98]. Specific surface patterning combined from micro- and nanostructures, with adjusted optical and electrical performance of the materials, it is important for functionalization of material surface [99]. Fabrication of micro-/nano-2D periodic structures is a promising approach for producing colorized metals or semiconductors, anti-reflection surface in solar cells, surfaces with unique wettability, bio-inspired structures, and photonic crystals [98-101].

Designing periodic structures will be demonstrated in the case of laser processing of the Ti/5×(Al/Ti)/Si multilayer structure with picosecond laser pulses under specific experimental conditions [102]. The influence of the Gaussian shape of the laser beam profile on the surface morphologies of the Ti/5×(Al/Ti)/Si sample, after irradiation at the fluence of 0.113 J cm^{-2}, is reflected by the non-uniform radial distribution of fluences: (i) at and near the centre of the irradiated zone, there is the highest local fluence (0.23 J cm^{-2}), and (ii) lower local fluence in wings of the laser beam profile, i.e., at rims of the irradiated zones. At a laser beam scanning velocity of 300 μm s^{-1} and steps of 300 μm, the whole area was irradiated with a total of 2670 spots, whereby the multiple overlap is most pronounced in the central part. The thermal relaxation time for metals after irradiation with picosecond pulses has a value in the interval of 10^{-10}-10^{-8} s [26]. The time between two consecutive pulses in these experimental conditions is approximately 11.1 ms, whereby this system was cooled and/or relaxed in the time between two pulses, without accumulation of heat. However, each subsequent pulse has come on the already modified material, and the multi-pulse irradiation certainly affects affected on the final modification.

Irradiation of the Ti/5×(Al/Ti)/Si system at given experimental parameters induced the creation of clearly regular rippled surface topography (Fig. 7.14). In the scanned area, the formation of ripples is reflected in the simultaneous occurrence of two LIPSS types, low and high spatial frequency LIPSS. It is interesting to note that the HSFLs (parallel to laser polarization) are created perpendicular to the top of each individual LSFL (normal to the direction of scanning). Using the AFM profiles, the periodicity of LSFLs is measured in the range of 1–1.1 μm which is almost the same a wavelength of the laser radiation (λ = 1064 nm). The depth between two consecutive ripples is estimated to be at about 15-25 nm, which means that only the first/top Ti layer took part in the formation of periodic structures. The periodicity of HSFLs is about 200 nm ($\lambda/5$) with the height in the range of 4-5 nm. Formation of LSFLs at higher values of local fluence is connected to optical interference of incident laser radiation and the scattered/diffracted light parallel to the surface, caused by some surface irregularities. Creation of HSFL is attributed to additional laser processing of already formed LSFL, because the spacing between two subsequent lines is smaller than the beam diameter. This additional laser treatment is responsible for HSFL from wings of the Gaussian beam and therefore with a lower value of the local fluence [85, 94]. It can be concluded that creation of co-existence periodic structures is a consequence of a lateral displacement of the sample during the irradiation together with multiple spot overlapping.

Fig. 7.14. AFM micrographs of the Ti/5×(Al/Ti)/Si sample after laser treatment in the scanning regime (150 ps, fluence 0.113 J cm^{-2}) with 3D view of the modified area [102].

7.4.2. Laser-induced Crater Formation

Many applications of laser processing of materials are inspired on the use of a focused laser beam used for the removal of material, which is a process known as laser ablation. This technique enables defined and selective removal of materials from very small volumes with excellent precision, therefore laser ablation is used for micro and nano-patterning. The characterization of crater morphology is necessary to predict the desired interaction between laser radiation and material determined by important parameters including ablation threshold, ablation rate, information about thermal effects in the material, and induced structures inside and around crater (redeposited material, rings, ripples, droplets, etc.) [103]. Control of the ablation rate and knowledge of the ablation threshold are a critical task for determining the production efficiency, dimensions, and the quality of laser induced structures in laser processing.

Laser processing of the material surface with ultra-short laser pulses resulted in well-defined ablated areas. This allows generation of well-defined areas or microstructures, especially at low laser fluence [104]. Selective laser ablation layer by layer can be achieved with careful selection of parameter for the femtosecond laser pulses and this can be seen nicely on the example of Ti/Zr multilayer system (Fig. 7.15). With

189

increasing of applied pulse energy, firstly it is noticed that number of concentric circles in the individual spots are increased (Fig. 7.15). These circles are connected with selective ablation of layers in multilayer Ti/Zr system, which are good recognize on the profiles. Maximum depth in the centre of spots and total roughness are gradually rising with pulse energy, but heights between ablated layers in these spots do not follow regularity. In fact, the height of the ejected material does not match with the thickness of as-deposited layers, but these deviations are not significant. Specifically, differences in height for first and other steps have different values, which would most likely originate from a different distribution of energy in Gaussian profile for given pulse energy. The width between ablated layers can be attributed to the consequence of Gaussian profile. Some physical mechanisms are investigated to evaluate how the fluence/pulse energy affects the thermal response (melting, evaporation, removal, etc.) of the specific material-metallic multilayer thin films.

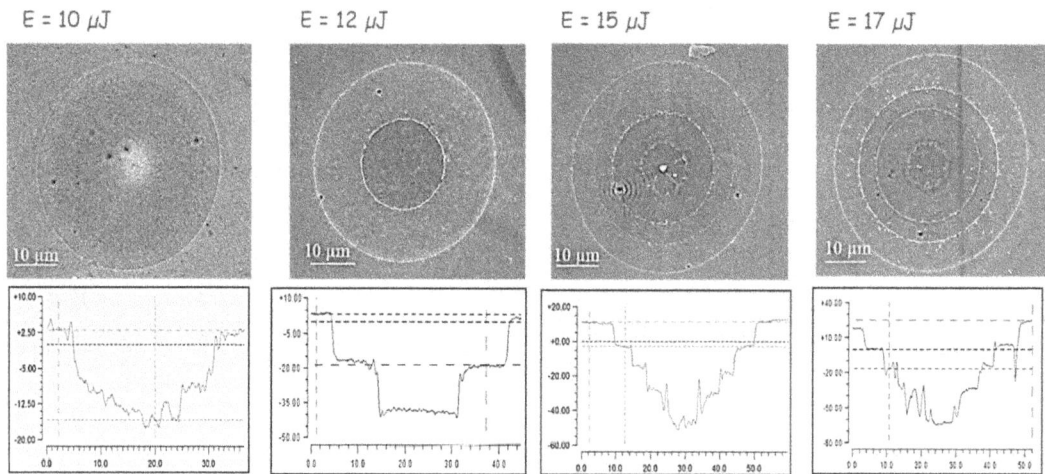

Fig. 7.15. SEM microphotographs and profiles for the craters obtained in Ti/Zr multilayer thin films induced by femtosecond pulses at different values of pulse energy.

Certainly, a first step towards description of the ablation in multilayer materials and the formation of craters, includes the consideration of fluid transport (or thermoplastic effects) and how phase transitions (or thermo-mechanical effects) influence the morphological changes in the irradiated material [26].

In laser processing very small portion of material can be removed by ablation process and crater is formed after irradiation, providing accurate depth control. The laser ablation strongly depends on laser parameters (pulse duration, energy per pulse, wavelength etc.), ambient conditions (vacuum, air, different liquids etc.) and thermo-physical properties of the material [105, 106]. At higher laser intensities ($I > 10^6$ W cm^{-2}), the crater formation is results of significant mass removal (Fig. 7.16). The size and geometry of the crater depends on composition and microstructure of ablated materials and the characteristics of the crater are controlled by setting parameters for the micromachining. [107-110]. Different mechanisms are involved in a laser ablation process, for laser pulses longer than few tenths of picoseconds, the laser induced processes are considered as thermally activated. Interaction with nanosecond laser pulses starts with absorption of pulse energy, leading to temperature increase and subsequent melting and vaporization of material, and ejection of material is appeared if pulse energy is quite high. The depth of energy deposition, the thermal diffusion and the material vaporization play important role in heat conduction from the laser exposure zone to the surrounding material. Increasing ablation efficiency by laser systems is achieved with combining parameters, such as simultaneous action of two separate laser systems or single system at a dual-wavelength [111]. Dual-wavelength system for sure improve ratio of crater diameter and ablation depth which is very important result for micromachining [112, 113]. Also, with dual-wavelength laser pulses can create deep craters with rather smooth morphology of their cavities.

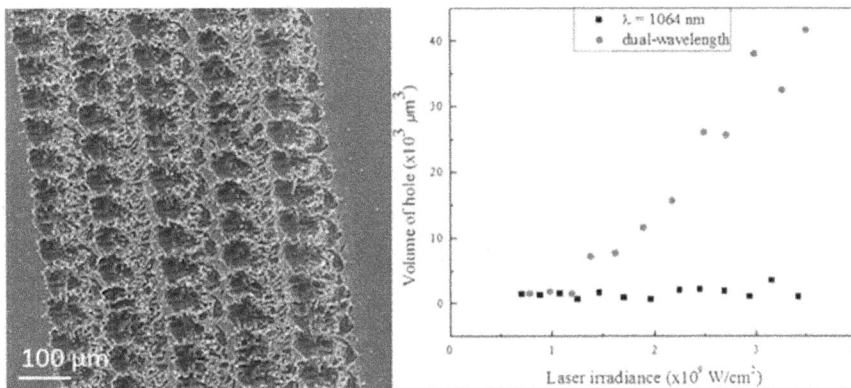

Fig. 7.16. SEM micrograph of the specific crater structures at dual-wavelength after irradiations by single laser pulse with 0.8×10^9 W cm^{-2} laser irradiance and volume of crater created at the center of the laser beam as a function of the laser irradiance at the different wavelengths [114].

The effect of increasing ablation by applying single- and dual-wavelength laser system is illustrated on a complex multilayer thin film including ten (Al/Ti) bilayers deposited on Si(100) substrate [114]. Single pulse laser irradiation of the sample was done at different laser irradiance in the range 0.25-3.5 $\times$ 10^9 W cm^{-2} and with the single-wavelength, either at 532 nm or 1064 nm or with both laser light simultaneously in the ratio of 1:10 for energy per pulse between second harmonic and 1064 nm. Most of the absorbed laser energy was rapidly transformed into heat, producing intensive modifications of composition and morphology on the sample surface. Profilometric measurements showed that volume of formed craters depends, for a given wavelength, on the applied laser irradiance (Fig. 7.16). At lower values of laser irradiance, volumes of craters are almost the same for both single- and dual-wavelengths. At higher energy, volumes of craters are drastically smaller when irradiated 1064 nm wavelengths. However, in the case of the dual-wavelength, increasing energy per pulse, at a single pulse action, tends to radically modify the 10×(Al/Ti)/Si system and create greater and/or deeper craters. From the applied energy point of view, laser energy absorption for dual-wavelength can be insufficient to induce modification compared with the monochromatic laser pulse [115]. However, small

191

fraction of the second harmonic mixing with the fundamental one is an important factor that can strengthen and improve the ablation due to more efficient laser-surface coupling [115]. More precisely, the absorption of a laser radiation leads to the ionization of the material which results in the formation of plasma and induces a large increase in temperature. At a wavelength of 1064 nm the generation of free electrons in silicon (substrate) is mainly caused by two non-linear processes, multi-photon ionization (two-photon ionization) and impact ionization [114]. Photon energy at 1064 nm (1.17 eV) is slightly above the band gap of silicon (1.12 eV), which means that the linear absorption is negligible and two-photon absorption is dominant process for starting of ablation.

On the other hand, at 532 nm photon energy is sufficient for linear absorption which is more efficient process then two-photon absorption. When electrons are excited into conduction band they are accelerated by electric field of laser and can produce additional free electrons by impact excitation. Dual-wavelength configuration indicates the existence of a coupling effect between ionization mechanisms, where linear ionization at 532 nm can initiate impact ionization by providing seed electrons required for engaging electronic avalanche [116]. During the irradiation with dual-wavelength laser pulses, an avalanche is responsible for permanent modifications of multilayer $10\times$(Al/Ti)/Si thin film structure.

7.4.3. Laser-assisted Formation of Mosaic Structure

The formation of the mosaic structure is closely related to the polycrystalline structure of the laser treated material. The polycrystalline nature of the material stimulates the morphological arrangement in certain crystallographic directions, which are dominant in the observed material. Depending on the laser parameters and the microstructure of the modified material, various physical and chemical processes can be activated, which initiate different mechanisms of the mosaic structure formation. In this section, two different mechanisms for creating mosaic structures will be shown on metallic multilayer systems.

At lower values of the laser fluence, when the ablation of the material is omitted, the formation of the mosaic structure occurs due to the cracking of the surface layer. Such a situation is illustrated in Ti/$5\times$(Al/Ti)/Si multilayer system after laser processing with picosecond (150 ps) pulses at the fluence of 0.057 J cm^{-2} (Fig. 7.17), where the modification of materials is based on the high temperature created at the surface during very short laser pulses.

If the pulse duration is longer than a few tenths of picoseconds, laser-material interaction can be considered as thermally activated [26]. For such a case with pulse longer than a few tenths of picoseconds, laser–material interaction is considered as thermally activated. Using the one-dimensional thermal model and COMSOL multiphysics simulation software, a time evolution of temperature in the Ti/$5\times$(Al/Ti)/Si multilayer system for the first applied laser pulse was calculated, (Fig. 7.17), [114, 117]. The surface temperature reaches a value of 1500 K, which is lower than the melting temperature of titanium (1941 K), but higher than melting point of aluminium (933 K – indicated by the dotted line on the graph). The temperature at 36 nm from the sample surface, within the first Al layer, reaches a maximum of 1053 K. This predicts that the first applied laser pulse induces melting of this aluminium layer and facilitates interdiffusion with the surrounding titanium layers. these differences in the melting temperatures for aluminium and titanium in the investigated system caused that the second Al layer to be found in the liquid state, while the surface Ti layer remained in solid state. Due to the expansion of the liquid phase in the Al sublayer, the internal stresses in the surface Ti solid layer is appeared and the final consequence was the appearance of micro-cracks. For the as-deposited Ti/5×(Al/Ti)/Si multilayer system, it has been found that the microstructure of Ti layer consists of the hpc Ti phase with the diffraction peaks Ti (111) and Ti (002), so that the resulting micro-cracks are expected to be oriented dominantly in these crystallographic directions.

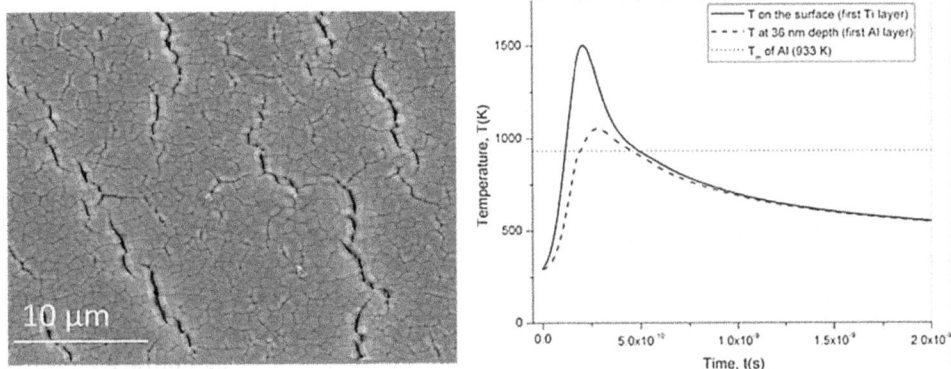

Fig. 7.17. SEM microphotograph laser irradiated Ti/5×(Al/Ti)/Si multilayer system and time evolution of temperature at the sample surface (Ti) and at depth of 36 nm (within the first Al layer beneath the sample surface) for the first applied laser pulse [117].

Another mechanism of the mosaic structure creation includes strong ablation of materials during the laser processing at relatively high values of fluences. In the case of irradiation of tungsten-titanium thin film on the silicon substrate with picosecond laser pulses with 16 J cm^{-2}, different mosaic surface structures were recorded (Fig. 7.18) [118]. At the micrometer level, in the central part of spot, the complete exfoliation of the WTi thin film and melting of the Si substrate occurred. The specific mosaic structure was formed as a consequence of cooling and resolidification of the molten silicon.

Whereby, the mosaic structure is existed in a rectangular form edged with micro-cracks oriented in the direction corresponding to the silicon (100) orientation. The dimensions of the rectangular area are several tens μm in the center of the spot and reducing to values under 10 μm at the periphery of the ablated area. In the same ablated area, a slightly different mosaic structure made of nanoparticles arranged in rectangular form. Redeposited and/or condensed nanoparticles, which were previously ejected during the laser irradiation, occupy positions in certain directions that are conditioned by the crystallographic directions of the Si substrate.

193

Fig. 7.18. SEM analysis of WTi/Si system after irradiation
with the picosecond Nd:YAG laser beam [118].

7.5. Applications and Perspective

Surface modification of materials using short and ultra-short laser pulses is a flexible, reproducible and suitable tool for creating a wide range of multiscale textures accompanied with changes of composition and properties of lot of metallic materials. Important application of laser modification of materials is laser surface texturing by formation of organized surface structures in the nanosecond to femtosecond time domain, which is dedicated to improve surface properties of different kinds of materials, including: extraordinary surface wettability, reduction of friction and wear, improve corrosion resistance, colorization of metallic surface, improve solar cell performance and activation of biomaterials. Laser surface texturing has the capability of controlling the wettability of different biological fluids and other liquids as well as the cell behaviour on the material surface including cell adherence, cell spreading, osteoblastic functions and bone matrix mineralisation. On the other hand, laser-induced surface modification on the material surfaces can control their optical and tribological properties, and may find applications in photodetectors, decorative materials and mechanical parts in different industrial fields.

Acknowledgment

This research was sponsored by the Ministry of Education, Science and Technological Development of the Republic Serbia through projects No. OI 171023 and III 45016.

References

[1]. G. A. Somorjai, Y. Li, Impact of surface chemistry, *PNAS*, Vol. 108, Issue 3, 2011, pp. 917-924.
[2]. H. Bubert, H. Jenett, Surface and Thin Film Analysis: Principles, Instrumentation, Applications, *Wiley-VCH Verlag GmbH*, 2002.
[3]. H. C. Gatos, Structure and properties of solid surfaces, in Surfaces and Interfaces of Glass and Ceramics (V. D. Frechette, W. C. La Course, V. L. Burdick, Eds.), Vol. 7, *Springer*, Boston, MA, 1974.

[4]. S. Monti, V. Carravetta, C. Battocchio, G. Iucci, G. Polzonetti, Peptide/TiO$_2$ Surface interaction: A theoretical and experimental study on the structure of adsorbed ALA-GLU and ALA-LYS, *Langmuir*, Vol. 24, Issue 7, 2008, pp. 3205-3214.

[5]. K. S. Birdi, Surface and Colloid Chemistry: Principles and Applications, *CRC Press*, 2009.

[6]. J. S. Hayes, E. M. Czekanska, R. G.Richards, The cell-surface interaction, in Tissue Engineering III: Cell – Surface Interactions for Tissue Culture (Advances in Biochemical Engineering/Biotechnology) (C. Kasper, F. Witte, R. Pörtner, Eds.), *Springer*, Berlin, Heidelberg, 2011.

[7]. K. Wandelt, Surface and Interface Science, *Wiley-VCH*, Weinheim, 2012.

[8]. A. Nagl, S. R. Hemelaar, R. Schirhagl, Improving surface and defect center chemistry of fluorescent nanodiamonds for imaging purposes – A review, *Analytical and Bioanalytical Chemistry*, Vol. 407, Issue 25, 2015, pp. 7521-7536.

[9]. W. D. Sproul, New routes in the preparation of mechanically hard films, *Science*, Vol. 273, 1996, pp. 889-892.

[10]. S. PalDey, S. C. Deevi, Properties of single layer and gradient (Ti, Al)N coatings, *Mater. Sci. Eng. A-Struct.*, Vol. 361, Issue 1-2, 2003, pp. 1-8.

[11]. F. M. Mwema, O. P. Oladijo, S. A. Akinlabi, E. T. Akinlabi, Properties of physically deposited thin aluminium film coatings: A review, *Journal of Alloys and Compounds*, Vol. 747, 2018, pp. 306-323.

[12]. P. A. Dowben, The metallicity of thin films and overlayers, *Surface Science Reports*, Vol. 40, 2000, pp. 151-247.

[13]. S. Mahdis, C. Youngjae, An overview of thin film nitinol endovascular devices, *Acta Biomaterialia*, Vol. 21, 2015, pp. 20-34.

[14]. D. Peruško, M. Mitrić, V. Milinović, S. Petrović, M. Milosavljević, The effects of pre-implantation of steel substrates on the structural properties of TiN coatings, *Journal of Materials Science*, Vol. 43, Issue 8, 2008, pp. 2625-2630.

[15]. G. Han, S. Zhang, P. P. Boix, L. H. Wong, L. Sun, S. Y. Lien, Towards high efficiency thin film solar cells, *Progress in Materials Science*, Vol. 87, 2017, pp. 246-291.

[16]. S. Petrović, D. Peruško, B. Gaković, M. Mitrić, J. Kovač, A. Zalar, V. Milinović, I. Bogdanović-Radović, M. Milosavljević, Effects of thermal annealing on structural and electrical properties of sputtered W-Ti thin films, *Surface & Coatings Technology*, Vol. 204, 2010, pp. 2099-2102.

[17]. D. Peruško, S. Petrović, M. Stojanović, M. Mitrić, M. Čizmović, M. Panjan, M. Milosavljević, Formation of intermetallics by ion implantation of multilatered Al/Ti nano-structures, *Nuclear Instruments and Methods in Physics Research B*, Vol. 282, 2012, pp. 4-7.

[18]. A. S. Ramos, M. T. Vieira, L. I. Duarte, M. F. Vieira, F. Viana, R. Calinas, Nanometric multilayers: A new approach for joining TiAl, *Intermetallics*, Vol. 14, 2006, pp. 1157-1162.

[19]. P. Panjan, A. Drnovšek, J. Kovač, Tribological aspects related to the morphology of PVD hard coatings, *Surface and Coatings Technology*, Vol. 343, 2018, pp. 138-147.

[20]. S. A. Shabalovskaya, G. C. Rondelli, A. L. Undisz, J. W. Anderegg, T. D. Burleigh, M. E. Rettenmayr, The electrochemical characteristics of native Nitinol surface, *Biomaterials*, Vol. 30, 2009, pp. 3662-3671.

[21]. S. Petrović, D. Peruško, M. Mitrić, J. Kovač, G. Dražić, B. Gaković, K. P. Homewood, M. Milosavljević, Formation of intermetallic phase in Ni/Ti multilayer structure by ion implantation and thermal annealing, *Intermetallics*, Vol. 25, 2012, pp. 27-33.

[22]. P. Bhatt, V. Ganeshan, V. R. Reddy, S. M. Chaudhari, High temperature annealing effect on structural and magnetic properties of Ti/Ni multilayers, *Applied Surface Science*, Vol. 253, 2006, pp. 2572-2580.

[23]. A. Kruusing, Underwater and water-assiste laser processing: Part 2 – etching, cutting and rarely used methods, *Optics and Lasers in Engineering*, Vol. 41, 2004, pp. 329-352.

195

[24]. I. Bozsoki, B. Balogh, P. Gordon, 355-nm nanosecond pulsed Nd:YAG laser profile measurement, metal thin film ablation and thermal simulation, *Optics and Laser Technology*, Vol. 43, 2011, pp. 1212-1218.

[25]. L.L Sartinska, B. Barchikovski, N. Wagenda, B. M.Rut, I. I. Timofeeva, Laser induced modification of surface structures, *Applied Surface Science*, Vol. 253, 2007, pp. 4296-4299.

[26]. D. Bauerle, Laser Processing and Chemistry, *Springer*, Berlin, 2000.

[27]. J. de Damborenea, Surface modification of metals by high power lasers, *Surface and Coating Technology*, Vol. 100-101, 1998, pp. 377-382.

[28]. R. F. Haglund, R. Kelly, Electronic Processes in Sputtering by Laser Beams, in *Proceedings of the Fundamental Processes in Sputtering of Atoms and Molecules Symposium (SPUT 92)*, Copenhagen, 1992, pp. 527-592.

[29]. M. von Allmen, Laser Beam Interaction with Materials, *Springer-Verlag*, 1987.

[30]. J. de Damborenea, Surface modification of metals by high power lasers, *Surf. Coat. Technol.*, Vol. 100-101, 1998, pp. 377-382.

[31]. B. S. Shin, J. Y. Oh, H. Sohn, Theoretical and experimental investigations into laser ablation of polyimide and copper film with 355-nm Nd:YVO$_4$ laser, *J. Mater. Process. Technol.*, Vol. 187-188, 2007, pp. 260-263.

[32]. S. A. Popov, A. N. Panchenko, A. V. Burachenko, F. N. Lubchenko, V. V. Mataibaev, Experimental study of the laser ablation plasma flow from the liquid Ga-In target, *IEEE Trans. Plasma Sci.*, Vol. 39, 2011, pp. 1412-1417.

[33]. E. Gyorgy, A. Perez del Pino, P. Serra, J. L. Morenza, Influence of the ambient gas in laser structuring of the titanium surface, *Surf. Coat. Technol.*, Vol. 187, 2004, pp. 245-249.

[34]. C. Tekmen, K. Iwata, Y. Tsunekawa, M. Okumiya, Influence of methane and carbon dioxide on in-flight particle behavior of cast iron powder by atmospheric plasma spraying, *Mater. Lett.*, Vol. 63, 2009, pp. 2439-2441.

[35]. S. D. Lu, Z. B. Wang, K. Lu, Enhanced chromizing kinetics of tool steel by means of surface mechanical attrition treatment, *Mater. Sci. Eng. A*, Vol. 527, 2010, pp. 995-1002.

[36]. B. S. Yilbas, C. Karatas, A. F. M. Arif, B. J. Abdul Aleem, Laser control melting of alumina surfaces and thermal stress analysis, *Opt. Laser Technol.*, Vol. 43, 2011, pp. 858-865.

[37]. X. G. Cui, C. Y. Cui, X. N. Cheng, X. J. Xu, J. Z. Lu, J. D. Hu, Y. M. Wang, Microstructure and tensile properties of the sub-micro and nano-structured Al produced by laser surface melting, *Mater. Sci. Eng. A*, Vol. 527, 2010, pp. 7400-7406.

[38]. M. Adamiak, L. A. Dobrzanski, Microstructure and selected properties of hot-work tool steel with PVD coatings after laser surface treatment, *Appl. Surf. Sci.*, Vol. 254, 2008, pp. 4552-4556.

[39]. P. Stefanov, N. Minkovski, I. Belchev, I. Avramova, N. Sabotinova, Ts. Marinova, XPS studies of short pulse laser interaction with copper, *Appl. Surf. Sci.*, Vol. 253, 2006, pp. 1046-1050.

[40]. L. L. Sartinski, S. Barchikovski, N. Wagenda, B. M. Rud, I. I. Timofeeva, Laser induced modification of surface structures, *Appl. Surf. Sci.*, Vol. 253, 2007, pp. 4295-4299.

[41]. A. K. Mondal, S. Kumar, C. Blawent, N. B. Dahotre, Effect of laser surface treatment on corrosion and wear resistance of ACM720 Mg alloy, *Surf. Coat. Technol.*, Vol. 202, 2008, pp. 3187-3198.

[42]. V. Antonov, I. Iordanova, S. Gurkovsky, Investigation of surface oxidation of low carbon sheet steel during its treatment with Nd:Glass pulsed laser, *Surf. Coat. Technol.*, Vol. 160, 2002, pp. 44-53.

[43]. E. Stratakis, Nanomaterials by ultrafast laser processing of surfaces, *Sci. Adv. Mater.*, Vol. 4, Issue 3-4, 2012, pp. 407-431.

[44]. M. Lampimaki, K. Lohtonen, M. Hirsimaki, M. Valden, Nanoscale oxidation of Cu(100): oxide morphology and surface reactivity, *J. Chem. Phys.*, Vol. 126, 2007, pp. 034703-034706.

[45]. M. H. Wong, F. T. Cheng, G. K. H. Pang, H. C. Man, Characterization of oxide film formed on NiTi by laser oxidation, *Mater. Sci. Eng. A*, Vol. 448, 2007, pp. 97-103.

[46]. F. Villermaux, M. Tabrizian, L. H. Yahia, M. Meunier, D. L. Piron, Excimer laser treatment of NiTi shape memory alloy biomaterials, *Appl. Surf. Sci.*, Vol. 109-110, 1997, pp. 62-66.

[47]. S. Petrovic, N. Bundaleski, D. Perusko, M. Radovic, J. Kovac, M. Mitric, B. Gakovic, Z. Rakocevic, Surface analysis of the nanostructured W-Ti thin film deposited on silicon, *Appl. Surf. Sci.*, Vol. 253, 2007, pp. 5196-5202.

[48]. S. Petrović, D. Peruško, J. Kovač, Z. Siketić, I. Radović-Bogdanović, B. Gaković, B. Radak, M. Trtica, Laser-induced surface oxidation of (Ni/Ti)/Si system with picosecond laser pulses, *Materials Chemistry and Physics*, Vol. 143, 2014, pp. 530-535.

[49]. S. Petrović, D. Peruško, J. Kovač, M. Panjan, B. Gaković, B. Radak, Lj. Janković-Mandić and M. Trtica, Laser treatment of nanocomposite Ni/Ti multilayer thin films in air, *Surface and Coatings Technology*, Vol. 211, 2012, pp. 93-97.

[50]. E. B. Simsek, E. Ozdemir, U. Beker, Zeolite supported monoand bimetallic oxides: promising adsorbents for removal of As(V) in aqueous solutions, *Chem. Eng. J.*, Vol. 220, 2013, pp. 402-411.

[51]. Y. Yang, D. Yan, Y. Dong, X. Chen, L. Wang, Z. Chu, J. Zhang, J. He, Influence of oxides addition on the reaction of Fe_2O_3-Al composite powders in plasma flame, *J. Alloys Compd.*, Vol. 579, 2013, pp. 1-6.

[52]. P. Branković, A. Milutinović-Nikolić, Z. Mojović, N. Jović-Jovičić, M. Perović, V. Spasojević, D. Jovanović, Synthesis and characterization of bentonites rich in beidellite with incorporated Al or Al-Fe oxide pillars, *Microporous Mesoporous Mater.*, Vol. 165, 2013, pp. 247-256.

[53]. S. Petrović, B. Gaković J. Kovač, P. Panjan, E. Stratakis, M. Trtica, C. Fotakis, B. Jelenković, Synthesis of ultra-thin oxide layer in laser-treated 3×(Al/Fe)/Si multilayer structure, *Journal of Materials Science*, Vol. 49, Issue 22, 2014, pp. 7900-7907.

[54]. S. Lee, H. Jeon, Characteristics of an Al_2O_3 thin film deposited by a plasma enhanced atomic layer deposition method using N_2O plasma, *Electron Mater. Lett.*, Vol. 3, Issue 1, 2007, pp. 7-21.

[55]. S. Nolte, C. Momma, H. Jacobs, A. Tunnermann, B. N. Chichkov, B. Wellegehausen, H. Welling, Ablation of metals by ultrashort laser pulses, *J. Opt. Soc. Am. B*, Vol. 14, Issue 10, 1997, pp. 2716-2722.

[56]. F. Di Niso, C. Gaudiuso, T Sibillano, F. P. Mezzapesa, A. Ancona, P. M. Lugara, Role of heat accumulation on the incubation effect in multi-shot laser ablation of stainless steel at high repetition rates, *Opt. Express*, Vol. 22, Issue 10, 2014, pp. 12200-12210.

[57]. N. K. Das, T. Shoji, An atomic study of hydrogen effect on the early stage oxidation of transition metal surfaces, *Int. J. Hydrogen Energy*, Vol. 38, 2013, pp. 1644-1656.

[58]. R. W. Pasco, P. J. Ficalora, Work function changes during H_2 adsorption on polycrystalline Fe and hydrogen embrittlement, *Scr. Metall.*, Vol. 14, 1980, pp. 667-671.

[59]. T. Kurzynowski, I. Smolina, K. Kobiela, B. Kuznicka, E. Chlebus, Wear and corrosion behaviour if Inconel 718 laser surface alloyed with rhenium, *Materials and Design*, Vol. 132, 2017, pp. 349-359.

[60]. B. Manne, H. Thiruvayapati, S. Bontha, R. M., Rangarasaiah, M. Das, V. K. Balla, Surface design of Mg-Zn alloy temporary orthopaedic implants: Tailoring wettability and biodegradability using laser surface melting, *Surface and Coatings Technology*, Vol. 347, 2018, pp. 337-349.

[61]. S. Bontha, N. W. Klingbeil, P. A. Kobryn, H. L. Fraser, Thermal process maps for predicting solidification microstructure in laser fabrication of thin-wall structures, *J. Mater. Process. Technol.*, Vol. 178, 2006, pp. 135-142.

[62]. Y. Chi, G. Gu, H. Yu, C, Chen, Laser surface alloying on aluminium and its alloys: A review, *Optics and Lasers in Engineering*, Vol. 100, 2018, pp. 23-37.

[63]. Y. N. Picard, D. N. Adams, J. A. Palmer, S. M. Yalisove, Pulsed laser ignition of reactive multilayer films, *Appl. Phys. Lett.*, Vol. 88, 2006, 144102.

[64]. E. Ma, C. V. Thompson, L. A. Clevenger, K. N. Tu, Self-propagating explosive reactions in Al/Ni multilayer thin films, *Appl. Phys. Lett.*, Vol. 57, 1990, 1262.

[65]. A. S. Rogachev, A. E. Grigoryan, E. V. Illarionova, I. G. Kanel, A. G. Merzhanov, A. N. Nosyrev, N. V. Sachkova, V. I. Khvesyuk, P. A. Tsygankov, Gasless combustion of Ti-Al bimetallic multilayer nanofoils, *Combust. Explos. Shock Waves*, Vol. 40, 2004, pp. 166-171.

[66]. B. Boettge, J. Braeuer, M. Wiemer, M. Petzold, J. Bagdahn, T. Gressner, Nanostructures design and fabrication for magnetic storage applications, *J. Micromech. Microeng.*, Vol. 20, 2010, 064018.

[67]. R. Gupta, M. Gupta, S. K. Kulkarni, S. Kharrazi, A. Gupta, S. M. Chaudhari, Thermal stability of nanometer range Ti/Ni multilayers, *Thin Solid Films*, Vol. 515, 2006, pp. 2213-2219.

[68]. M. Kohl, D. Dittmann, E. Quandt, B. Winzek, S. Miyazaki, D. M. Allen, Shape memory microvalves based on thin films or rolled sheets, *Materials Science and Engineering: A*, Vol. 273-275, 1999, pp. 784-788.

[69]. S. Petrović, B. Radak, D. Peruško, P. Pelicon, J. Kovač, M. Mitrić, B. Gaković, M. Trtica, Laser-induced surface alloying in nanosized Ni/Ti multilayer structures, *Applied Surface Science*, Vol. 264, 2013, pp. 273-279.

[70]. S. A. Shabalovskaya, G. C. Rondelli, A. L. Undisz, J. W. Anderegg, T. D. Burleigh, M. E. Rettenmayr, The electrochemical characteristics of native Nitinol surfaces, *Biomaterials*, Vol. 30, 2009, pp. 3662-3671.

[71]. G. Rondelli, B. Vicentini, Localized corrosion behaviour in simulated human body fluids of commercial Ni-Ti orthodontic wires, *Biomaterials*, Vol. 20, 1999, pp. 785-792.

[72]. H. C. Man, Z. D. Cui, T. M. Yue, Corrosion properties of laser surface melted NiTi shape memory alloy, *Scripta Materialia*, Vol. 45, 2001, pp. 1447-1453.

[73]. A. S. Ramos, M. T. Vieira, J. Morgiel, J. Grzonka, S. Simoes S, M. F. Vieira, Production of intermetallic compounds from Ti/Al and Ni/Al multilayer thin films – A comparative study, *J. Alloys Compounds*, Vol. 484, 2009, pp. 335-340.

[74]. J. C. Gachon, A. S. Rogachev, H. E. Grigoryan, E. V. Illarionova, J. J. Kuntz, D. Y. Kovalev, A. N. Nosyrev, N. V. Sachkova, P. A. Tsygankov, On the mechanism of heterogeneous reaction and phase formation in Ti/Al multilayer nanofilms, *Acta Mater.*, Vol. 53, 2005, pp. 1225-1231.

[75]. A. B. Mann, A. J. Gavens, M. E. Reiss, D. Van Heerden, G. Bao, T. P. Weihs, Modeling and characterizing the propagation velocity of exothermic reactions in multilayer foils, *J. Appl. Phys.*, Vol. 82, 1997, 1178.

[76]. D. Peruško, M. Čizmović, S. Petrović, Z. Siketić, M. Mitrić, P. Pelicon, G. Dražić, J. Kovač, V. Milinović, M. Milosavljević, Laser irradiation of nano-metric Al/Ti multilayers, *Laser Phys.*, Vol. 23, 2013, 036005.

[77]. D. Peruško, J. Kovač, S. Petrović, M. Obradović, M. Mitrić, V. Pavlović, B. Salatić, G. Jakša, J. Ciganović, M. Milosavljević, Selective Al-Ti reactivity in laser-processed Al/Ti multilayers, *Materials and Manufacturing Processes*, Vol. 32, Issue 14, 2017, pp. 1622-1627.

[78]. S. Nishiguchi, S. Fujibayashi, H. M. Kim, T. Kokubo, T. Nakamura, Biology of alkali- and heat-treated titanium implants, *Journal of Biomedical Materials Research A*, Vol. 67, 2003, pp. 26-35.

[79]. A. Arifin, A. B. Sulong, N. Muhamad, J. Syarif, M. I. Ramli, Material processing of hydroxyapatite and titanium alloy (HA/Ti) composite as implant materials using powder metallurgy: A review, *Mater Design*, Vol. 55, 2014, pp. 165-175.

[80]. F. Rubino, A. Astarita, P. Carlone, S. Genna, C. Leone, F. M. C. Minutolo, A. Squillace, Selective laser post-treatment on titanium cold spray coatings, *Materials and Manufacturing Processes*, Vol. 31, 2016, pp. 1500-1506.

[81]. A. Y. Vorobyev, C. L. Guo, Direct femtosecond laser surface nano/microstructuring and its applications, *Laser Photon. Rev.*, Vol. 7, 2013, pp. 385-407.

[82]. A. M. Kietzig, M. N. Mirvakili, S. Kamal, P. Englezos, S. G. Hatzikiriakos, Laser-patterned super-hydrophobic pure metallic substrates: Cassie to Wenzel wetting transitions, *J. Adhes. Sci. Technol.*, Vol. 25, 2011, pp. 2789-2809.

[83]. V. D. Ta, A. Dunn, T. J. Wasley, J. Li, R. W. Kay, J. Stringer, P. J. Smith, E. Esenturk, C. Connaughton, J. D. Shephard, Laser textured surface gradients, *Applied Surface Science*, Vol. 371, 2016, pp. 583-589.

[84]. E. Skoulas, A. Manousak, C. Fotakis, E. Stratakis, Biomimetic surface structuring using cylindrical vector femtosecond laser beams, *Scientific Reports*, Vol. 7, 2017, 45114.

[85]. G. D. Tsibidis, M. Barberoglou, P. A. Loukakos, E. Stratakis, C. Fotakis, Dynamics of ripple formation on silicon surfaces by ultrashort laser pulses in subablation conditions, *Phys. Rev. B*, Vol. 86, 2012, 115316.

[86]. J. Wang, C. Guo, Ultrafast dynamics of femtosecond laser-induced periodic surface pattern formation on metals, *Appl. Phys. Lett.*, Vol. 87, Issue 1, 2005, 251914.

[87]. X. C. Wang, G. C. Lim, W. Liu, C. B. Soh, S. J. Chua, Effects of 248 nm excimer laser irradiation on the properties of Mg-doped GaN, *Appl. Surf. Sci.*, Vol. 252, 2005, pp. 2071-2077.

[88]. F. Costache, S. Kouteva-Arguirova, J. Reif, Sub-damage-threshold femtosecond laser ablation from crystalline Si: surface nanostructures and phase transformation, *Appl. Phys. A*, Vol. 79, 2004, pp. 1429-1432.

[89]. J. Bonse, J. Kruger, Pulse number dependence of laser-induced periodic surface structures for femtosecond laser irradiation of silicon, *J. Appl. Phys.*, Vol. 108, 2010, 034903.

[90]. M. Barberoglou, D. Grey, E. Magoulakis, C. Fotakis, P. A. Loukakos, E. Stratakis, Controlling ripples' periodicity using temporally delayed femtosecond laser double pulse, *Opt. Express*, Vol. 21, Issue 15, 2013, pp. 18501-18508.

[91]. J. Bonse, J. Kruger, S. Hohm, A. Rosenfeld, Femtosecond laser-induced periodic surface structures, *J. Laser Appl.*, Vol. 24, 2012, 042006.

[92]. G. D. Tsibidis, E. Stratakis, K. E. Aifantis, Thermoplastic deformation of silicon surfaces induced by ultrashort pulsed lasers in submelting conditions, *J. Appl. Phys.*, Vol. 111, 2012, 053502.

[93]. A. Y. Vorobyev, C. Guo, Colorizing metals with femtosecond laser pulses, *Appl. Phys. Lett.*, Vol. 92, 2008, 041914.

[94]. S. M. Petrović, B. Gaković, D. Peruško, E. Stratakis, I. Bogdanović-Radović, M. Čekada, C. Fotakis, B. Jelenković, Femtosecond laser-induced periodic surface structure on the Ti-based nanolayered thin films, *Journal of Applied Physics*, Vol. 114, 2013, 233108.

[95]. A. Y. Vorobyev, V. S. Makin, C. Guo, Periodic ordering of random surface nanostructures induced by femtosecond laser pulses on metals, *J. Appl. Phys.*, Vol. 101, 2007, 034903.

[96]. J. Bonse, A. Rosenfeld, J. Kruger, Implications of transient changes of optical and surface properties of solids during femtosecond laser pulse irradiation to the formation of laser-induced periodic surface structures, *Appl. Surf. Sci.*, Vol. 257, 2011, pp. 5420-5423.

[97]. J. Eichstadt, G. R. B. E. Romer, A. J. Huis in 't Veld, Determination of irradiation parameters for laser-induced periodic surface structures, *Appl. Surf. Sci.*, Vol. 264, 2013, pp. 79-87.

[98]. J. Stašić, M. Trtica, B. Gaković, S. Petrović, D. Batani, T. Desai, P. Panjan, Surface modification of AISI 1045 steel created by high intensity 1064 and 532 nm picosecond Nd:YAG laser pulses, *Applied Surface Science*, Vol. 255, 2009, pp. 4474-4478

[99]. Y. Huang, S. Liu, W. Li, Y. Liu, W. Yang, Two-dimensional periodic structure induced by single-beam femtosecond laser pulses irradiating titanium, *Opt. Express*, Vol. 17, 2009, pp. 20756-20761.

[100]. A. Pan, J. Si, T. Chen, C. Li, X. Hou, Fabrication of two-dimensional periodic structures on silicon after scanning irradiation with femtosecond laser multi-beams, *Appl. Surf. Sci.*, Vol. 368, 2016, pp. 443-448.

[101]. T. Kondo, S. Juodkazis, V. Mizeikis, H. Misawa, S. Matsuo, Holographic lithography of periodic two-and three-dimensional microstructures in photoresist SU-8, *Opt. Express*, Vol. 14, 2006, pp. 7943-7953.

[102]. S. Petrović, D. Peruško, J. Kovač, P. Panjan, M. Mitrić, D. Pjević, A. Kovačević, B. Jelenković, Design of co-existence parallel periodic surface structure induced by picosecond laser pulses on the Al/Ti multilayers, *Journal of Applied Physics*, Vol. 122, Issue 11, 2017, 115302.

[103]. D. Bigoni, M. Milani, R. Jafer, C. Liberatore, S. Tarazi, L. Antonelli, et al., Influence of mechanical and thermal material properties on laser-produced crater-morphology and their study by focused ion beam & scanning electron microscope imaging, *J. Laser Micro. Nanoeng.*, Vol. 5, 2010, pp. 169-174.

[104]. A. Casal, R. Cerrato, M. P. Mateo, G. Nicolas, 3D reconstruction and characterization of laser induced craters by in situ optical microscopy, *Applied Surface Science*, Vol. 374, 2016, pp. 271-277.

[105]. M. H. Mahdieh, M. Nikbakht, Z. E. Moghadam, M. Sobhani, Crater geometry characterization of Al targets irradiated by single pulse and pulse trains of Nd:YAG laser in ambient air and water, *Appl. Surf. Sci.*, Vol. 256, 2010, pp. 1778-1783.

[106]. D. N. Patel, R. P. Singh, R. K. Thareja, Craters and nanostructures with lase rablation of metal/metal alloy in air and liquid, *Appl. Surf. Sci.*, Vol. 288, 2014, pp. 550-557.

[107]. Y. X. Yea, T. Xuana, Z. C. Liana, Y. Y. Fenga, X. J. Hua, Investigation of the crater-like microdefects induced by laser shock processing with aluminum foil as absorbent layer, *Appl. Surf. Sci.*, Vol. 339, 2015, pp. 75-84.

[108]. S. H. Alavi, S. P. Harimkar, Melt expulsion during ultrasonic vibration-assisted laser surface processing of austenitic stainless steel, *Ultrasonics*, Vol. 59, 2015, pp. 21-30.

[109]. A. B. Gojani, J. J. Yoh, J. H. Yoo, Extended measurement of crater depths for aluminum and copper at high irradiances by nanosecond visible laser pulses, *Appl. Surf. Sci.*, Vol. 255, 2008, pp. 2777-2781.

[110]. P. Mahanti, M. S. Robinson, D. C. Hummb, J. D. Stopar, A standardized approach for quantitative characterization of impact crater topography, *Icarus*, Vol. 241, 2014, pp. 114-129.

[111]. W. Zhao, W. Wang, X. Mei, G. Jiang, B. Liu, Investigations of morphological features of picosecond dual-wavelength laser ablation of stainless steel, *Opt. Laser Technol.*, Vol. 58, 2014, pp. 94-99.

[112]. S. Zoppel, R. Merz, J. Zehetner, G. A. Reider, Enhancement of laser ablation yield by two color excitation, *Appl. Phys. A*, Vol. 81, 2005, pp. 847-850.

[113]. B. Tan, K. Venkatkrishnan, N. R. Sivakumar, G. K. Gan, Laser drilling of thick material using femtosecond pulse with a focus of dual-frequency beam, *Opt. Laser Technol.*, Vol. 35, 2003, pp. 199-202.

[114]. B. Salatić, S. Petrović, D. Peruško, M. Čekada, P. Panjan, D. Pantelić, B. Jelenković, Single- and dual-wavelength laser pulses induced modification in 10×(Al/Ti)/Si multilayer system, *Applied Surface Science*, Vol. 360, 2016, pp. 559-565.

[115]. N. M. Bulgakova, V. P. Zhukov, A. R. Collins, D. Rostohar, T. J.-Y. Derrien, T. Mocek, How to optimize ultrashort pulse laser interaction with glass surfaces in cutting regimes?, *Appl. Surf. Sci.*, Vol. 336, 2015, pp. 364-374.

[116]. S. Rwyne, G. Duchatean, J. Y. Natoli, L. Lamaignere, Competition between ultraviolet and infrared nanosecond laser pulses during the optical breakdown of KH_2PO_4 crystals, *Appl. Phys. B*, Vol. 109, 2012, pp. 695-706.

[117]. D. Peruško, J. Kovač, S. Petrović, M. Obradović, M. Mitrić, V. Pavlović, B. Salatić, G. Jakša, J. Ciganović, M. Milosavljević, Selective Al-Ti reactivity in laser processed Al/Ti multilayers, *Materials and Manufacturing Processes*, Vol. 32, Issue 14, 2017, pp. 1622-1627.

[118]. S. Petrović, D. Peruško, D. Milovanović, J. Kovač, M. Čekada, P. Panjan, B. Gaković, M. Trtica, Effect of pulse duration in laser modification of nano-sized WTi films on Si-substrate, Journal of Optics, Vol. 12, 2010, 075602

Chapter 8
Low Level Laser Therapy

Priscila Daniele de Oliveira Perrucini,
Deise Aparecida Almeida Pires-Oliveira,
Rodrigo Franco de Oliveira,
Stheace Kelly Fernandes Szezerbaty,
Flávia Beltrão Pires, Jéssica Lúcio da Silva,
Larissa Dragonetti Bertin, Regina Célia Poli-Frederico

8.1. History

In 1917, through his research, Albert Einstein, in order to know more about the physical principles of the stimulated emission of light, demonstrated and proved his theoretical bases giving rise to the acronym LASER – Light Amplification By Stimulated Emission of Radiation stimulated radiation) that expresses exactly how light is produced [1-3].

Some years later, between 1920 and 1930, some scholars obtained biochemical responses with stimulations by light with wavelengths at the ends of the visible spectrum (red and blue) [4-6]. However, only in 1960 in the United States, Theodore Maiman built the first Laser apparatus with wavelength of 694.3 nm [7]. Then, he published the pioneering work on the stimulated emission of light using a synthetic ruby crystal in the red band, of the electromagnetic spectrum emerging the laser energy [8-10].

A few years later, in 1964, Patel et al. [11] developed the carbon dioxide laser in the infrared range and in the 1970s Mester and colleagues were the first to publish studies with the association of this therapy with tissue healing. His research topics included effects of the 488 nm and 515 nm Argon laser. Subsequent was introduced the Helium-Neon (HeNe), laser that emits red light with wavelength of 632.8 nm, replaced by a device of reduced price, more powerful, the diode laser, with wavelength of 660-950 nm [12].

Priscila Daniele de Oliveira Perrucini
University Anhanguera – UNIDERP (Universidade para o Desenvolvimento do Estado e da Região do Pantanal), Campo Grande-Mato Grosso do Sul, Brazil

In this way, the investigations related to the most different parameters, such as intensity, wavelength, density of energy, biological tissue types and modes of application were leveraged, providing a diversity of results. However, even with cautious acceptance by the scientific community about the issue until the 1980s, mainly due to the incipient molecular and cellular understanding of the mechanisms of electromagnetic radiation [13, 14].

In the second case, in the cytoplasm and cell nucleus [15, 16], it is important to understand the existence of a photobiomodulatory (laser) mechanism in the cellular respiratory chain to questions not yet elucidated [17].

Thus, since the creation of this first prototype, the laser has undergone several modifications and adaptations, with the development of numerous equipment that cover the use of this resource not only in medicine, but also in physiotherapy, dentistry and various other devices, barcode scanners [18].

8.2. Laser

The laser device consists of three primordial elements: The first element, a laser medium, which may be carbon dioxide, Argon, Helium-Neon (He-Ne), Yttrium Aluminium Garnet (YAG), eximers of dyes, rubies and semiconductor diodes, such as Gallium Arsenite (GaAs) and Gallium and Aluminum Arsenide (GaAlAs), among others; the second, a source of excitation, which may be a flash lamp or an electric arc, and finally by two mirrors. Thus, the functioning of these three elements results in the emission of light [19].

An electric current is able to excite atoms, which, once excited, begin to release photons along the axis of the cylinder, stimulating the emission of other identical photons, always in the same direction. Most of the photons are reflected by the mirrors, reinforcing the stimulated emission process, always producing a lot of identical photons. The emitted photons that are not parallel to the axis, leave the lateral walls of the cylinder, not participating in the stimulated reflection and emission process [1].

As previously mentioned, there is a great variety of lasers available in the literature in order to obtain different biomodulating effects among them: Rubi, He-Ne, AsGa and AsGaAl, among others [8, 20].

The AsGa laser has a wavelength in the region of 904 nm and the pulse duration generally of 100-200 nanoseconds allows for deep penetration without undesirable effects such as thermal effects, allowing shorter treatment times [21]. The AsGaAl laser has been used to accelerate the primary regenerative process, improving the quality of the bone structure and showing a more compact bone [22].

According to Andrade et al. [23] and Neves et al. [24], laser radiation presents unique properties that differentiate it from other light sources, such as, monochromaticity, coherence and collimation or unidirectionality. These characteristics are due in part to a structural sophistication, that is, a resonant optical cavity that allows the light that is produced by the passage of electric current in the device to be emitted in a stimulated way [13, 17].

According to power, laser radiation can be classified into: high-power lasers, known as High Level Laser Therapy (HLLT) characterized by its destructive effects on the tissues being used to cut, coagulate and cauterize, and the low lasers power, or therapeutic lasers, known as Low Level Laser Therapy (LLLT), which are commonly applied in tissue repair processes such as muscle, joint, nervous, bone, cutaneous traumatisms, being biostimulators, that is, acting primarily as accelerators in scarring processes [2, 19, 25].

It is known that the effectiveness of laser therapy and its respective biological effects are dependent on wavelength, power, dose and time [2]. The most used parameters are found in the low fluences, because it is believed that high doses can inhibit the repair process [26], causing aggressive effects at the cellular level; however, some studies have demonstrated the use of high doses with satisfactory results in the repair process [3, 27].

LLLT has become a widely used method in most countries and has been studied worldwide, one of the reasons for its popularity is related to its less invasive, athermic, painless, low cost [28] and time [29], with its energy and wavelengths capable of penetrating tissues [25], therefore, being used clinically to stimulate the tissue regeneration [18, 30-33].

8.3. Low Power Properties

A wide diversity of lasers is available to obtain different biomodulating effects, with variations in the depth of penetration, wavelength, duration of treatment, type of application, generation of thermal or athermic effects and, finally, in the objective of the treatment [8, 20-22].

Radiation is characterized by electromagnetic waves, visible or not, in which the applied energy effects work on the tissue area to be treated. It presents as a purpose the photoactivation of cellular mechanisms that aid in the rehabilitation of damaged areas [31].

Laser radiation presents unique properties that differentiate it from other light sources, such as [24, 39]:

A) Monochromaticity: The laser light is composed of photons that have only a single wavelength, all of them of the same color, therefore, presenting a pure light.

B) Consistency: The ordered displacement of the waves in relation to time with their equal amplitudes, since the depressions and peaks of the emitted waves combine perfectly in time and space, all light emitted photons have the same wavelength and fit perfectly and their displacement is structured with equal frequency in phase and emission.

C) Collimation: it shows that the light beam or photon is always parallel and the wavelength is always uniform, concentrating high intensity rays and energy, which rarely present radiation divergence with distance [2, 24, 31].

When it affects the tissue, a part of the light is not penetrated, it is reflected and the light that penetrates the tissue will be divided into three parts: one will be absorbed, one will be spread and another part will be transmitted [19]. The differences between the laser beams are given by the wavelength (λ), which is determined exclusively by the medium inside the optical resonance chamber [22], determining the mode of laser-tissue interaction and the process of absorption of the radiation, defining the depth of penetration and its effects [34].

Depending on the wavelength the light can still be classified as visible or red (380 to 750 nm) and invisible or infrared (over 750 nm) but there is divergence according to the literature, but it is known that each wavelength behaves differently from each tissue [30].

As for the energy dispersion of the laser, it is inversely proportional to the wavelength, the longer the wavelength, the deeper the energy penetration into the tissue [19].

Another important factor is the energy density (Ed), the amount of energy per unit area delivered to the tissues [19] which is given in J/cm² which designates the irradiation time, and its Watts (W) [20, 32]. Still, finally, laser radiation can be classified into: high-power lasers, known as High Level Laser Therapy (HLLT) and the low lasers power, or therapeutic lasers, known as Low Level Laser Therapy (LLLT) [2, 19, 25].

8.4. Indications and Contraindications

When it comes to indications of Low Level Laser Therapy (LLLT), we can high light several possibilities for its applicability. Its main indication is in pathological conditions where it is desired to obtain better quality and faster repair process, such as postoperative, soft tissue, bone and nervous repair [35, 36].

In the different stages of wound healing, including: the inflammatory phase, in which the immune cells migrate to the site of the tissue lesion; the proliferative phase, which includes stimulation of fibroblasts and macrophages and other repair components and the remodeling phase, consisting of collagen deposition and extracellular matrix reconstruction at the site of the wound [37, 38].

LLLT is also indicated to significantly reduce inflammation and prevent fibrosis. In addition, when properly administered, it is able to relieve pain caused by different musculoskeletal disorders [38, 39], such as tendinopathies, rheumatoid arthritis, temporomandibular disorders, and carpal tunnel syndrome [37, 38].

Regarding contraindications, studies indicate that LLLT should not be applied in [40, 41]:

- Eyes: Do not point laser beams at the eyes and all patients should wear appropriate safety glasses because of the risk of damage to the retina.

- Cancer: Do not treat the site of any known primary carcinoma or secondary metastasis unless the patient undergoes chemotherapy when LLLT can be used to reduce side effects, such as mucositis, so that therapy can be considered in patients a form of palliative relief.

- Pregnancy: Do not treat the developing fetus directly.

- Epileptics: Be aware that low-frequency pulsed visible light (<30 Hz) may trigger a seizure in photosensitive epileptic patients.

- Cardiac: Treatment directly on the chest in cardiac patients should be avoided, along with those who have a pacemaker.

- Other cases: Thyroid gland, growing child and diabetic patients.

8.5. Physiological Effects of Laser

The physiological low level laser therapy can be divided into: biochemical, bioelectric and bioenergetic. Laser has the potential to accelerate reactions such as: the release of preformed proteins (histamine, bradykinin and serotonin), and the modification as the enzymatic pathways (inhibiting or stimulating them). Studies have shown that laser irradiation may stimulate the production of adenosine triphosphate (ATP) within cells by accelerating mitosis [42, 43].

Once the laser radiation provides an increase in the production of ATP, the efficiency of the sodium and potassium pump is improved, with this, the difference of electric potential existing between the inside and outside of the cell is maintained with greater efficiency. Thus, the bioelectric effect of laser radiation is summarized in the maintenance of membrane potential [44].

The laser also has the potential to accelerate fibroblastic activities, collagen synthesis, neovascularization, increased leukocyte and phagocytic activity [45].

8.6. Therapeutic Effects of Laser

As a consequence of its physiological effects, low-power laser irradiation provides the following therapeutic effects: anti-inflammatory, analgesic, anti-oedematous and stimulant of trophism of the tissues thus leading to an accelerated and ordered tissue repair [44].

Low level laser therapy controls the production of substances released in the phenomena of pain and inflammation, such as prostaglandins, prostacyclins and serotonins. In this way, confirming both the anti-inflammatory and analgesic properties of low level laser therapy [46].

The circulatory effect of low level laser therapy as demonstrated by Miró [47] in his study on the action of laser in the microcirculation. These researchers concluded that the increase in both microcirculation and vasodilation is a result of elevated tissue metabolism and normalization of homeostasis, thus favoring the reduction of edema.

Lievens [48] demonstrated an anti-edema laser effect when irradiating animals with low level laser therapy after a surgical procedure. The author verified that the activation of the lymphatic flow of the animals belonging to the treated group occurred in comparison to the control group due to the regeneration of the lymphatic vessels.

With the increase of ATP production, the mitotic velocity is high, which provides a higher healing speed and also improves tissue tropism [44].

8.7. Effects of Low Power Laser

Genovese [49] reports that for the radiation emitted by the laser to have some effect on the tissue, it must penetrate and cause specific reactions in the cells and their structures. After the application and absorption of this energy by the human body it becomes another type of energy or biological effect that are the primary effects and classified into:

- Biochemical effect: The laser is able to control the release of substances linked to pain and inflammation, such as prostaglandins, prostacyclins, histamines, etc. In addition to modifying normal enzymatic reactions, exciting or inhibiting them. Radiation still exerts stimulus on the production of adenosine triphosphate (ATP) inside the cells, thus accelerating the process of mitosis.

- Bioelectric effect: the radiation helps to normalize the membrane potential, acting as a rebalancing and normalizer of the functional activity of the cell, because directly, the laser acts on the ionic mobility, and indirectly, increasing the amount of ATP produced by the cell.

- Bioenergetic effect: the laser provides the cells, tissues and organisms, a valid energy that stimulates, at all levels, the trophism, normalizing the deficiencies and balancing the inequalities.

8.8. Low Level Laser Therapy (LLLT)

Low level laser therapy is defined as the therapeutic application of relatively low-power lasers for the treatment of diseases and injuries using dosages that do not cause any detectable heating in irradiated tissues [50].

This therapy acts by photobiomodulating important functions at the cellular level, but the biological mechanisms involved after this type of irradiation have not yet been fully clarified, since their effects depend on the combination of some factors such as: wavelength, dose, potency, time and number of irradiations, optical properties of the tissues, types of irradiated cells, besides the physiological characteristics of the cells at the time of irradiation [29, 51, 52].

It is known that LLLT exerts important anti-inflammatory effects on healing processes such as reduction of chemical mediators, cytokines, edema, migration of inflammatory cells and increase of growth factors [51]. Additionally, it promotes neovascularization and proliferation of different cell types [29, 49, 52, 53].

According to Trainer [54], the first time low-power laser irradiation was performed in 1960 by Maimann, and since then many investigations have been carried out not only with the purpose of elucidating its mechanism of action, but also with the perspective of identify the appropriate dose to optimize the effect of this therapy.

The high costs of treating the lesions increase the importance of studies in search of adjunctive treatments that are able of interacting with the injured tissue and accelerate the rehabilitation process. In this perspective, the use of LLLT is increasingly investigated, since the laser-induced photobiomodulation is a non-pharmacological therapeutic resource that contributes to accelerate the recuperation of biological tissues [55, 56].

8.9. Effects of Low Level Laser Therapy on Healing and Fibroblasts

The principle of biostimulation promoted by the therapeutic laser was introduced more than 20 years ago, being applied primarily in dermatology, especially in the process of repairing cutaneous wounds. Soon after, it was suggested that biostimulation could also be useful for accelerating wound healing [57, 58].

The healing process is complex, involving three phases: a) inflammatory phase, where leukocyte and platelet cell migration occurs, b) proliferative phase, where there is an increase in fibroblastic cells and c) remodeling phase, where the fibroblasts participate in the process of restructuring of the extracellular matrix and collagen deposition. However, the main difficulty in the healing process occurs in the initial stages of the repair process, where an increase of edema, decrease of vascular proliferation and a significant reduction of leukocytes, macrophages and fibroblasts is observed [59]. Collagen synthesis is a fast and harmonic process that begins with interstitial injury and extends to the end of the healing phase when tissue remodeling occurs [60-62].

The wound healing process depends on several events involving cell proliferation and migration, aimed at restocking the injured area [63], and these events occur in response to the coordinated local release of growth factors and cytokines [58].

Among the existing physical resources, LLLT is more effective in stimulating healing [44]. Araújo et al. [64] demonstrated that TLBP can influence the production of

209

inflammatory mediators such as cytokines, important during the healing process by activating cell proliferation after irradiation of 1 J/cm^2 with the He-Ne laser (λ 632.8 nm) in periods of 8, 15 and 22 days.

In addition to growth factors, LLLT may also promote tissue repair, increase cell proliferation and expression of growth factors, and also modulate the synthesis of inflammatory cytokines by various cell types, including fibroblasts [26, 65].

Among the reports on the biomodulatory effects of LLLT, we highlight the formation of new blood capillaries through the release of growth factors. The release of cytokines and growth factors from injured cells is an important stage of wound healing, since it stimulates the formation of epithelial cells, new fibroblasts, increased collagen production and neovascularization [66].

Fibroblasts secrete multiple growth factors during the re-epithelization of the lesion and actively participate in the formation of granulation tissue and in the formation of the extracellular matrix. Thus, it is clear the important role played by the fibroblast during the process of tissue repair. The proliferation and increase of the cellular metabolism determine the success of the repair, since the collagen produced by the same is responsible for the strength and integrity of the new tissue formed [67, 68].

Among the most important functions of fibroblasts are extracellular matrix deposition (EMD), regulation of epithelial differentiation, regulation of inflammation, and involvement in healing. This cell type synthesizes many components of the fibrillar extracellular matrix, such as collagen type I, type III and type V (responsible for forming a fibril network) and fibronectin, contributing to the formation of basement membranes through the secretion of collagen type IV and laminin [69].

In all lesions, the blood supply of the fibroblasts responsible for the synthesis of collagen comes from an intense growth of new vessels, characterizing second intention healing and granulation tissue. The new vessels are formed from solid endothelial shoots, which migrate from the periphery to the center, on a fibrin mesh deposited in the wound bed [70]. This process is promoted by the expression of the Vascular Endothelial Growth Factor (VEGF) gene, stimulating the proliferation and migration of endothelial cells [71]. Rocha Jr et al. [72] observed a greater amount of fibroblasts in irradiated cells, evidencing a significant increase in fibroblastic proliferation and reduction of inflammatory infiltrate, concluding that LLLT accelerates the process of tissue repair.

There are several mechanisms by which the low power laser can induce the mitotic activity of fibroblasts. It stimulates the production of the basic fibroblast growth factor (FGFb), which is a multifunctional polypeptide, secreted by the fibroblasts themselves, capable of inducing not only proliferation but also fibroblast differentiation, and affects the immune cells secreting cytokines and other regulatory factors of growth for fibroblasts [29]. In vitro studies using cells from the macrophagic lineage show that, under low-power laser radiation, such cells release soluble factors that promote fibroblastic proliferation. The maturation of fibroblasts and their locomotion through the matrix are also influenced by LLLT [73].

Studies have shown that in vitro stimulation of fibroblasts through LLLT presents significant results, such as that of Basso et al. [74], which demonstrated a statistically significant difference in relation to increased metabolism and number of viable cells after irradiation of the LLLT in relation to the control, as well as Azevedo et al. [75], who also identified an increase in the cellular proliferation of laser irradiated fibroblasts $\lambda = 660$ nm in relation to the non-irradiated control.

8.10. Laser in the Biological Medium and Applicability

Muscular injuries are frequent in sports mainly due to bruising and excessive muscular forces, resulting in the removal of athletes from training and competitions [76]. Several therapeutic techniques are used in clinical practices in the attempt to accelerate and/or improve bone and muscular healing process, among them the LLLT [77]. Baroni et al [78] observed a reduction of muscle injury markers after using LLLT before eccentric exercise. Vieira et al. [79] reported a decrease in muscle fatigue after endurance training associated with LLLT. However, the greatest research challenge is to identify the ideal parameters for laser use [60].

LLLT presents a variety of applications in clinical practice such as: a) stimulation of various wounds types regeneration; b) treatment of various rheumatologic conditions; c) treatment of soft tissue lesions; d) pain relief; e) edema reduction, among other [31]. LLLT also stimulates biochemical, bioelectric and bioenergetic modifications, acting on increased metabolism and cell proliferation and maturation [29, 44]. The laser has the potential to accelerate biochemical reactions, releasing pre-forming substances such as bradykinin, histamine and serotonin. According Karu et al. [43], irradiation increases the synthesis of adenosine triphosphate (ATP), the synthesis and consumption of O_2 and mitochondrial membrane potential.

The primary effects of LLLT are directly caused by the absorption of the laser light, biochemical effects such as the release of preformed substances in the body, bioelectrical effects like the stabilization of membrane potential, and bioenergetic effects which provide cells with cellular bioenergy normalization [29]. The secondary or indirect effects originate from the primaries and act in the stimulation of microcirculation and cell trophism. Through its side effects, LLLT also exerts therapeutic effects in the tissue where it is applied.

The laser phototherapy effect on the mitochondria triggers the activation of ATP production by increasing the mitotic process due to the excitation of cellular respiration and endogenous porphyrins [26]. With the increase of ATP production, the efficiency of sodium and potassium pump is improved, maintaining the difference of electrical potential existing between the interior and exterior of the cell in a more effectively manner [20, 44].

Because of physiological factors, low-power laser irradiation provides the following effects: stimulation to microcirculation, anti-inflammatory and analgesic effects, edema reduction and induction of tissue repair [80], which accelerate and organize the tissue repairing process [81, 82].

According to Andrade et al. [2], the increase in collagen production occurs through mechanisms of photostimulation, in which certain laser powers, or doses, can act, thus modulating cell proliferation and raising the amount of fibroblasts growth factors.

Studies carried by Alfredo et al. [83] and Sakurai et al. [84] demonstrated that LLLT resulted in pain relief and increment of range of motion and functionality. The authors suggests as possible effects of the laser the anti-inflammatory and analgesic action by endogenous pain modulation via serotonin, where anti-inflammatory action can occur due to changes in cyclooxygenase pathway of the arachidonic acid metabolism, in addition to promoting the reabsorption of inflammatory exudates, favoring the elimination of substances such as bradykinin, histamine and acetylcholine.

At vascular level, Lins et al. [29] and Walsh et al. [73] observed that LLLT stimulates the proliferation of endothelial cells, resulting in the formation of numerous blood vessels, as well as an increased production of tissue granulation, stimulating the relaxation of the smooth vascular musculature. An improvement in microcirculation and stimulation in fibroblastic cells along with the neatly production of elastic fibers and collagens are factors that determine a better healing pattern [85].

Houreld et al. [86], 48 hours after irradiation with LLLT (λ 660 Nm) with 5 J/cm^2, observed a significant increase in the expression of genes involved in collagen production (COL14A1, COL4A1, COL4A3 and COL5A3), cell adhesion and growth (ITGA2, ITGA5, ITGB1, and ATG5), remodeling enzymes (CTSG, CTSL2, F13A1, F3, FGA, MMP7 and MMP9), cytoskeleton proteins (ACTC1 and RAC1) and inflammatory cytokines and chemokines (CD40LG, CXCL11, CXCL2, IFNg, IL10, IL2 and IL4).

Similar to a wavelength of 660 nm and dose of 5 J/cm^2, the study by Szezerbaty et al. [87] presented a biostimulatory effect in time 72 hours with verification of increased reticular activity, better organization and distribution of the cytoplasm and greater decondensation of chromatin. Additionally, these parameters were shown to be more effective in increasing the expression of the Endothelial Growth Factor (VEGF) gene and decrease in the transcripts of the Interleukin-6 (IL-6) gene.

On the other hand, the results of Martignago et al. [88] with 2 and 3 J/cm^2 demonstrated that the LLLT (λ 904 nm) in the parameters presented was able to stimulate the expression of VEGF and Collagen, type 1, alpha 1 (COL1α1) genes in mouse fibroblast cells in vitro. Thus, the dose of 2 J/cm^2 gave the cells an increase in the gene expression of the VEGF and COL1α1 genes, whereas at irradiation at 3 J/cm^2 they modulated only VEGF gene expression.

8.11. Laser and Bone Tissue

Several therapies related to bone healing have been studied in order to reduce bone regeneration time, as well as to accelerate the recovery time. Once there is a diversity of these therapies, LLLT becomes one therapeutic options, since it stimulates bone tissue cells and the expression of osteogenic factors [89], also by its action in the cell membrane permeability, leading to a photobiostimulation Through a chain of factors, both by enzymatic pathwayas by endogenous cell components, increasing cell metabolism and respiration [90], regulation of mitochondrial respiration [91], local circulation, collagen synthesis and a quick response to inflammation and other cellular effects [92].

According Oliveira et al. [93] the effects and properties of LLLT stimulate an increase of bone formation. The process of bone healing occurs in the body apart of external factors, however, due to pathologies or complications it can be affected by those factors. The LLLT works by assisting the bone metabolism on growth factors, biomodulation, in cell structure and in molecules present in this process [94, 95], as seen previously.

In accordance with Medeiros et al. [96], one of the resources (technique) that can help accelerate the bone healing process is LLLT, a therapy that has been used for its direct action in bone formation. This process is complex and necessary in several cases such as bone fracture and bone expansion, among others, which depend on the functionality of this process. With the interaction of several growth factors, it has its functionality related to each specific area of bone tissue, according to its need [97].

When LLLT is applied at low doses, an extra amount of Calcium (Ca^{+2}) is transported into the cytoplasm through a process that promotes cellular mitosis, RNA and DNA synthesis, and cell proliferation. However when a high dose of LLLT is applied, high doses of Calcium (Ca^{+2}) are produced, generating an increased activity of ATP, calcium pumps, which leads to a degradation of the ATP that is in the cell, inhibiting all cell metabolism [98, 99].

Bouvet-Gerbettaz [91] report that LLLT has been presenting positive mechanisms such as increased regulation of mitochondrial respiration, stimulation of cell proliferation, increased local circulation, collagen synthesis and a rapid response to inflammation and other cellular effects [100]. As a result of the development of the bone marrow, a new biostimulator for tissue repair and edema resolution [89] was proposed.

Parentiet al. [101] reported that the application of LLLT to osteoblastic cells had a biomodulating effect on cell viability when applied at a dose of 10 J/cm^2 and an inhibitory response when applied at 20 and 50 J/cm^2. Favoring positive responses for clinical utility when at lower doses.

Thus, it is understood that the low-level laser therapy reproduces positive effects in relation to osteoblastic cellular biomodulation [92]. For the clinical therapeutic scope, this is of extreme importance for several methodologies, which can be applied to the patient. It is assumed that LLLT may benefit acceleration of the bone regeneration process [102].

Studies have described the application of LLLT as a biomodulator of morphological characteristics, among them, the increase in cell proliferation and influence of the maturation process in osteogenesis [103], effects on bone formation, when using the laser Arseneto gallium and aluminum (AsGaAl) with wavelength of 830 nm at a dose of 3 J/cm^2 [104] in the cell growth, later this same laser with doses of 1 and 3 J/cm^2 [105] and in the process of regeneration and bone repair when irradiated at 808 nm at a dose of 5 J/cm² [106].

The importance of the studies applied to the cell culture derives from the fact that scientific studies are feasible so that the therapy can be applied in the clinical scope, and that parameters such as dose and wavelength, are exposed in a clear way, as evidenced by these experiments [31].

Henriques et al. [80] describes from a literature review, effects of LLLT on changes in cellular biochemical activity, acting on the regulation of cellular pH, causing structural alterations in the cell and increased cellular metabolism.

According to Silva et al. [107], LLLT stimulates cell proliferation, bone nodule formation, alkaline phosphatase (AF) activity, and expression of osteocalcin (OC) and others genes in the proliferative phase, since LLLT exerts a stimulatory effect on osteoblastic activity and that the irradiation favors the proliferation of the osteoblasts being responsible for the formation of the bone.

Already Jawad et al. [108] indicated that laser irradiation has a biomodulating effect on the healing process of the bone, increasing the vascularization, causing the formation of the trabecular osteoid tissue, resulting in a faster metabolism, and increasing the bone callus reaction, which leads to acceleration of bone regeneration.

In addition, Pires-Oliveira et al. [26] report that, when applying LLLT, 904 nm (AsGa) at energy densities of 6 J/cm² and 50 mJ/cm², identified intense mitochondrial and reticular activity by increasing protein synthesis, when LLLT was applied in fibroblast cells.

Some markers of bone formation are essential for the bone regeneration process, the transcription factor of Runt related transcription factor 2 (RUNX2) [109], essential for the formation of osteoblasts. In addition, there are biochemical markers that characterize bone formation and resorption, as in this case the protein present in the extracellular matrix [110].

The organism is composed of a regulatory system, in which several metabolic events promote the development of tissues, one of these events is the characterization of genes, correlated with the growth, proliferation and differentiation of bone cells. An element considered to initiate this process is the RUNX2 gene, also known as Core Binding factor 1-CBFA1, and previously as Acute Myeloid Leukemia 3 (AML3) [109].

During the bone repair process, growth factors and biomarkers are metabolically present and together with active or repressed transcriptional factors, they actively participate in the dynamics of osteoblastic cell interaction involving all stages of cell differentiation. The RUNX2 gene plays a key role in this metabolic process [111].

The action of RUNX2 is essential in the initial phase, in which the transition of the mesenchymal cells to the osteoblastic lineage occurs. RUNX2 is considered an essential gene for the entire process of osteoblastic determination, a process that depends on multiple steps to regulate the extracellular matrix [111].

Osteocalcin is an important extracellular matrix protein specific for bone tissue [112]. It is a protein that acts essentially in bone mineralization [97] and is expressed in a late stage of the process of bone regeneration and occurs during the process of differentiation of mature osteoblastic cells [110].

However, the LLLT is the subject of several studies in the health area, it reproduces beneficial effects in relation to osteoblastic cellular biomodulation [92] being an important component for clinical applicability, but they are not elucidated, all cellular and biochemical responses such as its effects to the bone system [113].

It is of great importance for the clinical practice data such as those described above where, it has demonstrated beneficial factors after the use of low level laser therapy in different cells present in the human organism, therefore, it is proposed that the use of this therapy in certain wavelengths can, when used by the professional satisfactory results in the early stages can be obtained [114].

8.12. Conclusions

Low Level Laser Therapy has long been used in clinical practice to accelerate soft and hard tissue repair processes. It has several biomodulating effects on cells and tissues, activating or inhibiting physiological and biochemical processes. Although this resource has positive effects, there is a great diversity in relation to the dosimetry used. Therefore, it is extremely important that the dose-effect relationship be standardized so that the effects of the therapy result in favorable stimuli to the organism.

References

[1]. L. Davidovich, Os quanta de luz e a ótica quântica, *Revista Brasileira de Ensino de Física*, Vol. 37, Issue 4, 2015, pp. 4205-4212.

[2]. F. S. S. D. Andrade, R. M. O. Clark, M. L. Ferreira, Effects of low-level laser therapy on wound healing, *Revista do Colégio Brasileiro de Cirurgiões*, Vol. 41, Issue 2, 2014, pp. 129-133.

[3]. S. B. Tajali, J. C. Macdermid, P. Houghton, R. Grewal, Effects of low power laser irradiation on bone healing in animals: A meta-analysis, *Journal of Orthopaedic Surgery and Research*, Vol. 5, Issue 1, 2010, pp. 1-10.

[4]. G. Verdonk, L'antagonismebiologique in vitro entre diferentes bandes visible du spectresolaire, *CompteRendus des Séances de la Société de Biologie*, Vol. 124, 1937, pp. 258-261.

[5]. K. L. Poliakov, G. M. Margolin, V. L. Fedder, La modification de la chronoxie (N. ishiadici) d'une grenouille a l'action sur son corps des rayons lumineux d'une différente longueur d'onde, *The J. of Physiol.*, Vol. 18, 1935, pp. 1012-1019.

[6]. H. Kustner, Die biologische Wirkung von Strahlenverschiedenen Wellenlängen, *Zentralblatt fur Gyndkologie*, Vol. 55, 1931, pp. 2986-31.

[7]. T. Watson, Eletroterapia: Prática Baseada em Evidências, 12 Ed., *Elsevier,* Rio de Janeiro, 2009.

[8]. T. Kushibiki, T. Hirasawa, S. Okawa, M. Ishihara, Regulation of miRNA expression by Low-Level Laser Therapy (LLLT) and Photodynamic Therapy (PDT), *International Journal of Molecular Sciences*, Vol. 14, 2013, pp. 13542-13558.

[9]. J. P. Silva, M. A. Silva, A. P. Almeida, I. Lombardi Junior, A. P. Matos, Laser therapy in the tissue repair process: A literature review, *Photomedicine Laser Surgical*, Vol. 28, Issue 1, 2010, pp. 17-21.

[10]. L. A. Lopes, Análise in vitro da proliferação celular de fibroblastos de gengiva humana tratados com laser de baixa potência, PhD Thesis, *Universidade de São Paulo*, 1999.

[11]. C. K. N. Patel, Continous-wave action on vibration-rotational transitions of CO_2, *Physical Review Online Archive*, Vol. 136, 1964, pp. 1187-1193.

[12]. C. Lucas, R. W. Stanborough, C. L. Freeman, R. J. Haan, Efficacy of low level laser therapy on wound healing in human subjects: A systematic review, *Lasers in Medical Science*, Vol. 15, Issue 2, 2000, pp. 84-93.

[13]. T. I. Karu, Photobiological fundamentals of low-power therapy, *IEEE J. of Quantum Electronics*, Vol. 23, Issue 10, 1987, pp. 1703-1717.

[14]. T. I. Karu, Low-power laser therapy, in Biomedical Photonics Handbook (T. Vodinh, Ed.), Vol. 48, *CRC Press*, Boca Raton, 2003, pp. 1-25.

[15]. T. I. Karu, S. F. Kolyakov, Exact action spectra for cellular responses relevant to phototherapy, *Photomedicine and Laser Surgical*, Vol. 23, Issue 4, 2005, pp. 355-361.

[16]. E. F. Schubert, Light Emitting Diodes, 2nd Ed., *Cambridge*, England, 2006.

[17]. I. F. L. Dias, C. P. C. M. Siqueira, D. O. Toginho Filho, J. L. Duarte, E. Laureto, L. M. Lima, *Effects of Light on Biological Systems*, Vol. 30, Issue 1, 2009, pp. 33-40.

[18]. R. M. Medina-Huertas, F. J. M. Moreno, E. L. Bertos, J. R. Torrecillas, O. G. Martínez, C. Ruiz, The effects of low-level diode laser irradiation on differentiation, antigenic profile, and phagocytic capacity of osteoblast-like cells (MG-63), *Lasers Medical Science*, Vol. 29, Issue 4, 2014, pp. 1479-1484.

[19]. T. M. Cavalcanti, R. Q. A. Barros, M. H. C. V. Catão, A. P. A. Feitosa, R. D. A. U. Lins, Knowledge of the physical properties and interaction of laser with biological tissue in dentistry, *Anais Brasileiros de Dermatologia*, Vol. 86, Issue 5, 2011, pp. 955-960.

[20]. M. C. T. Heckler, D. J. Barberini, R. M. Amorim, Laserterapia de baixa potência em cultivos celulares. *Jornal Brasileiro de Ciência Animal*, Vol. 7, 2014, pp. 541-565.

[21]. J. T. Hashmi, Y. Y. Huang, S. K. Sharma, D. B. Kurup, L. Taboada, J. D. Carroll, M. R. Hamblin, Effect of pulsing in low-level light therapy, *Lasers Surgical Medicine*, Vol. 42, Issue 6, 2010, pp. 450-466.

[22]. D. S. Blaya, A. L. Freddo, M. G. Oliveira, Análise comparativa da terapia laser sobre processos de reparo ósseo, *Revista Brasileira de Ortopedia*, Vol. 4, Issue 1, 2015, pp. 1-10.

[23]. A. G. Andrade, C. F. Lima, A. K. B. Albuquerque, Efeitos do laser terapêutico no processo de cicatrização das queimaduras: Uma revisão bibliográfica, *Revista Brasileira Queimaduras*, Vol. 9, Issue 1, 2010, pp. 21-30.

[24]. L. S. Neves, C. M. S. Silva, J. F. C. Henriques, R. H. Cançado, R. P. Henriques, G. Janson, A utilização do laser em ortodontia, *Revista Dental Press Ortodon Ortop Facial*, Vol. 10, Issue 5, 2005, pp. 149-156.

[25]. H. R. S. Caetano, E. A. C. Zanuto, Ação da laserterapia de baixa intensidade sobre parâmetros bioquímicos, Uma revisão de literatura, *Colloquium Vitae*, Vol. 5, 2013, pp. 63-69.

[26]. D. A. A. P. Pires-Oliveira, R. F. Oliveira, S. U. Amadei, C. P. Soares, R. F. Rocha, Laser 904 nm action on bone repair in rats with osteoporosis, *Osteoporos Int.*, Vol. 21, 2010, pp. 2109-2114.

[27]. A. C. M. Renno, P. A. Mcdonnell, M. C. Crovace, E. D. Zanotto, L. Laakso, Effect of 830 nm laser phototherapy on osteoblasts grown in vitro on biosilicate scaffolds, *Photomedicine Laser Surgical*, Vol. 28, Issue 1, 2010, pp. 131-133.

[28]. P. Oliveira, K. R. Fernandes, E. F. Sperandio, F. A. C. Pastor, O. Nonaka, N. A. Parizotto, A. C. M. Renno, Comparative study of the effects of low-level laser and low-intensity ultrasound associated with Biosilicate on the process of bone repair in the rat tibia, *Revista Brasileira Ortopedia*, Vol. 47, Issue 1, 2012, pp. 102-107.

[29]. R. D. A. U. Lins, E. M. Dantas, K. C. R. Lucena, M. C. V. Catão, A. F. G. Garcia, L. G. C. Neto, Biostimulation effects of low-power laser in the repair process, *Anais Brasileiro Dermatologia*, Vol. 85, Issue 6, 2010, pp. 849-855.

[30]. G. Catorze, Laser: Fundamentos e indicações em dermatologia, *Surgical and Cosmetic Dermatology*, Vol. 37, Issue 1, 2009, pp. 5-27.

[31]. S. Kitchen, Eletroterapia: Prática Baseada em Evidências, 11 Ed., *Manole*, São Paulo, 2003.

[32]. E. H. Loreti, V. L. W. Pascoal, B. V. Nogueira, L. V. Silva, D. F. Pedrosa, Use of laser therapy in the healing process: A literature review, *Lasers Surgical Medicine*, Vol. 33, Issue 2, 2015, pp. 104-116.

[33]. R. M. Silva, P. R. Andrade, A laserterapia na osteogênese: Uma revisão de literatura, *Revista Brasileira Ciências Saúde*, Vol. 10, Issue 34, 2012, pp. 56-62.

[34]. E. C. P. Leal Júnior, R. A. B. Lopes-Martins, B. M. Baroni, T. Marchi, R. P. Rossi, D. Grosselli, Comparison between single-diode low-level laser therapy and LED multi-diode (cluster) therapy (LEDT) applications before high-intensity exercise, *Photomedicine and Laser Surgery*, Vol. 27, Issue 4, 2009, pp. 617-623.

[35]. R. S. Machado, S. Viana, G. Sbruzzi, Low-level laser therapy in the treatment of pressure ulcers: Systematic review, *Lasers Medical Science*, Vol. 32, Issue 4, 2017, pp. 937-944.

[36]. S. Haslerud, L. H. Magnussen, J. Joensen, R. A. B. Lopes-Martins, J. M. Bjordal, The efficacyoflow-level laser therapy for shouldertendinopathy: a systematicreviewand meta-analysis of randomized controlled trials, *Physiotherapy Research Internacional*, Vol. 20, Issue 2, 2014, pp. 108-125.

[37]. R. J. Bensadoun, Photobiomodulation or Low-level Laser Therapy in the Management of Cancer Therapy-induced Mucositis, Dermatitis and Lymphedema, *Curr Opin Oncol.*, 30, 4, 2018, pp. 226-232.

[38]. R. S. Machado, S. Viana, G. Sbruzzi, Low-level laser therapy in the treatment of pressure ulcers: Systematic review, *Lasers Medical Science*, Vol. 32, Issue 4, 2017, pp. 937-944.

[39]. Z. Yu. Huang, J. Ma, J. Chen, B. Shen, F. X. Pei, V. B. Kraus, The effectiveness of low-level laser therapy for non specific chronic low back pain: A systematic review and meta-analysis, *Artrite Research Therapy*, Vol. 17, 2015, 360.

[40]. Y. T. Chen, H. H. Wang, T. J. Wang, Y. C. Li, T. J. Chen, Early application of low-level laser may reduce the incidence of postherpetic neuralgia (PHN), *J. Am. Acad. Dermatol.*, Vol. 75, Issue 3, 2016, pp. 572-577.

[41]. M. F. Cavalcanti, U. H. Silva, E. C. Leal-Junior, Comparative study of the physiotherapeutic and drug protocol and low-level laser irradiation in the treatment of pain associated with Temporo Mandibular Dysfunction, *Photomedical Laser Surgical*, Vol. 34, Issue 12, 2016, pp. 652-656.

[42]. A. Mester, Laser biostimulation, *Photomedicine Laser Surgical*, Vol. 31, Issue 6, 2013, pp. 237-239.

[43]. T. Karu, L. Pyatibrat, G. Kalendo, Irradiation with He-Ne laser increases ATP level in cells cultivated in vitro, *Journal of Photochemistry and Photobiology B: Biology*, Vol. 27, Issue 3, 1995, pp. 219-222.

[44]. M. C. Veçoso, Laser em Fisioterapia, 5th Ed., *Lovise*, São Paulo, 1993.

[45]. H. M. P. Demir, M. Kirnap, M. Calis, I. Ikizceli, Comparison of the effects of laser, ultrasound, and combined laser + ultrasound treatments in experimental tendon healing, *Lasers Surgical Medicine*, Vol. 35, Issue 1, 2004, pp. 84-89.

[46]. V. Campana, M. Moya, A. Gavotto, H. Juri, J. A. Palma, Effects of diclofenac sodium and He:Ne laser irradiation on plasmatic fibrinogen levels in inflammatory processes, *Journal Clinical Laser Medicine Surgical*, Vol. 16, Issue 6, 1998, pp. 317-320.

[47]. L. C. M. Miró, C. Charras, C. Jambom, J. M. Chevalier, Estudio capiloroscópico de la acción de un láser de AsGa sobre la microcirculación, *Inv. Clin. Laser*, Vol. 1, Issue 2, 1984, pp. 9-14.

[48]. P. C. Lievens, The effect of I.R. Laser irradiation on the vasomotricity of the lymphatic system, *Laser Medical Science*, Vol. 6, 1991, pp. 189-191.

[49]. W. J. Genovese, Laser de Baixa Intensidade. Aplicações Terapêuticas em Odontologia, *Lovise*, São Paulo, 2007.

[50]. A. C. G. Henrique, C. Cazal, J. F. L. De Castro, Ação da laserterapia no processo de proliferação e diferenciação celular: Revisão da literatura, *Revista do Colégio Brasileiro de Cirurgiões*, Vol. 37, Issue 4, 2010, pp. 295-302.

[51]. J. A. A. C. Piva, E. M. C. Abreu, V. S. Silva, R. A. Nicolau, Ação da terapia com laser de baixa potência nas fases iniciais do reparo tecidual: princípios básicos, *Anais Brasilero Dermatologia*, Vol. 86, Issue 5, 2011, pp. 947-954.

[52]. A. M. Rocha Júnior, R. G. Oliveira, R. E. Farias, L. C. F. Andrade, F. M. Aarestrup, Modulação da proliferação fibroblástica e da resposta inflamatória pela terapia a laser de baixa intensidade no processo de reparo tecidual, *Anais Brasileiro Dermatologia*, Vol. 81, Issue 2, 2006, pp. 150-156.

[53]. M. M. Marques, A. N. Pereira, N. A. Fujihara, F. N. Nogueira, C. P. Eduardo, Effect of low-power laser irradiation on protein synthesis and ultrastructure of human gingival fibroblasts, *Lasers Surgical Medicine*, Vol. 34, 2004, pp. 260-265.

[54]. M. Trainer, The 50th anniversary of the laser, *World Patent Information*, Vol. 32, 2010, pp. 326-330.

[55]. R. J. Mendonca, J Coutinho-Netto, Cellular aspects of wound healing, *Anais Brasileiro Dermatologia*, Vol. 84, Issue 3, 2009, pp. 257-262.

[56]. P. Moore, T. D. Ridgway, E. W. Higbee, E. W. Howard, M. D. Lucroy, Effect of wavelength on low-intensity laser irradiation stimulated cell proliferation in vitro, *Lasers in Surgery and Medicine*, Vol. 36, 2005, pp. 8-12.

[57]. F. G. Basso, D. G. Soares, T. N. Pansani, L. M. Cardoso, D. L. Scheffel, C. A. S. Costa, J. Hebling, Proliferation, migration, and expression of oral-mucosal-healing-related genes by oral fibroblasts receiving low-level laser therapy after inflammatory cytokines challenge, *Lasers Medical Science*, Vol. 48, Issue 10, 2016, pp. 1006-1014.

[58]. T. N. Pansani, F. G. Basso, A. P. S. Turrioni, D. G. Soares, J. Hebling, C. A. S. Costa, Effects of low-level laser therapy and epidermal growth factor on the activities of gingival fibroblasts obtained from young or elderly individuals, *Lasers Medical Science*, Vol. 32, Issue 1, 2016, pp. 45-52.

[59]. P. T. C. Carvalho, N. Mazzer, F. A. Reis, A. C. G. Belchior, I. S. Silva, Analysis of the influence of low-power HeNe laser on the healing of skin wounds in diabetic and non-diabetic rats, *Acta Cir. Bras.*, Vol. 21, Issue 3, 2006, pp. 177-183.

[60]. A. M. R. Júnior, L. C. F. Andrade, R. G. de Oliveira, F. M. Aarestrup, R. E. Farias, Modulation of fibroblast proliferation and inflammatory response by low-intensity laser therapy in tissue repair process, *Anais Brasileiro Dermatologia*, Vol. 81, Issue 2, 2006, pp. 150-156.

[61]. D. W. Thomas, I. D. Oneill, K. G. Harding, Cutaneous wound healing: a current perspective, *J. Oral Maxillofac. Surg.*, Vol. 53, Issue 4, 1995, pp. 442-447.

[62]. S. Rochkind, M. Rousso, M. Nissan, Systemic effects of low power laser irradiation on the peripheral and central nervous system, cutaneous wounds, and burns, *Lasers Surgical Medicine*, Vol. 9, Issue 2, 1989, pp. 174-182.

[63]. J. M. Reinke, H. Sorg, Wound repair and regeneration, *European Surgical Research*, Vol. 49, 2012, pp. 35-43.

[64]. C. E. Araújo, M. S. Ribeiro, R. Favaro, D. M. Zezell, T. M. Zorn, Ultrastructural and autoradiographical analysis show a faster skin repair in He-Ne laser-treated wounds, *Journal Photochem. Photobiol. Biological*, Vol. 86, Issue 2, 2007, pp. 87-96.

[65]. T. N. Pansani, F. G. Basso, A. P. Turirioni, C. Kurachi, J. Hebling, C. A. de Souza Costa, Effects of low-level laser therapy on the proliferation and apoptosis of gingival fibroblasts treated with zoledronic acid, *International Journal Oral Maxillofacial Surgical*, Vol. 43, 2014, pp. 1030-1034.

[66]. N. K. Khoo, M. A. Shokrgozar, I. R. Kashani, A. Amanzadeh, E. Mostafavi, H. Sanati, L. Habib, In vitro therapeutic effects of low level laser at mRNA level on the release of skin growth factors from fibroblasts in diabetic mice Avicenna, *Journal Medicine Biotech.*, Vol. 6, Issue 2, 2014, pp. 113-118.

[67]. F. G. Basso, C. F. Oliveria, C. Kurachi, J. Hebling, C. A. S. Costa, Biostimulatory effect of low-level laser therapy on keratinocytes in vitro, *Lasers Medical Science*, Vol. 28, 2012, pp. 367-374.

[68]. R. F. Oliveira, D. A. A. Pires-Oliveira, W. Monteiro, R. A. Zangaro, M. Magini, C. Pacheco Soares, Comparison between the effect of low-level laser therapy and low-intensity pulsed ultrasonic irradialtion in vitro, *Photomedine Laser Surgical*, Vol. 26, Issue 1, 2008, pp. 6-9.

[69]. J. J. Tomasek, G. Gabbiani, B. Hinz, C. Chaponnier, R. A. Brown, Myofibroblasts and mechanoregulation of connective tissue remodeling, *Nature Revista*, Vol. 3, 2002, pp. 349-363.

[70]. M. F. G. S. Tazima, Y. Vicente, T. Moriya, Biologia da ferida e cicatrização, *Medicina*, Vol. 41, Issue 3, 2008, pp. 259-264.

[71]. F. R. P. Lopes, B. C. Lisboa, F. Frattini, F. M. Almeida, M. A. Tomaz, P. K. Matsumoto, F. Langone, S. Lora, P. A. Melo, et al., Enhancement of sciatic nerve regeneration after Vascular Endothelial Growth Factor (VEGF) gene therapy, *Neuropathol. Appl. Neurobiol.*, Vol. 37, Issue 6, 2011, pp. 600-612.

[72]. A. M. Rocha Júnior, B. J. Vieira, L. C. F. Andrade, F. M. Aarestrup, Effects of low-level laser therapy on the progress of wound healing in humans: the contribution of in vitro and vivo experimental studies, *Journal Vascular Brasileiro*, Vol. 6, Issue 3, 2007, pp. 257-265.

[73]. L. J. Walsh, The current status of low level laser therapy in dentistry, Part 1, Soft tissue applications, *Austr. Dental Journal*, Vol. 42, 1997, pp. 247-254.

[74]. F. G. Basso, T. N. Pansani, A. P. Turrioni, V. S. Bagnato, J. Hebling, C. A. S. Costa, In vitro wound healing improvement by low-level laser therapy application in cultured gingival fibroblasts, *Int. Journal Dent.*, Vol. 2012, 2012, 719452.

[75]. L. H. Azevedo, F. P. Eduardo, M. S. Moreira, C. P. Eduardo, M. M. Marques, Influence of different power densities of LILT on cultured human fibroblast growth – A pilot study, *Lasers Medical Science*, Vol. 21, 2006, pp. 86-89.

[76]. T. A. H. Jarvinen, M. Kaariainen, M. Jarvinen, H. Kalimo, Muscle strain injuries, *Curr. Op. Rheumatol.*, Vol. 12, 2000, pp. 155-161.

[77]. J. A. M. Marino, C. Taciro, J. A. S. Zuanon, C. Benatti Neto, N. A. Parizotto, Efeito do laser terapêutico de baixa potência sobre o processo de reparação óssea em tíbia de rato, *Revista Brasileira Fisioterapia*, Vol. 7, Issue 2, 2003, pp. 167-173.

[78]. B. M. Baroni, E. C. P. Leal Júnior, T. De Marchi, A. L. Lopes, M. Salvador, M. A. Vaz, Low level laser therapy before eccentric exercise reduces muscle damage in humans, *Eur. J. Appl. Physiol.*, Vol. 110, Issue 4, 2010, pp. 789-796.

[79]. W. H. B. Vieira, C. Ferraresi, S. E. A. Perez, V. Baldissera, N. A. Parizotto, Effects of low level laser therapy (808 nm) on isokinetic muscle performance of young women submitted to endurance training: a randomized controlled clinical trial, *Lasers Medical Science*, Vol. 27, Issue 2, 2012, pp. 497-504.

[80]. A. C. G. Henriques, C. Cazal, J. F. L. Castro, Ação da laserterapia no processo de proliferação e diferenciação celular: Revisão da literatura, *Ver. Col. Brasileiro Cirurgia*, Vol. 37, Issue 4, 2010, pp. 295-302.

[81]. C. H. Santuzzi, D. F. Pedrosa, B. V. Nogueira, H. F. Buss, M. O. V. M. Freire, W. L. S. Gonçalves, Combined use of low level laser therapy and cyclooxygenase-2 selective inhibition on skin incisional wound reepithelialization in mice: A preclinical study, *Anais Brasileiro Dermatologia*, Vol. 86, Issue 2, 2011, pp. 278-283.

[82]. G. M. Assis, A. D. L. Moser, Laser therapy in pressure ulcers: limitations for evaluation of response in persons with spinal cord injury, *Nursing*, Vol. 22, Issue 3, 2013, pp. 850-856.

[83]. P. P. Alfredo, J. M. Bjordal, S. H. Dreyer, S. R. Meneses, G. Zaguetti, V. Ovanessian, et al., Efficacy of low level laser therapy associated with exercises in knee osteoarthritis: A randomized double-blind study, *Clinical Rehabil.*, Vol. 26, Issue 6, 2012, pp. 523-533.

[84]. Y. Sakurai, M. Yamaguchi, Y. Abiko, Inhibitory effect of low-level laser irradiation on LPS-stimulated Prostaglandin E2 production and cyclooxygenase-2 in human gingival fibroblasts, *Eur. Journal Oral. Sci.*, Vol. 108, Issue 1, 2000, pp. 29-34.

[85]. A. B. Junior, R. G. Villa, W. J. Genovese, Laser na Odontologia, *Pancast*, São Paulo, 1991.

[86]. N. N. Houreld, S. M. Ayuk, H. Abrahamnse, Expression of genes in normal fibroblast cells (WS1) in response to irradiation at 660 nm, *Journal Photochem Photobiological B*, Vol. 5, Issue 130, 2014, pp. 146-152.

[87]. S. K. F. Szezerbaty, R. F. Oliveira, D. A. A. Pires-Oliveira, C. Pacheco-Soares, D. Sartori, R. C. Poli-Frederico, The effect of low-level laser therapy (660 nm) on the gene expression involved in tissue repair, *Lasers in Medical Science*, Vol. 33, 2017, pp. 315-321.

[88]. C. C. Martignago, R. F. Oliveira, D. A. Pires-Oliveira, P. D. Oliveira, C. Pacheco Soares, P. S. Monzani, R. C. Poli-Frederico, Effect of low-level laser therapy on the gene expression of collagen and vascular endothelial growth factor in a culture of fibroblast cells in mice, *Lasers Medical Science*, Vol. 27, Issue 1, 2015, pp. 203-208.

[89]. R. Somaiya, G. Kaur, Future of bone repair, *Bone and Tissue Regen Ins.*, Vol. 6, 2015, pp. 1-7.

[90]. K. M. Alghamdi, A. Kumar, N. A. Moussa, Low-level laser therapy: a useful technique for enchancing the proliferation of various cultured cells, *Lasers Medical Science*, Vol. 27, Issue 1, 2011, pp. 237-249.

[91]. S. Bouvet-Gerbettaz, E. Merigo, J. P. Rocca, G. F. Carle, N. Rochet, Effects of low-level laser therapy on proliferation and differentiation of murine bone marrow cells into osteoblasts and osteoclasts, *Lasers Surgical Medicine*, Vol. 41, 2009, pp. 291-297.

[92]. M. Khadra, S. P. Lyngstadaas, H. R. Haanaes, K. Mustafa, Effect of laser therapy on attachment, proliferation and differentiation of human osteoblast-like cells cultured on titanium implant material, *Biomaterials*, Vol. 26, 2005, pp. 3503-3509.

[93]. F. A. Oliveira, A. A. Matos, S. S. Matsuda, M. A. R. Buzalaf, V. S. Bagnato, M. A. A. M. Machado, C. A. Damante, R. C. Oliveira, C. Peres-Buzalaf, Low level laser therapy modulates viability, alkalinephosphataseandmatrix metalloproteinase-2 activates of osteoblasts, *Journal Photochem. Photobiological*, Vol. 169, 2017, pp. 35-40.

[94]. A. R. Coombe, C.-T. G. Ho, C. C. Chapple, L. W. P. Yum, M. A. Darendeliler, N. Hunter, The effects of low level laser irradiation on osteoblastic cells, *Clinical Orthod. Research*, Vol. 4, 2001, pp. 3-14.

[95]. C. S. Santinoni, H. F. F. Oliveira, V. E. S. Batista, C. A. A. Lemos, F. R. Verri, Influence of low-level laser therapy on the healing of human bone maxillofacial defects: A systematic review, *Journal Photochem. Photobiol. Biological*, Vol. 169, 2017, pp. 83-89.

[96]. M. A. B. Medeiros, L. E. A. G. Nascimento, T. C. L. Lau, A. L. B. B. Mineiro, M. M. Pithon, E. F. Sant'anna, Effects of laser vs ultrasound on bone healing after distraction osteogenesis: A histomorphometric analysis, *Angle Orthodontist*, Vol. 85, Issue 4, 2015, pp. 555-561.

[97]. F. Judas, P. Palma, R. L. Falacho, H. Figueiredo, Estrutura Dinâmica do Tecido Ósseo, *Clínica Universitária de Ortopedia*, 2011.

[98]. D. T. Meneguzzo, T. Bavaresco, A. F. Lucena, Resultados da laserterapia em pacientes com feridas crônicas, *HCPA*, Porto Alegre, 2017.

[99]. F. G. Antunes, Efeito do laser de baixa intensidade na atividade biológica de células-tronco da polpa de dentes decíduos humanos, PhD Thesis, *Universidade Federal do Rio Grande do Norte*, 2017.

[100]. P. V. Peplow, T. Y. Chung, G. D. Baxter, Laser photobiomodulation of proliferation of cells in culture: A review of human and animal studies, *Photomedical Laser Surgical*, Vol. 28, Issue 1, 2010, pp. S3-S40.

[101]. S. I. Parenti, L. Checci, M. Fini, M. Tschon, Different doses of low-level laser irradiation modulate the in vitro response of osteoblast-like cells, *Journal Biomed. Opt.*, Vol. 19, Issue 10, 2014.

[102]. J. Nissan, D. Assif, M. D. Gross, A. Yaffe, I. Binderman, Effect of low intensity laser irradiation on surgically created bony defects in rats, *Journal Oral Rehabil.*, Vol. 33, 2006, pp. 619-624.

[103]. N. Bloise, G. Ceccarelli, P. Minzioni, M. Vercellino, L. Benedetti, M. G. C. Angelis, M. Imbriani, L. Visai, Investigation of low-level laser therapy potentiality on proliferation and differentiation of human osteoblast-like cells in the absence/ presence of osteogenic factors, *Journal Biomedical*, Vol. 18, Issue 12, 2013, 128006.

[104]. M. T. Pagin, F. A. Oliveira, R. C. Oliveira, A. C. P. Sant'ana, M. L. R. Rezende, S. L. A. Greghi, C. A. Damante, Laser and light-emitting diode effects on pre-osteoblast growth and differentiation, *Lasers Medical Science*, Vol. 29, 2014, pp. 55-59.

[105]. J. R. Zortéa, G. S. Coura, L. A. Savi, C. M. O. Simões, R. S. Magini, Protocolo preliminar de cultura de fibroblastos gengivais humanos, *Revista Brasileira Implantodont Prótese Implant*, Vol. 12, 2005, pp. 190-196.

[106]. M. H. Aras, Z. B. Bozdag, T. Demir, R. Oksayan, S. Yanik, O. Sokucu, Effects of Low-level laser therapy on changes in inflammation and in the activity of osteoblasts in the expanded premaxillary suture in an ovariectomized rat model, *Photomedical. Laser Surgical*, Vol. 33, Issue 3, 2015, pp. 136-144.

[107]. A. P. R. B. Silva, A. D. Petri, G. E. Crippa, A. S. Stuani, A. L. Rosa, M. B. S. Stuani, Effect of low-level laser therapy after rapid maxillary expansion on proliferation and differentiation of osteoblastic cells, *Lasers Medical Science*, Vol. 27, 2011, pp. 777-783.

[108]. M. M. Jawad, A. Husein, A. Azlina, M. K. Alam, R. Hassan, R. Shaari, Effect of 940 nm low-level laser therapy on osteogenesis in vitro, *Journal of Biomedical Optics*, Vol. 18, Issue 12, 2013, pp. 128001-128006.

[109]. H. Drissi, Q. Luc, R. Shakoori, S. C. S. Lopes, J. Y. Choi, A. Terry, M. Hu, S. Jones, J. C. Neil, J. B. Lian, J. L. Stein, A. J. Wijnen, G. Stein, Transcriptional autoregulation of the bone related CBFA1/RUNX2 gene, *Journal Cell. Physiol.*, Vol. 184, 2000, pp. 341-350.

[110]. A. Neve, A. Corrado, F. P. Cantatore, Osteocalcin: Skeletal and extra-skeletal effects, *Journal Cell Physiol.*, Vol. 228, 2013, pp. 1149-1153.

[111]. T. Nakasa, M. Yoshizuka, M. A. Usman, E. E. Mahmoud, M. Ochi, MiRNAs and bone regeneration, *Curr. Genomics*, Vol. 16, 2015, pp. 441-452.

[112]. E. F. Pípi, Efeitos do laser terapêutico e do ultrassom pulsado de baixa intensidade na expressão de genes relacionados à diferenciação celular durante o processo de reparo ósseo, Dissertação ao Programa de Pós Graduação em Biotecnologia, *Universidade Federal de São Carlos*, Brazil, 2010.

[113]. G. T. Nogueira, R. A. M. Ferrari, M. D. Martins, S. K. Bussadori, T. D. Silva, K. P. S. Fernandes, Efeito da laserterapia de baixa potência sobre o tecido ósseo – Revisão de literatura, *ConScientiae Saúde*, Vol. 8, Issue 4, 2009, pp. 671-676.

[114]. F. B. Pires, Influência da Terapia Laser de Baixa Potência em células osteoblásticas OFCOL-II e sua ação no perfil da expressão do MIR-148A e MIR-21 e dos genes RUNX2 e Osteocalcina associados as reparo ósseo, PhD Thesis, *Unopar*, Londrina, Brasil, 2018.

Chapter 9
Fiber-end Integrated Micro- and Nano-Structures for Sensor Applications

Xinping Zhang

9.1. Introduction

Integration of photonic structures onto the end surface of optical fibers to achieve sensor devices developed new approaches for special detections and have attracted extensive research and engineering interests. Remarkable advances have been made in this field, which also pushed the development of lab-on-fiber techniques, fiber sensors, micro- and nano-photonics, biosensors, photoelectronics, and micro- and nano-fabrication methods. The advantages of fiber sensors and micro-/nano-photonic sensors are also integrated together in such fiber-end integration techniques. Possibly long-range and flexible detection using optical fibers enable easy reaching of the target samples by the sensor devices, which are difficult to reach or to detect using conventional techniques. This feature becomes extremely important if the samples to be detected are located in a space with extreme conditions, e. g. , in a space with extremely high or low temperatures, in the space filled with poison or easily flammable gas, inside or underneath liquids, or in a location simply not reachable or difficult to reach by conventional detection methods.

However, multiple challenges have to be overcome to realize micro-/nano-scale photonic structures on the fiber end surface. These challenges mainly lie in the small area of the fiber end, the large aspect ratio of the device using an optical fiber as its substrate, the difficulties to manage or control a system based on a very thin optical fiber. They will directly bring difficulties to the fabrication, characterization, and application of the sensor devices. A variety of techniques have been explored to solve such challenges. Non-periodical structures, including micro cantilevers [1], subwavelength silicon-on-insulator waveguides [2], microlens and woodpiles [3] have been achieved on optical fiber ends by focused ion beam etching or scanning electron beam lithography. Three-dimensional writing using femtosecond lasers has produced complicated nanostructures, including metallic structures for SERS application [4]. In these structures, the optical fibers function both as a waveguide for delivering the excitation and the signal

Xinping Zhang
College of Applied sciences, Beijing University of Technology, Beijing 100124, P. R. China

light of the sensor device and as a mechanically controlled arm for positioning the sensor device on the fiber tip.

If photonic devices are to be integrated on the fiber end surface, large-scale periodical structures have to be patterned on the optical fiber end. Thus, the in-coupling light will first excite the resonance modes in the photonic structures propagating in the plane parallel to the end surface of the fiber, which will be coupled back to the fiber body and delivered by the fiber to the photoelectronic or spectroscopic detection system. Dielectric, metallic, and hybrid photonic structures have been integrated on the fiber end surface [5-12]. In particular, metallic photonic structures introduce additional photophysical processes with surface plasmon resonance and its coupling with the involved photonic resonance modes. Ion-beam etching, electron-beam etching, mechanical contact, nano-imprinting, and photolithography have been utilized in accomplishing these fabrications. Hybrid dielectric and metallic structures with a period of 900 nm have been produced on the fiber end surface [5]. Refractive-index sensor and acoustic detection functions have been explored for this device. A sensitivity of about 125 nm/RIU has been demonstrated for the refractive-index sensing measurement. Plasmonic photonic structures consisting of periodically arranged metallic nanostructures constitute a large portion of the fiber-end-based sensor devices. Two-dimensional arrays of metallic nanostructures have been patterned on the cleaved facets of optical fibers using nanoskiving [7], which can be applied in sensors based on localized surface plasmon resonances and surface-enhanced Raman scattering (SERS), in optical filters, and in diffraction gratings. Optical fiber metatip was proposed by integrating a phase-gradient plasmonic metasurface on an optical fiber tip [9]. This enables flexible light-manipulation in optical fibers, which are potentially applicable in optical communications, fiber-based signal processing, imaging and sensing. Nanosphere lithography was recently demonstrated for the patterning of the fiber tips [10]. The fabrication was based on the self-assembly of polystyrene nanospheres and additional processes of thermal evaporation, plasma etching, and sonication. The produced structures show excellent performance in SERS detection of low-concentration molecules. Furthermore, metallic nanostructures fabricated using e-beam lithography and lift-off processes were transferred first to a soft thin film through mechanical contact, then easily to the end surface of an optical fiber [12]. Above examples summarized some of the typical methods and recent advances in micro-/nano-patterning on fiber-end surface.

This chapter focuses its topics on the fabrication methods of nanophotonic structures on the end facets of optical fibers, as well as possible applications of the produced structures/devices on refractive-index or molecular concentration sensors.

9.2. Fiber-end Integration of Photonic Structures Using Interference Lithography for Multi-parametric Sensors

Interference lithography is a straightforward and convenient method for patterning photonic structures onto fiber end surface, although this is a very limit area. The challenge lies mainly in coating the recording medium or photoresist onto such a small area with a diameter in microns or hundreds of microns. A most simple and effective approach is to dip the fiber tip into the solution of photoresist (PR), so that the droplet may be held by the surface tension to the fiber end surface. Using high-pressure air to blow the fiber tip, the excess photoresist can be removed and a homogeneous thin film was produced on the end surface, as shown in Fig. 9.1(a). The disadvantage of this method is that it is difficult to control the thickness of the PR layer.

An alternative method is demonstrated in Fig. 9.1(b), multiple fibers are bundled together to form a large flat surface, so that spin-coating of PR may be employed to produce controllable thin films on the large-area common surface of the fiber ends. It is understandable that the length of the fiber is very limited for the fabrication in Fig. 9.1(b). Multi-beam interference lithography is then employed to pattern two-dimensional grating structures into the PR film, as shown in Fig. 9.1(c). For practical applications, a subsequent fiber fusion splicing procedure is necessary to connect the short fiber to the main body of the fiber sensing system, as shown in Fig. 9.1(d).

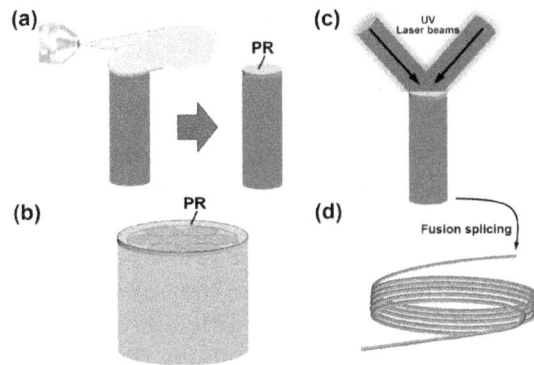

Fig. 9.1. (a) Producing a thin layer of photoresist on the fiber end surface by high-pressure air blowing; (b) Spin-coating of photoresist onto the flat surface of a fiber bundle; (c) Two-beam interface lithography into the photoresist layer on the fiber end; (d) Fusion splicing of the end-structured short fiber segment onto the long fiber constituting the sensing system.

As an example, we fabricated a waveguide grating on the end surface of an optical fiber with a diameter of 600 μm [13], where the waveguide is a 430-nm ZnO layer deposited by radio-frequency magnetron sputtering. Then, a PR thin film was produced by the method in Fig. 9.1(a). Interference lithography using UV laser beams at 355 nm produced a PR grating with a period of about 425 nm and a modulation depth of about 200 nm. Certainly, two-dimensional grating can be produced if a second exposure is performed after the fiber is rotated by 90 degrees about its axis. High-quality gratings can be produced, as verified by the diffraction pattern in Fig. 9.2(b) and the SEM image in Fig. 9.2(c). The green and red light was diffracted into the space on the backside of the fiber end surface for a grating period of 425 nm, where the diffraction was based on the equation:

$$n\Lambda(\sin\theta_i + \sin\theta_d) = \lambda, \tag{9.1}$$

where n is the refractive index of the fiber material, Λ is the grating period, θ_i and θ_d are the angles of incidence and diffraction, λ is the wavelength of light.

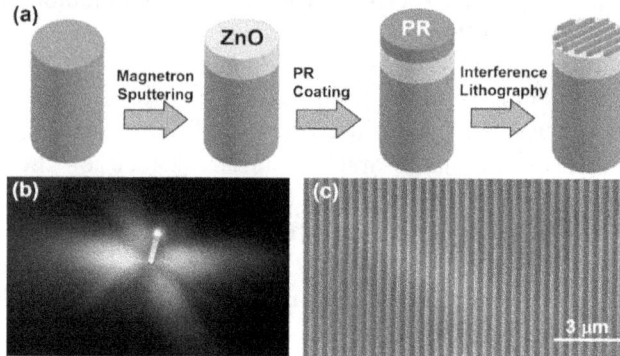

Fig. 9.2. (a) Fabrication procedures for the waveguide grating on the end facet of an optical grating; (b) Diffraction pattern by the PR grating on the fiber end into the backward space; (c) SEM image of the grating.

The application of the fiber-end-integrated waveguide grating in refractive-index sensors is based on the sensitivity of the waveguide resonance mode to the change in the environmental refractive index. The theory for the waveguide grating structures has been investigated extensively [14-16]. The waveguide resonance mode is very sensitive to the angles of incidence and diffraction. Therefore, in the transmission space, due to the large divergence angle of the output of the fiber, different spectroscopic response can be measured at different output angles with respect to the fiber axis. For the convenience of measurement, in the practical geometry, we may measure the output spectrum by locating the detection head at different distance from the center axis, as shown in Fig. 9.3 (a) by X. At $X = 0$, the waveguide resonance modes based on the +1 and -1 orders of diffractions are degenerate, so that one optical extinction peak is observed. With increasing the value of X, the resonance spectrum split into two branches, which evolve into opposite directions, as shown in Fig. 9.3(b).

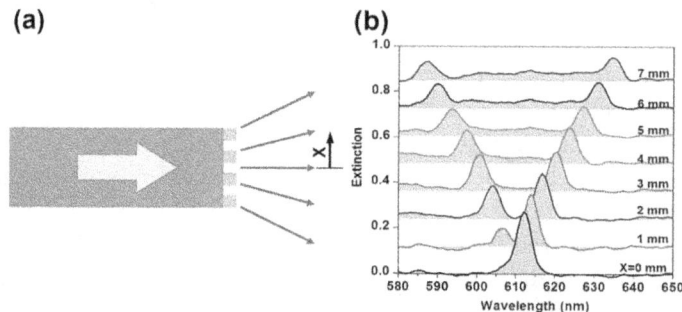

Fig. 9.3. (a) Geometry for the measurement on spectroscopic response of the waveguide grating on the fiber end surface; (b) Transmissive optical extinction spectrum as a function of the displacement from the center axis of the fiber.

Due to the large divergence of the output of the fiber end, there is a large spatial and angle dispersion of the output spectrum. The separation between the two branches is as large as 50 nm at $X = 7$ mm, corresponding to a sensitivity of about 7 nm/mm. Therefore, this fiber-end integrated device is intrinsically a displacement and angle sensor. The equivalent angle sensitivity is about 4.5 nm per degree. Furthermore, the change in the environmental refractive index also modifies the resonance condition of the waveguide mode. Thus, the fiber-end-integrated device is also a sensitive refractive-index sensor. In above considerations, a multi-functional sensor for the detection of the changes of various physical parameters has been achieved, including displacement, tilting angles, molecular concentrations, and environmental temperatures.

9.3. Electron-beam Lithography and Reactive Ion-beam Etching of Metallic Nanostructures onto Fiber End Facets as Refractive-index Sensors

Although optical methods, including interference lithography and mask-based photolithography, are convenient for the fabrication of large-area micro- and nano-structures, they are not suitable for achieving specially shaped structures like rings. Electron-beam lithography becomes advantageous in such cases. Complicated patterns in micro- or nano-scale can be easily produced by a graph generator, interaction between focused electron beams with polymeric e-beam resist enables writing of the designed pattern into the photoresist film. The photoresist pattern can be used as a master for further fabrication, e. g., of metallic plasmonic structures.

The practical fabrication procedures for gold nanoring structures on the end surface of an optical fiber are illustrated in Fig. 9.4 [17]. In order to overcome electrostatic charging of the fiber facets, a 30-nm-thick gold film was first evaporated onto the fiber end surface. Then, a layer of poly(methyl methacryate) (PMMA) was coated on the top of the gold film and used as e-beam resist. The sample was exposed with an electron beam with dose of 150 μC/cm². After development, 5-nm Cr and 30 nm Au were evaporated sequentially on the top surface of the fiber. After lift-off in acetone, the fiber end was etched for 90 s with Ar-ions accelerated by a voltage of 0. 7 kV and a beam current of 70 mA. It needs to be noted that for the convenience of fabrication, multiple fiber segments were fixed into a brass mount [17], so that all fabrication procedures were carried out on a large flat surface, instead of on a single fiber tip.

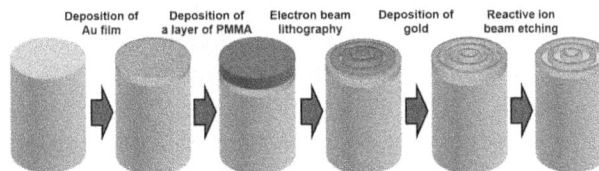

Fig. 9.4. Fabrication procedures for the gold nanorings on the end facet of an optical fiber [17].

Fig. 9.5 shows the fabrication results. The SEM image in Fig. 9.5 (a) presents high-quality gold nanorings with excellent homogeneity, smooth edges and surfaces, high contrast and regularity. The separation between the adjacent rings is constantly 900 nm. Fig. 9.5(b) shows the diffraction pattern photographed in a plane facing the output of the fiber. Strong radial dispersion of the color can be observed in Fig. 9.5(b), however, each single color forms a symmetric ring, implying diffraction by a symmetric ring grating.

The sensing performance of the gold nanorings is based on the diffraction anomaly by the ring grating, which can be characterized by:

$$n\Lambda(1 \pm \sin\theta_i) = m\lambda, \qquad (9.2)$$

where n is the environmental refractive index, Λ is the grating period, θ_i is the angle of incidence, m is the order of diffraction, and λ is the wavelength of the diffracted light. According to equation (9.2), the sensitivity of the sensor to the environmental change in refractive index can be evaluated by:

$$d\lambda/dn = \Lambda(1 \pm \sin\theta_i) \qquad (9.3)$$

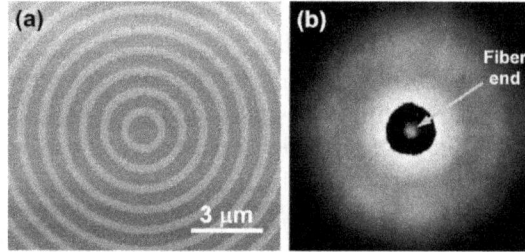

Fig. 9.5. (a) SEM image of the gold concentric nanoring structures [17]. (b) Diffraction pattern recorded in a plane facing the output of the gold-nano-ring structured fiber end.

Thus, at $\theta_i = 0$, we have $d\lambda/dn = \Lambda$, implying that the sensitivity is proportional to the grating period. For a grating period of 900 nm, we may achieve a sensitivity of 900 nm/RIU. This can be verified by the measurement results in Fig. 9.6, where we compare a series reflective optical extinction measurements on different environmental media, including air, alcohol, isopropanol, and butylamine. The refractive index increases from 1 to 1. 357, 1. 378, and 1. 402, when the environment was changed from air to alcohol, isopropanol, and butylamine. A linear spectral shift of the Rayleigh anomaly can be observed in Fig. 9.6. A spectral shift as large as 320 nm was measured when the fiber end was dipped into alcohol from air, verifying a Rayleigh anomaly nature of the sensor signal, where $\Delta n = 0. 357$, $\Lambda = 900$ nm, $\theta_i \approx 0$, and $\Delta\lambda \approx 900 \times 0. 357 = 321. 3$ nm.

A further mechanism is the splitting of the sensor signal resulted from Rayleigh anomaly, as shown in Fig. 9.6, which are guided by the dashed lines in magenta. Each spectral peak as the sensor signal splits into two branches, which actually resulted from the possibly deviation of the angle of incidence ($\theta_i \neq 0$) from normal incidence ($\theta_i = 0$). This is easily understood by considering light-propagation in an optical fiber before reaching the ring grating, where the fiber modes allow different output angles at the fiber ends. If assuming

$\lambda_0 = n\Lambda$ for $\theta_i = 0$, then the dependence of resultant sensor signal on the angle of incidence can be evaluated by:

$$\lambda = \lambda_0 \pm n\Lambda\sin\theta_i, \tag{9.4}$$

and

$$\Delta\lambda = |\lambda - \lambda_0| = n\Lambda\sin\theta_i \tag{9.5}$$

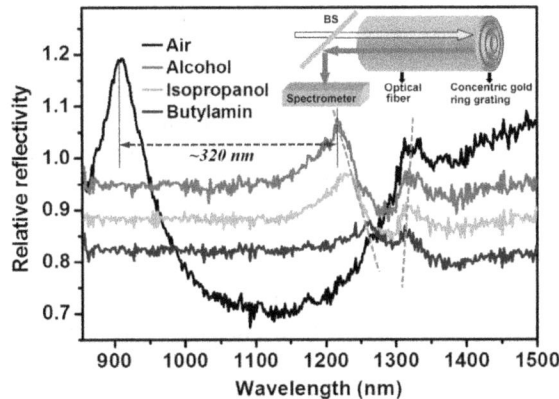

Fig. 9.6. Refractive-index sensing measurements on different liquids with comparison to that in air. Inset: geometry for the measurement.

Therefore, we have two branches for the +1 and -1 orders of diffraction. Moreover, the separation between the two branches is a function of the refractive index, the grating period, and the angle of incidence. To evaluate the dependence of this separation on the refractive index of the environmental medium, we carry out such a derivation:

$$d\Delta\lambda/dn = \Lambda\sin\theta_i + n\Lambda\cos\theta_i \cdot d\theta_i/dn \tag{9.6}$$

We need to note that the diffraction has to take place after the light is refractive out of the fiber and incident onto the grating. Therefore, θ_i reduces with increasing the value of n, or we have $d\theta_i/dn < 0$. As a result, equation (6) becomes:

$$d\Delta\lambda/dn = -\Lambda(n\cos\theta_i \cdot |d\theta_i/dn| - \sin\theta_i) \tag{9.7}$$

Considering the conditions of our experiments, we may have $d\Delta\lambda/dn < 0$. This explains the reduced separation between the two branches in Fig. 9.6, as highlighted by the dashed lines in magenta. Actually, this mechanism can be also utilized as a new criterion to characterize the sensor performance, which seems even more sensitive than the spectral shift of one of the branches. As shown by a comparison in Fig. 9.7, we plotted the spectral "shift" and "separation" as a function of refractive indices of the environmental solvent (alcohol, isopropanol, abd butylamine). Clearly, the variation of the spectral "separation" exhibits larger slope, implying higher sensitivity, as shown more clearly in Fig. 9.7(b).

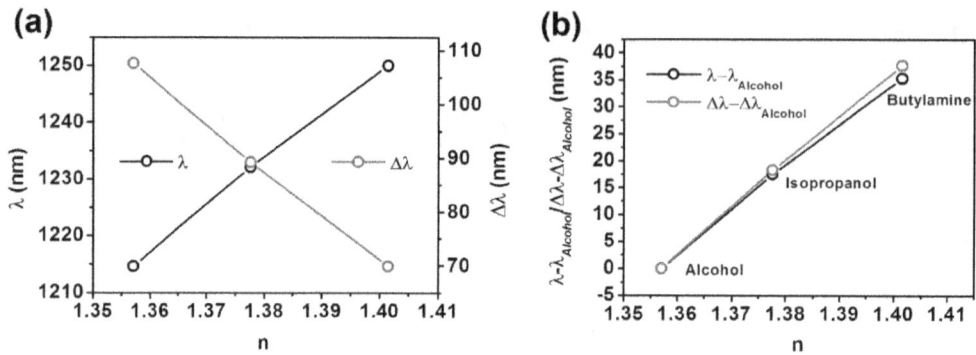

Fig. 9.7. (a) Center wavelength of the sensor signal (black) and spectral separation between the two branches of the sensor signal (red) as a function of refractive indices of alcohol, isopropanol, and butylamine. (b) Relative spectral shift with reference to alcohol (black) and relative spectral separation with reference to alcohol (red) as a function of refractive index.

Additionally, localized surface plasmon resonance (LSPR) of the gold nanoring can also be utilized to achieve more sensor functions. The LSPR is located not only in each ring as a whole, but also in the gold strip constituting the ring structure. It is more interesting for optimizing the structures such that the LSPR may couple with the Rayleigh anomaly through Fano coupling, which will introduce new mechanisms for the sensor device and may enhance the sensitivity further.

Apparently, the gold nanoring structures are symmetric about the center axis, thus, there is no polarization dependence of the sensor signals. This a unique feature and advantage of such a sensor device. Thus, the advantages of this fiber-end integrated sensor device lie in the miniaturization of the sensor device onto the fiber end surface with a size of 600 μm in diameter, possibly remote and flexible sensing utilizing the advantages of optical fibers, and polarization independence as a best fitting of the sensor to fiber end facet.

9.4. Direct Laser Writing of Plasmonic Nanostructures onto Fiber Tips

Chemically synthesized solution-processed colloidal gold nanoparticles supply multifold possibilities to produce various plasmonic nanostructures [18-20], The most convenient fabrication is the annealing of the spin-coated film of the colloidal gold nanoparticles on planar substrates, where randomly distributed gold nanoparticles can be obtained with strong localized surface plasmon resonance [21-24]. This is based on the low melting point of the gold nanoparticles smaller than 10 nm. The ligands covering the gold nanoparticles, which are important for the dispersity of the gold nanoparticles in organic solutions, have a sublimation temperature as low as 150 °C, whereas, the gold nanoparticles become molten when being heated to about 200 °C. By controlling the concentration of the colloidal solution, the annealing temperature, or by adjusting the spin-coating speed, the plasmonic nanostructures can be tuned easily in their morphological or structural properties and their optical spectroscopic response [18].

The low melting temperature of the small gold nanoparticles also allows direct writing of metallic micro- and nano-structures by laser beams. Interference ablation has been demonstrated in the fabrication of one- and two-dimensional metallic gratings [25, 26] with excellent homogeneity in large-area structures. Interactions between plasmonic and photonic resonance modes induced strong optical response with narrow spectral bands, which can be applied practically in optical filters and sensor devices.

Making use of the intrinsic advantages of optical fibers, we can also achieve plasmonic nanostructures directly on the fiber tips, including the end facets and the side walls, where the laser light is delivered by the optical fiber to reach the fiber tip, so that it interacts with colloidal gold nanoparticles, inducing an effective annealing process. Consequently, randomly distributed gold nanostructures can be produced on both the end and side surfaces of the fiber tips. The basic fabrication procedures are illustrated in Fig. 9.8. In the first stage, the fiber tip was immersed into the colloidal solution of gold nanoparticles with an expected or designed depth, so that the colloidal gold nanoparticles coat the fiber surface with a suitable length, as shown in Fig. 9.8(a). The homogeneity, the thickness, and the distribution of this coating layer depend strongly on the concentration of the colloidal solution, the environmental conditions outside the solution, as well as on how the fiber was held during the drying process.

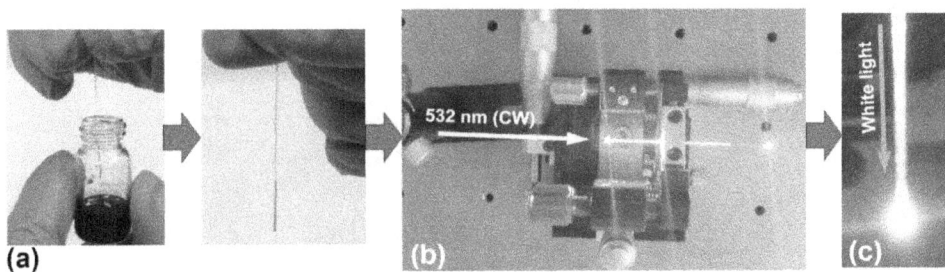

Fig. 9.8. (a) Coating the fiber tip with colloidal gold nanoparticles by immersing one of the fiber end into the colloidal solution; (b) Laser annealing using 200-mW laser power at 532 nm; (c) Demonstration of the plasmonic scattering of light by sending a white-light beam into the fiber from the unstructured end.

Then, the fiber was mounted onto an optical holder and direct laser annealing can be performed by sending the laser beam from the other end of the fiber without being coated with colloidal gold nanoparticles. In the fabrication illustrated in Fig. 9.8, we used a concentration of 100 mg/mL for the colloidal solution of gold nanoparticles. As shown in Fig. 9.8 (b), a continuous-wave green laser at 532 nm with a power of about 200 mW was focused into the fiber at end without gold-nanoparticle coating. Thus, the laser power was delivered to the other end to interact with the colloidal gold nanoparticles. After the annealing process, gold nanoparticles with random distributions in their sizes, shapes, and separation distances were produced on the fiber end and side surfaces. Strong light scattering by the gold nanoparticles can be observed when a white-light beam is coupled into the fiber. The bright orange color of the scattered light in Fig. 9.8 (c) indicates strong plasmonic scattering of light and excellent homogeneity of the gold nanoparticles.

Fig. 9.9 shows the microscopic characterization of the laser-annealed fiber tip. Fig. 9.9 (a) shows the photograph of the fiber tip captured under an optical microscope. The bright red color indicates strong plasmonic scattering in the red spectral range and the purity of the red color implies the homogeneity of the produced gold nanoparticles. As shown in Fig. 9.9 (a), the fiber has a diameter of 400 μm. The gold nanoparticle structures coat the fiber tip with a length longer than 5 mm.

Figs. 9.9 (b) and the 9.9 (c) show the scanning electron microscopic (SEM) characterization of the produced gold nanostructures on the fiber end facet and side wall, respectively. Gold nanoparticles with a diameter larger than 200 nm and roughly spherical shapes were produced at the end facet of the fiber. The separation between them is also mostly larger than 200 nm, as shown in Fig. 9.9 (b). However, the gold nanostructures become more irregular on the side surface when compared with those on the end, as shown in Fig. 9.9 (c). Moreover, they become smaller at larger distance from the end facet. This can be understood by considering that the annealing on the end facet is a direct interaction between the incident laser with the colloidal gold nanoparticles, however, that on the side wall was based on the annleaing with the reflected laser by the annealed end facet. The fiber-mode confined laser beam cannot interact directly with the gold nanoparticles on the side surface. Thus, the interaction becomes weaker at larger distance from the end facet. As a result, weak annealing process produced smaller gold nanoparticles.

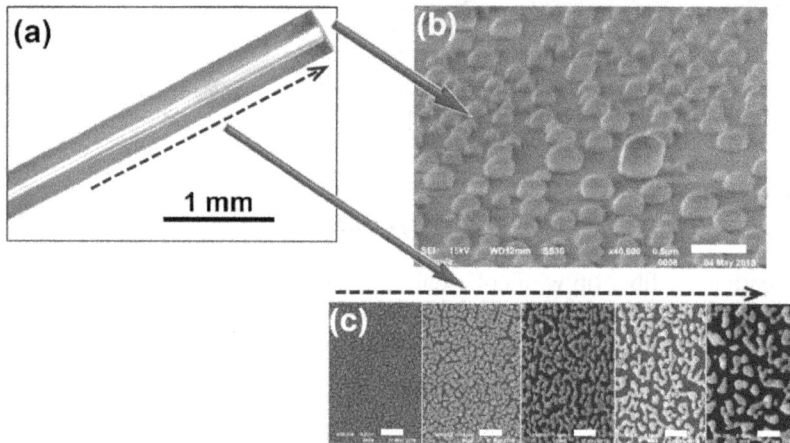

Fig. 9.9. (a) Observation of the fiber tip coated with gold nanoparticles under an optical microscope using transmission mode; (b) SEM image of the gold nanoparticles on the end facet of the fiber; (c) SEM images of the gold nanoparticles on the side surface of the fiber tip, showing distribution of the gold nanoparticles with the distance from the fiber end surface. All scale bars in Figs. 9.5 (b) and 9.5 (c) indicate 500 nm.

This laser annealing method enables efficient and homogeneous fabrication of plasmonic nanostructures on the fiber tips. Strong interaction between the laser and the gold nanoparticles facilitates strong attachment of the gold nanoparticles onto the surface of the fiber tip. By adjusting the concentration of the colloidal solution and the power of the annealing laser beam, it is convenient to tune the structural parameters of the gold nanostructures and consequently tune the spectroscopic and sensing performance. The randomly distributed gold nanoparticles on the fiber tip enable direct applications in refractive-index and SERS sensors, which is based on the strong localized surface plasmon and the induced local field enhancement.

9.5. Transfer of Metallic Photonic Structures onto Fiber-end Facets through Soft "Welding" [27]

A more favorable method to achieve sensor devicing on the end surface of an optical fiber is the transfer of the structures fabricated successfully on a large-area planar substrate to the fiber tip. This approach allows convenient fabrication of high-quality photonic structures, without being limited by the small area of the fiber end, by the management of a thin fiber into the fabrication system, and by the mounting of a miniaturized device with a large aspect ratio into the spectroscopic and sensor measurement schemes. In short, the transfer method may overcome all challenges that may be encountered in the direct fabrication on fiber tips.

The "soft welding" method is most convenient for transferring the large-area photonic structures on a planar substrate to the fiber tips. The basic idea is shown schematically by an example in Fig. 9.10, where metallic photonic crystals (MPCs) consisting of periodically arranged gold nanostructures to the end facet of an optical fiber. MPCs were firstly fabricated on a large-area glass substrate and the indium-tin-oxide (ITO) was used as the buffer layer between MPCs and the substrate. A droplet of the solution of polymethyl methacrylate (PMMA) in xylene with a concentration of 5 mg/ml was used as the solder for the welding, as shown in Fig. 9.10(a). After complete evaporation of the solvent, the whole device was immerse into the solution of hydrochloride (HCl) acid with a concentration of 20 %. It took less than 30 minutes for the ITO layer to be dissolved completely, and the MPC structures are transferred to the end surface of the fiber, as shown in Fig. 9.10(b). After being rinsed in pure water, the structures were dried and trimmed, so that the device can be used as plasmonic sensors, as shown in Fig. 9.10(c).

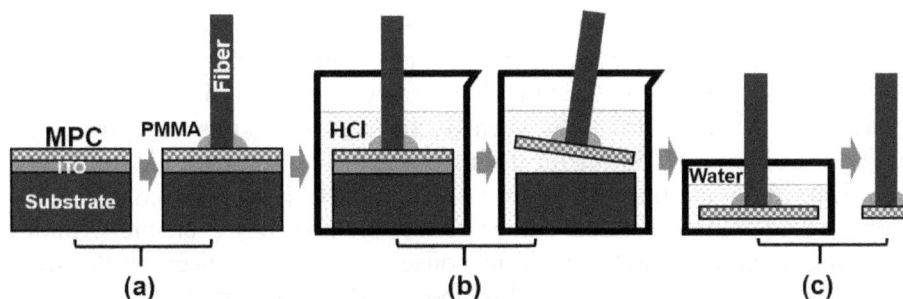

Fig. 9.10. Soft-welding procedures for the transfer of metallic photonic crystals (MPC) from a glass substrate to the fiber end surface with an ITO layer employed as a buffer layer. (a) Welding of the fiber end to the top surface of the MPC using a droplet of PMMA solution in chloroform; (b) Liftoff through etching the ITO in hydrochloride acid solution; (c) Rinsing in water and trimming the edges of the transferred device [27].

It needs to be noted that during the welding process, the end surface should be parallel to the top surface of the MPCs, so that close contact can be achieved between the surfaces of the fiber end and the MPCs. Thus, the tilting angle of the fibers should be able to be adjusted with respect to the normal of the substrate. Therefore, a sophisticatedly designed holder is required to ensure the relative positioning between the fibers and the MPCs, as shown in Fig. 9.11 (a). Furthermore, another important advantage of this method is that the large-area MPCs can be divided into multiple domains with each domain welded to an individual fiber. Thus, after the etching process, the transfer can be accomplished simultaneously to each individual fiber. As shown in Fig. 9.11 (b), the whole system, including the holder, will be moved together during the etching process in HCl solution. Fig. 9.11 (c) shows the photograph of a practical experimental result, where a thin-film MPC device was transferred to the end surface of an optical fiber with a diameter of 600 μm.

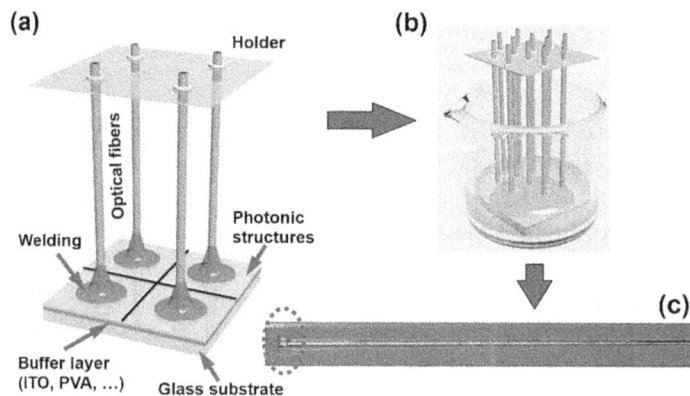

Fig. 9.11. (a) Separation of a large-area photonic structure into multiple divisions and welding of them separately to different fiber ends using a droplet of a polymer solution; (b) Liftoff of the photonic structures from the substrate by etching away the buffer layer in hydrochloride acid; (c) The photograph of the finished sample with a photonic structured thin film transferred to the end surface of an optical fiber.

The quality of the transferred MPC structure can be evaluated by the microscopic and spectroscopic performance. Fig. 9.12 (a) shows the observation of the end surface of the fiber covered with the MPC structures. Clearly, the MPC structures were strongly welded to the end surface. However, during the drying of the PMMA droplet, the welded thin film becomes wrinkled in multiple locations. Such wrinkles are unavoidable for such a fabrication method. This is a disadvantage of this method, however, the wrinkles seem not influencing much the spectroscopic response of the MPCs, because the size of the wrinkles is much larger than the period of the metallic nanostructures. The diffraction pattern in Fig. 9.12 (b) verifies the high quality of the photonic structures. The spectroscopic response with strong LSPR and Fano coupling [28, 29] make the device suitable for application in sensors. These sensors are important for the detection of environmental refractive index change, in particular in liquids.

Fig. 9.12. (a) Observation of the end facet of the optical fiber after being transferred with the thin-film MPC. (b) Diffraction pattern at the output end of the fiber when a white-light beam was coupled into the fiber from the other end.

9.6. Flexible Transfer of Metallic Photonic Structures [30]

As has been discussed in Section 9.5, the soft-welding method has a problem of producing wrinkles in the transferred structures. This method may be improved further by modifying the soft-welding into the flexible thin-film transferring procedures, as illustrated in Fig. 9.13: (a) Interference lithography was first employed to produce PR grating as a master structure on a ITO glass substrate; (b) Spin-coating of colloidal gold nanoparticles to fill the grating grooves; (c) Annealing at 200 °C to sublimate the ligands covering the gold nanoparticles and melt the small gold nanoparticles to make them aggregated; (d) Rinsing with acetone to remove the PR template grating; (e) Annealing at 350 °C to melt the gold nanoparticles and make them connect to form continuous gold nanowires; (f) Spin-coating of PMMA onto the top surface of the gold-nanowire grating; (g) Lift-off of the gold-nanowire grating and transfer onto a hollow plate by immersing the sample in HCl solution; (h) Decking of the thin-film gold-nanowire grating onto the end surface of an optical fiber; (i) Annealing at 200 °C and rinsing in chloroform to fix the gold nanowires onto the end surface of the fiber and to remove PMMA; (j) Deep annealing at 500 °C to produce more homogeneous gold nanowires and to sinter the gold nanowires onto the fiber end surface.

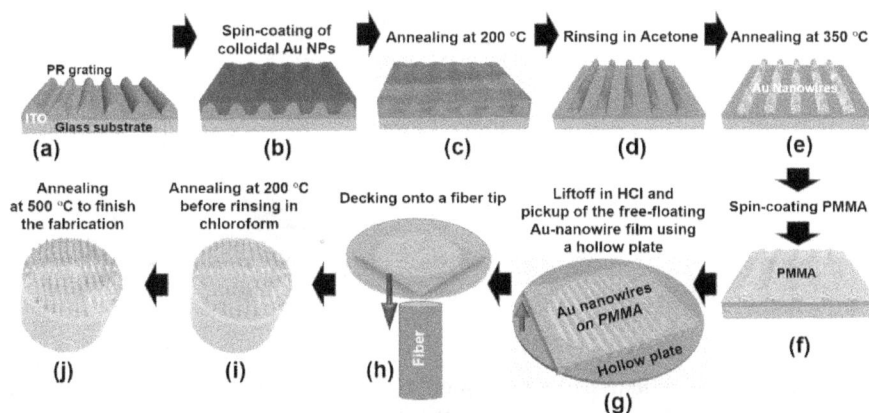

Fig. 9.13. Fabrication procedures for the transfer of gold nanowire gratings onto the end surface of optical fibers [30].

Homogeneous gold-nanowire gratings are produced on the fiber end surface. An example of the fabricated structures is shown by the SEM image in Fig. 9.14 (a) and the optical microscope image in Fig. 9.14 (b). The inset of Fig. 9.14 (b) shows the details of the gold nanowires. As seen in Fig. 9.14 (a), no wrinkles were produced and whole structures were transferred homogeneously onto the fiber end. In particular, due to the high-temperature annealing, no PMMA or PR was remaining with the gold-nanowire gratings. This favors excellent contacts between the gold nanowires and the fiber surface and between the sensor device and the liquid samples to be detected.

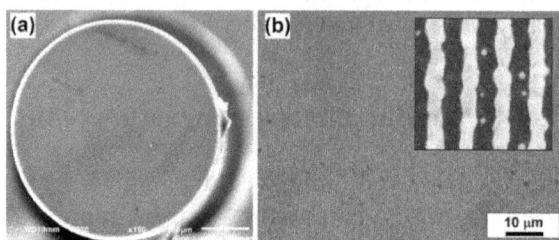

Fig. 9.14. (a) SEM image of the fiber end surface transferred with grating structures of gold nanowires; (b) Optical microscope image of the grating structures on the fiber end surface. Inset: enlarged view of the gold nanowires [30].

Fig. 9.15 shows an example of refractive index sensing measurement. Fig. 9.15(a) demonstrates how the fiber-end integrated sensor was managed in the experimental setup. For long-range detection, the fabricated sensor on a fiber segment needs to be connected to a long fiber by fusion splicing. At the detection end of the fiber, the reflected sensor signal was coupled into the spectrometer by discrete optical elements. Fig. 9.15(b) shows the sensor signal as a function of the sample solution prepared by dissolving glucose into pure water with a concentration of 3 %, 7 %, and 10 %. The sensor signal has been calculated by $-\log[I_s(\lambda)/I_0(\lambda)]$, where $I_s(\lambda)$ and $I_0(\lambda)$ are the reflection spectra when the structured fiber end was immerse in glucose/water solution and in pure water, respectively. The amplitude of the sensor signal is defined by S, as shown in Fig. 9.15(b).

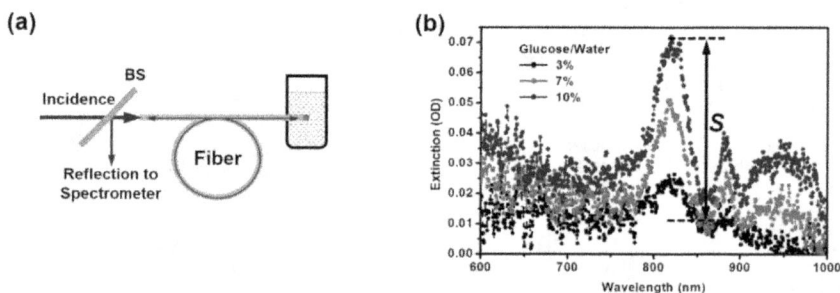

Fig. 9.15. (a) Reflective optical extinction spectra of the fiber-end-integrated gold-nanowire grating for TE and TM polarizations; (b) Refractive index sensor measurements on the glucose/water solution with different concentrations using the fiber-end-integrated gold-nanowire grating [30].

References

[1]. D. Iannuzzi, S. Deladi, J. W. Berenschot, S. de Man, K. Heeck, M. C. Elwenspoek, Fiber-top atomic force microscope, *Rev. Sci. Instrum.*, Vol. 77, 2006, 106105.

[2]. C. Pang, F. Gesuele, A. Bruyant, S. Blaize, G. Lerondel, P. Royer, Enhanced light coupling in sub-wavelength single-mode silicon on insulator waveguides, *Opt. Express*, Vol. 17, 2009, pp. 6939-6945.

[3]. H. E. Williams, D. J. Freppon, S. M. Kuebler, R. C. Rumpf, M. A. Melino, Fabrication of three-dimensional micro-photonic structures on the tip of optical fibers using SU-8, *Opt. Express*, Vol. 19, 2011, pp. 22910-22922.

[4]. Z. Xie, S. Feng, P. Wang, L. S. Zhang, X. Ren, L. Cui, . R. Zhai, J. Chen, Y. L. Wang, X. K. Wang, W. F. Sun, J. S. Ye, P. Han, P. J. Klar, Demonstration of a 3D radar-like SERS sensor micro- and nanofabricated on an optical fiber, *Adv. Opt. Mater.*, Vol. 2015, 3, pp. 1232-1239.

[5]. M. Consales, A. Ricciardi, A. Crescitelli, E. Esposito, A. Cutolo, A. Cusano, Lab-on-fiber technology: Toward multifunctional optical nanoprobes, *ACS Nano*, Vol. 6, 2012, pp. 3163-3170.

[6]. M. Pisco, F. Galeotti, G. Quero, A. Iadicicco, M. Giordano, A. Cusano, Miniaturized sensing probes based on metallic dielectric crystals self-assembled on optical fiber tips, *ACS Photon.*, Vol. 1, 2014, pp. 917-927.

[7]. D. J. Lipomi, R. V. Martinez, M. A. Kats, S. H. Kang, P. Kim, J. Aizenberg, F. Capasso, G. M. Whitesides, Patterning the tips of optical fibers with metallic nanostructures using nanoskiving, *Nano Lett.*, Vol. 11, 2011, pp. 632-636.

[8]. M. Principe, M. Consales, A. Micco, A. Crescitelli, G. Castaldi, E. Esposito, V. La Ferrara, A. Cutolo, V. Galdi, A. Cusano, Optical fiber meta-tips, *Light: Sci. Appl.*, Vol. 6, 2017, e16226.

[9]. V. Savinov, N. I. Zheludev, High-quality metamaterial dispersive grating on the facet of an optical fiber, *Appl. Phys. Lett.*, Vol. 111, 2017, 091106.

[10]. M. Piscol, F. Galeotti, G. Quero, G. Grisci, A. Micco, L. V. Mercaldo, P. D. Veneri, A. Cutolo, A. Cusano, Nanosphere lithography for optical fiber tip nanoprobes, *Light: Sci. Appl.*, Vol. 6, 2017, e16229.

[11]. N. Wang, M. Zeisberger, U. Hübner, M. A. Schmidt, Nanotrimer enhanced optical fiber tips implemented by electron beam lithography, *Opt. Mater. Express*, Vol. 8, 2018, pp. 2246-2255.

[12]. E. J. Smythe, M. D. Dickey, G. M. Whitesides, F. Capasso, A technique to transfer metallic nanoscale patterns to small and non-planar surfaces, *ACS Nano*, Vol. 3, 2009, pp. 59-65.

[13]. S. F. Feng, X. P. Zhang, H. M. Liu, Fiber coupled waveguide grating structures, *Appl. Phys. Lett.*, Vol. 96, 2010, 133101.

[14]. D. Rosenblatt, A. Sharon, A. A. Friesem, Resonant grating waveguided structures, *IEEE J. Quantum Electron.*, Vol. 33, 1997, pp. 2038-2059.

[15]. A. Sharon, D. Rosenblatt, A. A. Friesem, Narrow spectral bandwidths with grating waveguide structures, *Appl. Phys. Lett.*, Vol. 69, 1996, pp. 4154-4156.

[16]. S. F. Feng, X. P. Zhang, J. Y. Song, H. M. Liu, Y. R. Song, Theoretical analysis on the tuning dynamics of the waveguide-grating structures, *Opt. Express*, Vol. 17, 2009, pp. 427-436.

[17]. S. F. Feng, S. Darmawi, T. Henning, P. J. Klar, X. P. Zhang, A miniaturized sensor consisting of concentric metallic nanorings on the end facet of an optical fiber, *Small*, Vol. 8, 2012, pp. 1937-1944.

[18]. X. P. Zhang, H. M. Liu, S. F. Feng, Solution-processible fabrication of large-area patterned and unpatterned gold nanostructures, *Nanotechnology*, Vol. 20, 2009, 425303.

[19]. X. P. Zhang, B. Q. Sun, H. C. Guo, D. Nau, H. Giessen, R. H. Friend, Metallic photonic crystals based on solution-processible gold nanoparticles, *Nano Lett.*, Vol. 6, 2006, pp. 651-655.

[20]. X. P. Zhang, B. Q. Sun, H. C. Guo, N. Tetreault, H. Giessen, R. H. Friend, Large-area two-dimensional photonic crystals of metallic nanocylinders based on colloidal gold nanoparticles, *Appl. Phys. Lett.*, Vol. 90, 2007, 133114.

[21]. H. M. Liu, X. P. Zhang, T. R. Zhai, T. Sander, L. M. Chen, P. J. Klar, Centimeter-scale-homogeneous SERS substrates with seven-order global enhancement through thermally controlled plasmonic nanostructures, *Nanoscale*, Vol. 6, 2014, pp. 5099-5105.

[22]. T. R. Zhai, X. P. Zhang, Z. G. Pang, X. Q. Su, H. M. Liu, S. F. Feng, L. Wang, Random laser based on waveguide-plasmonic gain channel, *Nano Lett.*, Vol. 11, 2011, pp. 4295-4298.

[23]. X. P. Zhang, H. W. Li, Y. M. Wang, F. F. Liu, Stimulated emission within the exciplex band by plasmonic-nanostructured polymeric heterojunctions, *Nanoscale*, Vol. 7, 2015, pp. 5624-5632.

[24]. X. P. Zhang, C. Y. Huang, M. Wang, P. Huang, X. K. He, Z. Y. Wei, Transient localized surface plasmon induced by femtosecond interband excitation in gold nanoparticles, *Sci. Rep.*, Vol. 8, 2018, 10499.

[25]. Z. G. Pang, X. P. Zhang, Direct writing of large-area plasmonic photonic crystals using single-shot interference ablation, *Nanotechnology*, Vol. 22, 2011, 145303.

[26]. Y. H. Lin, X. P. Zhang, Flexible interference ablation using fibers to split and deliver laser pulses for direct plasmonic nanopatterning, *Appl. Phys. Lett.*, Vol. 105, 2014, 123102.

[27]. X. P. Zhang, F. F. Liu, Y. H. Lin, Direct transfer of metallic photonic structures onto end facets of optical fibers, *Front. Phys.: Opt. Photon.*, Vol. 4, 2016.

[28]. X. P. Zhang, S. F. Feng, J. Zhang, T. R. Zhai, H. M. Liu, Z. G. Pang, Sensors based on plasmonic-photonic coupling in metallic photonic crystals, *Sensors*, Vol. 12, 2012, pp. 12082-12097.

[29]. X. P. Zhang, X. M. Ma, F. Dou, P. X. Zhao, H. M. Liu, A biosensor based on metallic photonic crystals for the detection of specific bioreactions, *Adv. Funct. Mater.*, Vol. 21, 2011, pp. 4219-4227.

[30]. Y. Wang, F. F. Liu, X. P. Zhang, Flexible transfer of plasmonic photonic structures onto fiber tips for sensor applications in liquids, *Nanoscale*, Vol. 10, 2018, pp. 16193-16200.

Chapter 10
Multi-wavelength Interferometric Distance Sensors

Sucheta Sharma, Peter Eiswirt and Jürgen Petter

10.1. Introduction

Interferometric techniques are of great importance when it comes to accurate imaging of geometrical features of optical surfaces [1-5]. It is one of the most well-established methods for surface metrology, not only restricted to optics industry but also hugely preferred by the field of instrumentation and automotive engineering [6-7]. Such non-contact surface measurement techniques using interferometry cover a wide variety of applications from manufacturing of optical components to analyzing production of machined, etched, molded parts in different industrial areas including aerospace, medical devices and semiconductor-based technologies [8-11].

Interferometric measurement techniques work by using the phenomenon of wave interference between the reflected beam from the target with the beam reflected from a reference surface. For surface analysis they are much more advantageous compared to mechanical contact methods as they do not make any contact with the surface under test, thus cannot be a cause of any serious damage during measurement. However, despite giving good resolution and measurement accuracy, single wavelength phase shifting interferometry has a disadvantage which is commonly known as '2π phase ambiguity' [12-15]. This problem is inevitable because of their inherent feature of using only one wavelength for performing phase measurements when the test surface contains roughness with step heights more than half a wavelength [16-17]. Usage of multiple wavelengths extends the range of unambiguousness by generating synthetic or beat wavelengths [18-22]. Many approaches like - Two-Wavelength Interferometry (TWI), Multi-Wavelength Interferometry (MWLI), Scanning White Light Interferometry (SWLI) have helped to achieve a larger dynamic range [23-28].

In this chapter, we will focus on the MWLI technique and present how this method is employed in an interferometric measurement setup by using four wavelengths for

Sucheta Sharma
AMETEK GmbH BU Taylor Hobson/Luphos, Weiterstadt, Germany

numerous metrological applications based on optics industry [29]. The main emphasis will be given on the Electro-Optic (EO) phase modulation process which is being used for developing the prototype of a new generation MWLI phase shifting distance sensor. The EO sensor can perform faster phase modulation to increase the speed of distance measurement compared to conventional Piezo actuated mechanical phase modulation method [30-31]. The chapter will be initiated with a brief introduction of the primary concepts behind the MWLI process. We will also discuss about different phase shifting mechanisms and their applications in driving the sensor systems. The experimental setup of the current MWLI measurement system with the mechanical phase modulation driven distance sensor will be introduced. Finally, it will be followed by the main research part on the implementation of an EO distance sensor. The experimental results will show the potential of the linear EO effect for replacing the current mechanical phase shifting technique in order to achieve the goal of faster distance measurement.

10.2. Basic Theory

10.2.1. Single Wavelength and Multi-Wavelength Interferometry

Single wavelength homodyne interferometry refers to the method of analyzing the interference pattern of superposed reference and target waves where both have the same optical frequency [32-34]. The intensity distribution of the interference signal can be expressed as

$$I = I_{Ref} + I_{Target} + 2 \cdot \sqrt{I_{Ref} I_{Target}} \cdot \cos \varphi$$

$$= [I_{offset} + I_{amplitude} \cdot \cos \varphi] \tag{10.1}$$

Here, φ is the phase difference between two interfering beams from reference and target having I_{Ref} and I_{Target} as corresponding intensities [35-36]. Considering the round-trip Optical Path Difference (OPD), φ can be expressed as

$$\varphi = \frac{2\pi}{\lambda} OPD = \frac{2\pi}{\lambda} (2 \times n \times L), \tag{10.2}$$

where n is the refractive index of the medium and L is the distance between reference and target. φ carries the displacement information when the target is moved from the reference. The intensity of the detected pattern will vary sinusoidally if φ changes resulting in fringe movement. It can be expressed as $\varphi = [2\pi \cdot m + \varphi']$, where m is the interference fringe order and φ' is in between 0 and 2π. Now, while analyzing the fringes and simultaneously unwrapping the phase, the target needs a very slow but continuous movement from the reference. Any disturbance (which is of the order or more than $\lambda/2$) during the measurement might bring discontinuity in the phase calculation as the integer order (m) of φ will be lost. This problem is known as '2π phase ambiguity'. Single wavelength phase shifting interferometry works excellently for characterizing an optically smooth target where the surface discontinuity is less than $\lambda/2$. However, the problem of

2π phase ambiguity is unavoidable for such system and it mainly occurs because of its shorter range of unambiguousness.

The idea of multiple wavelengths helps to overcome the problem of a short unambiguity range of single wavelength interferometers by generating a synthetic wavelength (Λ_s). The concept can be explained for an interferometer operating with two wavelengths. In a TWI, two close wavelengths, λ_1 and λ_2, are used to evaluate the phase. These two wavelengths generate a longer beat wavelength which can be expressed as [37-41],

$$\Lambda_s = \frac{\lambda_1 \lambda_2}{|\lambda_1 - \lambda_2|} \qquad (10.3)$$

Unlike the case of a single wavelength interferometer, the Λ_s provides larger unambiguity range where the synthetic phase progresses slowly compared to the constituting wavelengths λ_1 and λ_2. Thus, usage of multiple wavelengths protects the measurement process from getting interrupted by any sudden disturbances which might lead to a phase error resulting in 2π phase ambiguity. Fig. 10.1 shows the measurement principle by using the concept of Λ_s [29].

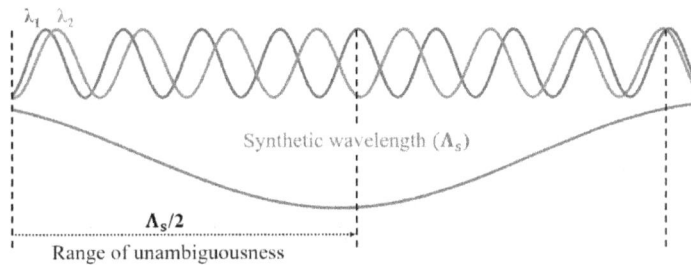

Fig. 10.1. Extended unambiguity range of synthetic/beat wavelength (Λ_s) in MWLI.

In this chapter, we will present a multi wavelength measurement scheme where four wavelengths are used instead of two. The phase is first evaluated at each of the operating wavelengths which are aligned on the same path of the interferometer then the synthetic phases are determined and included in the calculation of the distance measurement. The four laser sources have been carefully chosen for the MWLI system to design the Λ_s. By taking suitable combinations of these four wavelengths (let's say, λ_1, λ_2, λ_3 and λ_4), several synthetic or beat wavelengths can be generated like $-\Lambda_{12}, \Lambda_{23}, \Lambda_{34}, \Lambda_{14}, \Lambda_{123}$..etc. One of the advantages of using four source wavelengths is that it can produce such combinations to generate different Λ_s with varying unambiguity range and measurement uncertainty. The longest Λ_s produced in this manner, determines the range of unambiguousness but the uncertainty in measurement increases according to the ratio of the synthetic to the source wavelengths. Hence, the shortest Λ_s gives the smallest measurement uncertainty [15]. The largest range of unambiguousness which can be generated for the MWLI system by using the two closest wavelengths is 1.25 mm. It is more than 1000 times larger than the range of unambiguousness of a single wavelength interferometer which operates with $\lambda \sim 1550$ nm.

10.2.2. Phase Modulation Process

Phase shifting interferometry (PSI) is one of the best methods for both single and multiple wavelength interferometers for performing optical measurements with improved precision [42]. One of the common approaches of phase modulation in PSI is based on the introduction of sinusoidal optical path length variation for evaluating the φ values of the interference pattern [43-45]. The phase modulation process works completely under a machine-controlled mechanism which offers much better precision in measuring the phase difference between the interfering wave fields compared to the visual fringe counting method. If a sinusoidal phase shift of amplitude ψ, phase ξ and period $T = \frac{2\pi}{\omega}$ is added with the original interference phase (φ) then the Eq. (10.1) can be rewritten for the intensity distribution I of the detected phase modulated interference signal at some point (x, y) of the detector plane as [46],

$$I(x, y, t) = I_{offset}(x, y) +$$

$$+I_{amplitude}(x, y) \cdot cos[\varphi(x, y) + \psi sin(\omega t + \xi)] \qquad (10.4)$$

There are various ways available for achieving phase modulation in an interferometric setup. We will particularly focus on two approaches which are being used for realizing our sensor systems –(a) Mechanical phase modulation; (b) Electro-Optic (EO) phase modulation.

In a mechanical phase modulation process the reference or the target is made to execute a controlled periodic movement with the help of an attached Piezoelectric Transducer (PZT). The phase shifts are produced by the PZT to change the geometrical length of the optical path between the reference and the target. It is one of the most widely used techniques for obtaining phase modulation which is also used to drive our current MWLI sensor unit. The PZT needs to move the target or reference arm by an amount $\sim\lambda/4$ in order to achieve π phase modulation for an interferometric system where the light is passing the distance between the reference and the target (i.e. L) twice according to Eq. (10.2). However, the motion of the PZT can also have some non-linearity which might lead to improper reference/target movement resulting in erroneous phase shift [47-48]. The mechanical phase shifting method also suffers from other problems (for distance measurement applications using our current MWLI system) which include internal heat generation and vibration related misalignment when modulating frequency is increased. We will discuss this in detail in the Section 10.3.3.

The linear EO effect or Pockels effect is another technique for including the phase shift by changing the optical properties of the medium of propagation with the application of an applied voltage in a controlled manner [49-51]. The refractive index $n(E')$ of an EO medium varies with the applied electric field (E'). If E' is applied along the x axis of the EO crystal then the corresponding change in refractive index will be [52-53],

$$\Delta n_x \approx \frac{1}{2}n_x{}^3 rE', \qquad (10.5)$$

where r is the EO coefficient. Commonly used crystals which show Pockel's effect are LiNbO$_3$, LiTaO$_3$, NH$_4$H$_2$PO$_4$ (ADP) and CdTe [52]. Hence, a light wave can be phase modulated by linear EO effect which can be a great replacement for mechanical phase shifting method in order to achieve faster phase modulation as it is not associated with any mechanical vibration to influence the measurement stability.

EO modulators can be classified in two categories – longitudinal and transverse, depending on how E' is applied with respect to the direction of propagation of light [54-55]. In the first configuration, the applied E' is parallel to the wave vector direction where for the later the voltage is applied perpendicular to the direction of propagation of light. Now, for obtaining phase modulation, suppose voltage V has been given along the x axis of the crystal. Hence, the total amount of phase shift due to the applied V (following Eq. (10.2)) is

$$\varphi' = \frac{2\pi}{\lambda}[(n_x + \Delta n_x) \cdot (2L')] = (\varphi + \Delta \varphi), \tag{10.6}$$

where $\varphi = \frac{2\pi}{\lambda}[n_x(2L')]$ is the natural phase term and L' is the length of the modulator along which the light propagates. The multiplication of '2' with L' in the Eq. (10.6) suggests that the target beam has passed through the modulator twice before interfering with the reference beam (as discussed in Section 10.2.1) considering that the modulator has been positioned after the reference plane but before the target surface. $\Delta \varphi$ is the electrically induced phase shift which is given by: $\Delta \varphi = \frac{2\pi}{\lambda}[\Delta n_x(2L')]$ where Δn_x refers to the change in refractive index as given in the Eq. (10.5).

For a longitudinal modulator, $E' = V/L'$, and the phase shift is $\Delta \varphi \approx \frac{\pi}{\lambda}[2n_x^3 rV]$. The $\Delta \varphi$ is independent of L' and is directly proportional to the applied voltage V. Under transverse configuration, V is applied across the depth (d') of the crystal so that the light propagates perpendicular to E'; where, $E' = V/d'$. Hence, the resulting $\Delta \varphi \approx \frac{\pi}{\lambda}[2n_x^3 rV (L'/d')]$ is a function of the ratio (L'/d') and V. The voltage which can produce a phase shift of π is called the half-wavelength voltage (V_π). The expressions of V_π for longitudinal and transverse configurations are given in Eq. (10.7) and Eq. (10.8) respectively,

$$V_\pi = [\lambda/(n_x^3 r)], \tag{10.7}$$

$$V_\pi = [\lambda/(n_x^3 r)][(d'/L')], \tag{10.8}$$

Experiments have been carried out on a LiNbO$_3$ based transverse EO modulator as the V_π for the transverse configuration can be reduced by suitably choosing the crystal dimensions. In Section 10.5 of this chapter we have discussed in detail about how the concept of linear EO effect is being used for realizing the new MWLI distance sensor in order to replace the current PZT based mechanical phase shifter sensor.

10.2.3. Evaluation of Interferometric Phase

After the introduction of phase modulation, the intensity (I) of the interference pattern becomes a time-varying function $I(t)$ as explained in the Section 10.2.2 at Eq. (10.4). $I(t)$ represents the intensity distribution of the phase modulated interference signal. Now in this section, we have briefly discussed about one of the methods that shares similarity with the approach by which the interferometric phase (φ) is unwrapped for our MWLI distance measurement system. This method of interferometric phase measurement under sinusoidal phase modulation has been comprehensively described in the article by *A. Dubois* [46].

The $I(t)$ is integrated over four quarters of the modulation period T as shown in Fig. 10.2

$$v_p(x,y) = \int_{(p-1)T/4}^{pT/4} I(x,y,t)\, dt, \; p = 1,2,3,4 \tag{10.9}$$

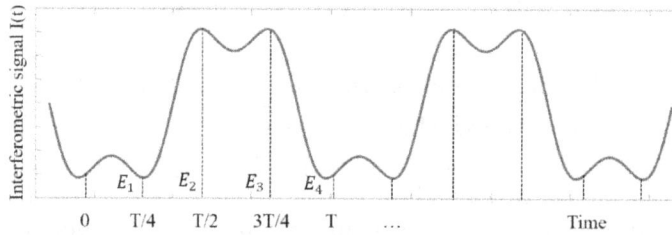

Fig. 10.2. An example of how I(t) is integrated over the four quarters of the time period T.

To calculate the integral in Eq. (10.9), the $I(t)$ is expressed as a sum of Fourier components by using the Bessel function of the first kind J_n:

$$I(t) = I_{offset} + AJ_0(\psi)cos\varphi + 2Acos\varphi \sum_{n=1}^{+\infty} J_{2n}(\psi) \cos[2n(\omega t + \xi)] -$$

$$- 2Asin\varphi \sum_{n=0}^{+\infty} J_{2n+1}(\psi)sin[(2n+1)(\omega t + \xi)] \tag{10.10}$$

By putting the expression of $I(t)$ [Eq. (10.10)] in Eq. (10.9) and by taking linear combinations of the integrated parts i.e. v_1, v_2, v_3 and v_4 it can be written

$$\Xi_s = -v_1 + v_2 + v_3 - v_4 = (4T/\pi)Y_s A \sin\varphi, \tag{10.11}$$

$$\Xi_c = -v_1 + v_2 - v_3 + v_4 = (4T/\pi)Y_c A \cos\varphi, \tag{10.12}$$

where

$$Y_s = \sum_{n=0}^{+\infty} (-1)^n \frac{J_{2n+1}(\psi)}{2n+1} \sin[(2n+1)\xi],$$

$$Y_c = \sum_{n=0}^{+\infty} \frac{J_{4n+2}(\psi)}{2n+1} \sin[2(2n+1)\xi], \tag{10.13}$$

The phase φ can be unwrapped by the equation:

$$\tan \varphi = \frac{Y_c}{Y_s} \frac{\Xi_s}{\Xi_c},$$ (10.14)

Hence, φ depends on the linear combinations of four frames Ξ_s and Ξ_c. It also depends on the modulation amplitude ψ and phase ξ through the parameters Y_s and Y_c. Using the Eq. (10.14) the value of φ can be evaluated for calculating the measured distance.

10.3. System Design

10.3.1. Working Principle of Fizeau Interferometer

The MWLI setup which will be discussed in this section is based on the concept of a Fizeau interferometer. Therefore, we give a short description of the measurement mechanism of a Fizeau interferometric process. The working principle is based on interference by the division of the amplitude where the original wave front is split into two or more parts and directed along different optical paths and then recombined to produce an interference pattern [56]. The Fizeau interferometer compares one optical surface to another by maintaining a small air gap between them where fringes of equal thickness are observed [57].

Along a fringe, the gap has constant thickness and adjacent fringes corresponds to change in thickness of half a wavelength. Fig. 10.3 shows the schematic diagram of a Fizeau interferometer. One of the two surfaces used here is called the reference plane or known surface. The other surface in the setup is the test/target surface which contains imperfections and they are compared with the reference surface during the process of surface analysis. The fringes are observed as a result of the interference from the direct reflection of the source by the reference and the test/target surfaces. Such two-dimensional measurement scheme has applications in testing optical elements [58]. However, in this study, we have used the one-dimensional Fizeau measurement method to scan the surface of an object under test with a point probe system.

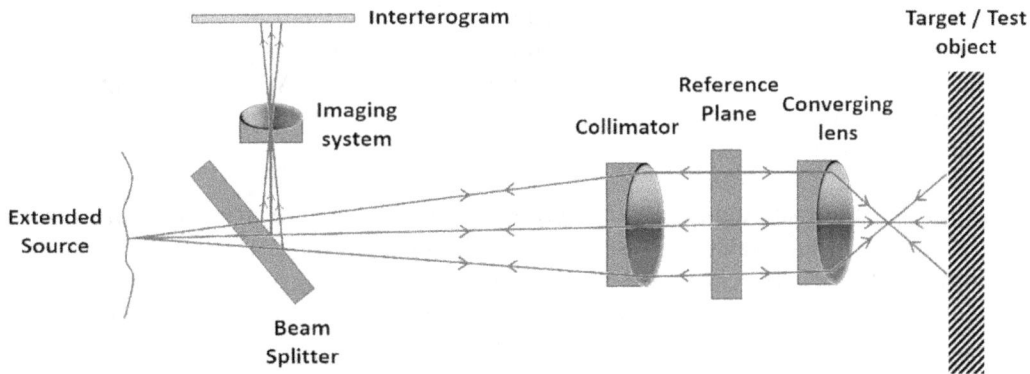

Fig. 10.3. Schematic diagram of Fizeau interferometric setup.

10.3.2. Experimental Setup of Multi-wavelength Interferometer

The working principle of the MWLI setup is based on the one-dimensional Fizeau-homodyne interferometry [30-31]. The four operating wavelengths range from 1530-1630 nm. Hence, as the source wavelengths are in the infrared region, the necessity of photodiodes and electronic data analysis arises for the MWLI system. However, one of the biggest advantages of using the infrared wavelengths from diode laser sources is that it allows the system to be entirely optical fibre based which reduces the alignment complexities and also helps for relatively easy assembling of optical and electrical components within the device. Fig. 10.4 shows the schematic diagram of the MWLI distance measurement system. It is currently being operated with a PZT actuated mechanical phase shifter sensor which will be explained in detail in Section 10.3.3.

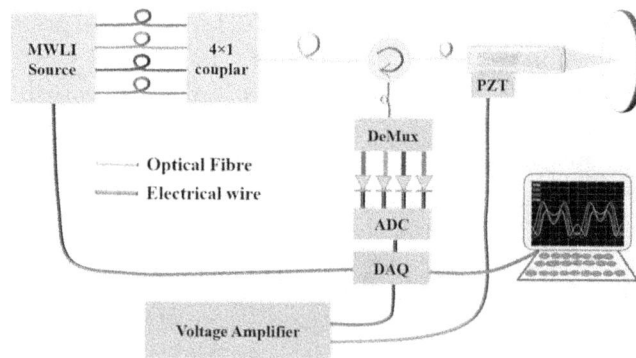

Fig. 10.4. Schematic diagram of MWLI setup (four colors represent four operating wavelengths).

The four wavelengths are coupled by a 4×1 coupler after being sent out by the MWLI source. After that the beam is directed towards the sensor where the reference and the target signals are collected and coupled back into the fibre ferule for creating the interference. The resultant interference intensities are then separated by the Demultiplexer (DEMUX) for the four operating wavelengths which are then individually detected by the four photodiodes. This process of detection converts the optical signal into an electrical output which then goes through Analog to Digital Conversion (ADC) before entering the Data Acquisition (DAQ) system. At the DAQ, the final process of phase unwrapping and simultaneous distance calculation take place.

The phase detection and data analysis section of the setup is able to retrieve and analyse under both – (a) Individual interferometric mode where four wavelengths work in the construction of four different folded Fizeau interferometers which separately collect data. Finally, the phase measurement is performed by taking the average of these separately calculated phases. This mode is especially suitable for testing optically smooth surfaces. (b) Interleave or synthetic wavelength mode which uses the concept of the beat wavelength for increasing the dynamic range e.g. for measuring surfaces with step heights exceeding $\sim\lambda/2$ of the source wavelengths as discussed in the Section 10.2.1. This approach is comprised of four simultaneous phase detection modes namely–Smooth,

Course, Rough and Aslope. The combinations of the operating wavelengths are chosen carefully to generate different synthetic wavelengths for which the dynamic range increases ascendingly from Smooth to Aslope. The highest unambiguity range (mentioned previously ~1.25 mm) is found at the Aslope mode which is used for measuring very rough surfaces having surface non uniformity comparable to millimetre range. However, all four modes run simultaneously while performing phase measurement with interleave. Depending on the types of applications and target/test objects, the mode of operation of the MWLI system can be selected for carrying out high precision surface analysis.

10.3.3. Mechanical Phase Modulation Sensor

This section focuses on the current MWLI distance sensor which is being operated with mechanical phase modulation. Fig. 10.5 shows the schematic diagram of the sensor. The components arrangement includes an optical fibre which releases the input four wavelengths from guided propagation through the optical fibres from MWLI source (as discussed in Section 10.3.2), to freely propagate in order to hit the target. The ferule is attached with a PZT which takes the driving voltage from the DAQ.

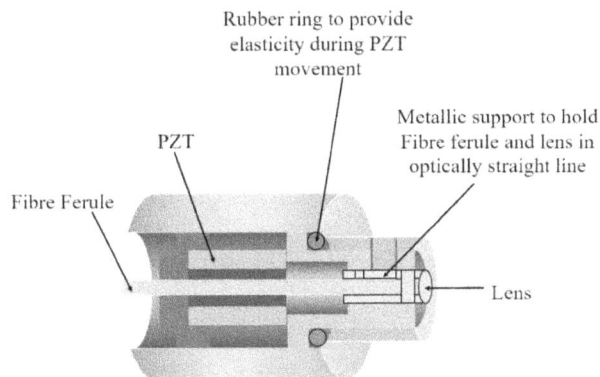

Fig. 10.5. Schematic diagram of mechanical phase modulation sensor.

The PZT makes the ferule to move back and forth in order to change the length ~$\lambda/4$ between the reference and the target for shifting the phase by π. The DAQ controls the amount of phase shift by controlling the driving voltage and simultaneously retrieves the modulated signal in order to send it to data analysis unit for unwrapping the phase. The mechanical phase shifting distance sensor is currently being operated with 1.25 kHz modulating frequency i.e. retrieval of one data point is possible for a given L within ~800 µs.

In the sensor some amount of light ~4 % is reflected back from the fibre end and is taken as the reference wave. The rest passes through it onto the target. The target wave gets the phase modulation by the PZT before being coupled back into the fibre to interfere with the reference beam. The sensor follows the same approach of the Fizeau interferometric process for receiving the reference and the target beam. The interference signal then goes through the process of detection and conversion to the electrical signal from optical input in the previously explained manner (at the Section10.3.2).

10.4. Applications

The PZT driven MWLI distance sensor (Section 10.3.3) has several applications. Among these e.g. high accuracy in 3D form measurement which is essential to manufacture optical surfaces mainly aspheric lenses [29, 59]. However, not only aspheres but it can also be used for measuring:

- Normal spherical lenses;

- Optical flats;

- Slight freeforms;

- Acylindrical mirrors;

- Plano – convex/concave, cylindrical lenses;

- Various types of prisms (equilateral / right-angle prisms);

- Larger aspheres and lenses for telescope optics;

- Surfaces with steep slopes with varying pitch directions and small surfaces like moulds for smartphone lenses.

Fig. 10.6 shows some examples of measurements on different types of object surfaces using the MWLI distance probe. Typical target applications include polished, fine-ground, segmented, rough and continuous/discontinuous surfaces. Here, especially the MWLI interferometer allows for the measurability of rough and discontinuous surfaces.

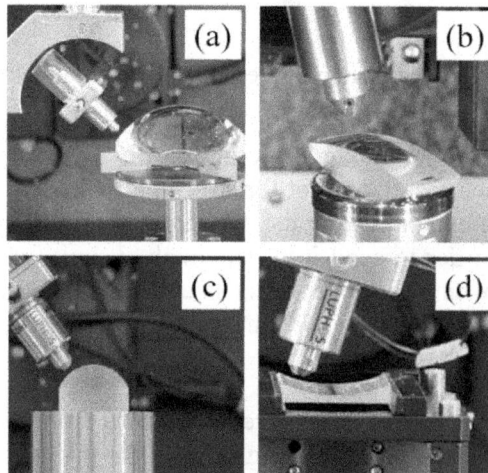

Fig. 10.6. Target surface measurement: (a) Aspheric; (b) Aspheric segmented; (c) Spherical; (d) Cylindrical concave.

Fig. 10.7 shows the schematic diagram of measuring an aspheric surface using the PZT based mechanical phase shifter sensor. The object probe performs a scanning point measurement. This is placed normal and equidistant to the surface and follows the design curvature while measuring the object distance. The design curvature can be estimated by the device if the object parameters are given as input. The object probe makes a controlled movement along the axes i.e. R and Z, while the rotary T-axis aligns the probe always perpendicular to the surface. During the scanning process the object under test is rotated by C-axis causing a spiral scanning path on the surface. The result of the measurement is expressed through a full 3D map which shows the deviation of the actual curvature from the theoretically estimated designed surface features. Apart from measuring surface profiles of optical components, the MWLI probes have several other applications, for example in roundness measurement and high precision alignment.

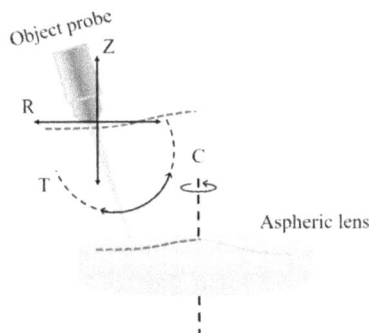

Fig. 10.7. Measurement mechanism of object probe on aspheric surface.

10.5. Electro Optic Phase Shifting Distance Sensors

10.5.1. The Need of Electro Optic Phase Modulation

The PZT based mechanical phase shifter probe works excellently at the current modulating frequency ~1.25 kHz as mentioned above. However, the objective is to speed up the measurement process. In order to achieve faster measurement using the PZT probe, the modulating frequency needs to be increased. Fig. 10.8 shows the result when the sensor is made to operate at higher modulating frequency [31, 60] where the distance drift with time is shown for different modulating frequencies. As depicted here, it indicates that the PZT is suffering from internally generated heat due to which the distance drift is occurring in the first few minutes after starting the experiment. The amount of drift increases with the modulating frequency resulting in more instability which is not suitable for performing any distance measurement until the system gets into a stable state. At 5 kHz modulating frequency the distance drift is ~300 nm within the first 10 minutes. Apart from the internally generated heat, the PZT also makes unbearable audible noise when it is operated at higher modulating frequency. Hence, the results suggest that the PZT probe will not be suitable for speeding up the measurement and it needs to be replaced with a different type of phase shifting process to overcome such limitations.

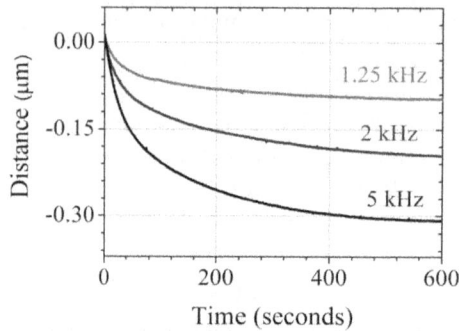

Fig. 10.8. Distance drift of the PZT based mechanical phase shifter sensor when the modulating frequency is increased from 1.25 kHz to 5 kHz.

Linear EO phase modulation process is considered as a good option for replacing the PZT based sensor. As described in the Section10.2.2 the EO crystal modulates the phase by changing the refracting index of the medium and as a result, the optical path length changes to shift the phase of the input light. Hence, mechanical movement or vibration related misalignment cannot occur under this type of phase modulation technique. Moreover, the required driving voltage V_π (which will be explained in detail in the following section) is also comparatively less than the PZT sensor. The performance at higher modulating frequency has also been checked for the EO sensor and the results which will be explained in detail at Section 10.5.3, did not show the initial distance drift and any dependency on the driving frequency like in the case of the PZT sensor. In the remaining part of the chapter we will discuss about the research on developing a MWLI distance sensor based on linear EO phase modulation process for achieving the goal of faster measurement.

10.5.2. Requirements

The experiments have been carried out on an EO crystal $LiNbO_3$ with a Ti-indiffussed waveguide. The length of the crystal is ~15 mm and the electrode separation is ~50 µm. The corresponding V_π has been calculated using the equation (10.8) which gave the value ~12-14 V. Hence, the driving voltage of the EO sensor is almost less than one fourth of what is required for the PZT probe. Fig. 10.9 shows the MWLI experimental setup under EO phase modulation.

The electrode connected $LiNbO_3$ crystal has been placed between the fibre ferule and the target. Like in the case of PZT sensor, the reference light is collected from the fibre ferule and the target beam passes through the EO crystal twice before interfering with the reference. In order to avoid extra reflexes from the end face of the waveguide, the crystal edge which is facing the target has been made slightly angled ~5°. Thus, the back reflected light follows a different path which has been indicated in the diagram of Fig. 10.10 and does not interfere with the target beam.

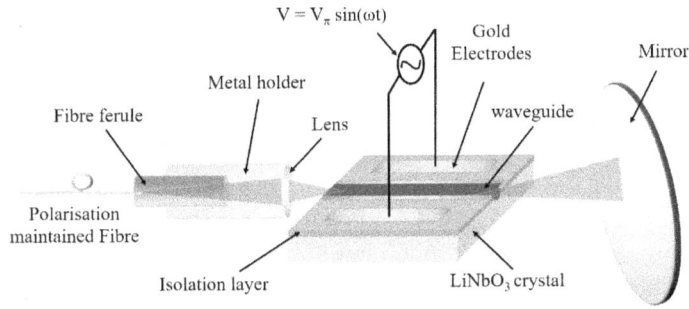

Fig. 10.9. Schematic diagram of experimental set up under EO phase modulation.

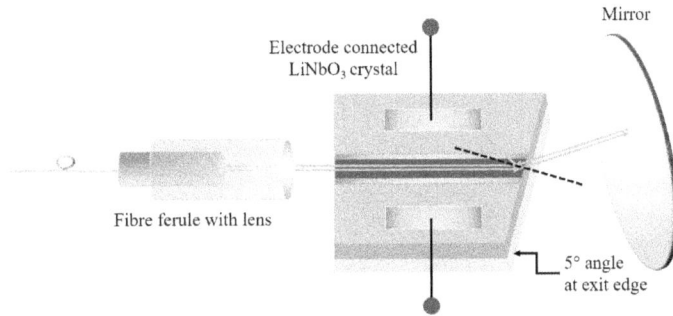

Fig. 10.10. Top view of the experimental setup with EO crystal LiNbO$_3$.

10.5.3. Comparison between Mechanical and Electro Optic Phase Modulating Sensors

In this section we present a comparison on the phase modulation efficiency between the PZT and the Electro Optic Modulator (EOM). Fig. 10.11(a) and (b) show the phase modulated interferometric signal governed by Eq. (10.4) under the modulating frequency 1.25 kHz for PZT and EOM respectively. The four operating wavelengths have been represented by four colors in the graph. The graphs show that the EO approach results in the phase modulation that is as effective as with the current PZT based system.

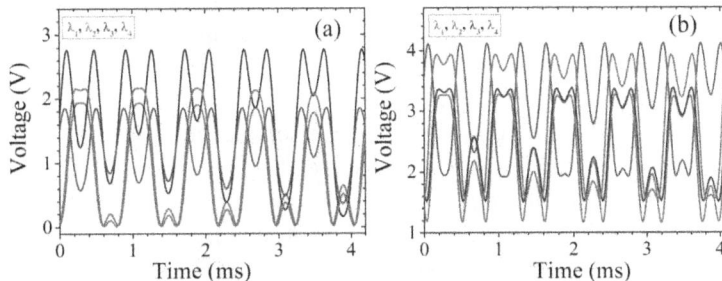

Fig. 10.11. Phase modulated interferometric signal from (a) PZT and (b) EO sensor.

The distance drift was checked for the EO sensor to compare with the PZT probe when the target was kept fixed. The results have been reported in detail in our article 'Electro optic sensor for high precision absolute distance measurement using multiwavelength interferometry' [30]. As depicted in Fig. 10.12, the drift under EO phase modulation is ~40 nm and for PZT sensor it is ~15 nm in 1200 seconds. However, it should be noted that the data was taken for the PZT when it attained stability after the initial drift of ~80 nm due to the internally generated heat as discussed in Section 10.5.1. The distance drift observed for the case of PZT (under stable condition) is comparatively less but the EO sensor is under experimentation which is more sensitive to environmental disturbances i.e.- the thermal expansion of rail within the setup or air turbulence. Details of these possible influences will be discussed in Section 10.5.4.4.

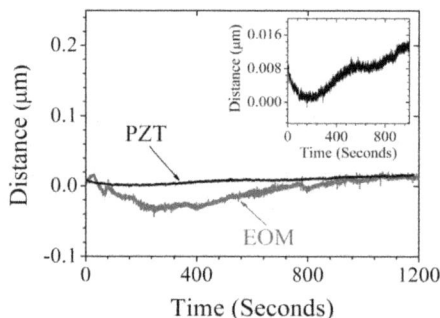

Fig. 10.12. Comparison on observed distance drift under the PZT and EO driven phase modulation over 1200 seconds for a fixed target position (Inset: Zoomed view of the distance drift for the PZT sensor).

Another test was done to check the ability of the EOM for measuring the target displacement. Unlike the previous experiment which was performed under static target condition, here the target mirror has been moved by introducing a controlled periodic mechanical oscillation using a Piezo. It was operated with a triangular voltage waveform from an external function generator under a comparatively low driving frequency ~0.3 Hz (to avoid measurement sensitivity due to vibration in the setup). The piezo actuated positioning of the mirror has been first monitored with the PZT probe and afterwards with the EO phase modulation process in order to draw a comparison between these two modulation techniques. The graph in Fig. 10.13 shows that the EO phase modulation-based system is capable of measuring the ~0.8 μm amplitude of target displacement similar to that of the PZT sensor.

As the main motivation behind choosing the EO phase modulation is to speed up the measurement process, hence experiments have been carried out for checking the performance at higher modulating frequency ~5 kHz. Fig. 10.14(a) shows the result on the study of distance drift for a fixed target position under both types of modulation processes. The distance measurement using the EOM contains mechanical noise due to external vibration, but the influence of internally generated temperature drift is almost negligible ~40 nm compared to the PZT which has shown ~300 nm distance change within

800 seconds. In Fig. 10.14(b), the phase modulated signal of the EOM at 5 kHz has been shown. In comparison to the previous signal at Fig. 10.11(b), it is relatively less smooth. The reason behind this is the reduction of the number of data points per cycle as the modulating frequency has been increased. The data receiving rate of the DAQ from the photodiodes has not been changed while switching to higher modulating frequency. The rate is 100 kHz for the current system which is designed to operate at 1.25 kHz. Hence, for each cycle of the modulated signal 80 data points are received by the DAQ from the photo diodes. Now, at this situation when the modulating frequency was increased to 5 kHz, the data rate per cycle decreased to 20 data points. As a result, the resolution became progressively worse. For future application the data rate of DAQ can be increased through proper software adjustments and all other electronic constituents of the device need to be synchronised. However, in order to check whether the system would be capable to perform at four times higher modulating frequency (5 kHz) the data rate has been kept constant for initial measurements but in future it will be adjusted to have better resolution.

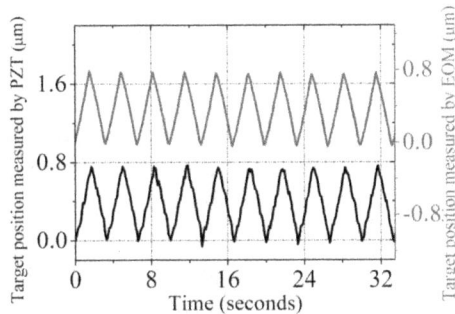

Fig. 10.13. Performance evaluation on measuring piezo actuated positioning of the target using PZT and EO phase modulation at modulating frequency 1.25 kHz.

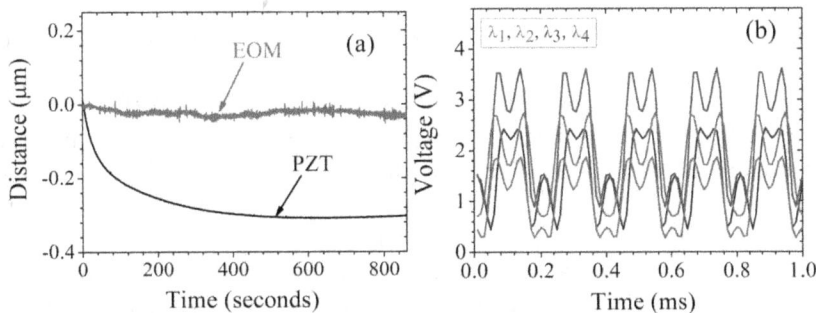

Fig. 10.14. (a) Comparison on interferometric stability at higher modulating frequency (~5 kHz) between PZT and EO phase modulation (b) Phase modulated interferometric signal using EOM at 5 kHz.

The test on displacement measurement under periodic target movement with EO phase modulation has been repeated for 5 kHz and compared with the PZT system under the same operating condition. The results have been shown in Fig. 10.15 where 0.8 μm Piezo activated target displacement has been measured by EOM at higher modulating frequency like the PZT based sensor. The EOM showed more stability during this measurement where the thermal drift experienced by the PZT (black curve in Fig. 10.15) is visible in the plot.

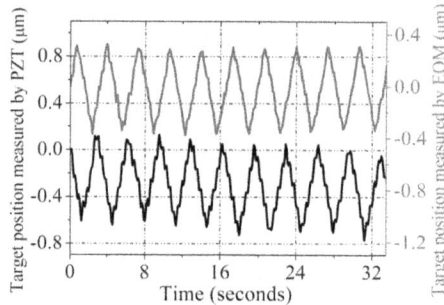

Fig. 10.15. Comparison on measurement of piezo actuated target displacement at 5 kHz between PZT and EO phase modulation.

The work which has been presented in this section is aimed at the development of an EO distance sensor for replacing the PZT based mechanical phase modulation. The results depicted the phase modulation ability of the EO sensor which showed equivalent efficiency similar to the present PZT probe for nanometre-scale absolute distance measurement. However, a slow drift in distance measurement was observed at both modulating frequencies 1.25 kHz and 5 kHz for the EO phase modulation during static interferometric measurement. Therefore, in the following sections, different possibilities will be discussed to attain better interferometric system stability for performing distance measurement with EO phase modulation.

10.5.4. Measurement Stability Study Using Electro Optic Distance Sensor

In order to investigate the sources behind the observed distance drift under static target condition for the EO sensor, a study on different influences like modulating frequency, operating wavelengths and external temperature will be presented here. Previously, the compatibility of EO phase modulation to replace the PZT distance sensor for performing faster distance measurement in an MWLI set up has been presented. In this section, the cause of the distance drift for a new prototype of the EO sensor having lower driving voltage requirement $V_\pi \sim 5V$ will be discussed. The results presented here have been reported in our article 'Study on Interferometric Stability Based on Modulating Frequency, Operating Wavelengths and Temperature using an Electro Optic Multi-Wavelength Distance Sensor' [31]. In this study, at first, experiments have been carried out to check the influence of modulating frequency when the target was kept fixed. Then the drift of the four source wavelengths was observed for 14 hours. The synchronicity in the distance measurement by each of the employed wavelengths was also tested for the EO sensor at both 1.25 kHz and 5 kHz. Finally, the dependence of the gradual drift in measured distance on the external temperature has been studied.

10.5.4.1. Influence of Modulating Frequency

As mentioned earlier, the prime cause behind replacing the PZT probe is its inability to perform at higher modulating frequency due to internal heat generation. Whereas the EO sensor does not show any rapid distance change like the PZT but it has a slow and continuous optical path length drift. In order to find any dependency of the distance drift on the modulating frequency, distance data have been recorded for 1.25 kHz and 5 kHz and then compared with the result of PZT.

Fig. 10.16 (a) and (b) depict the result on distance measurement when the target / mirror was kept fixed for 600 seconds where the total drift as monitored for the EOM was ~70 nm and ~100 nm for 1.25 kHz and 5 kHz respectively. The reason behind this drift can be an effect of external temperature variation which will be explained in Section 10.5.4.4. Particularly, the peak which can be observed in the distance vs. time graph at Fig. 10.16(a), is a result of mechanical instability caused due to such temperature variation. The PZT experienced a fast distance change with higher rate ~90 nm and ~260 nm at 1.25 kHz and 5 kHz respectively within 180 seconds after starting the experiment. The EOM showed much stable behavior with comparatively less amount of drift ~10-15 nm during that time frame. Hence, the results suggest that unlike the case of PZT, the distance drift observed under EO phase modulation is not due to the modulating frequency.

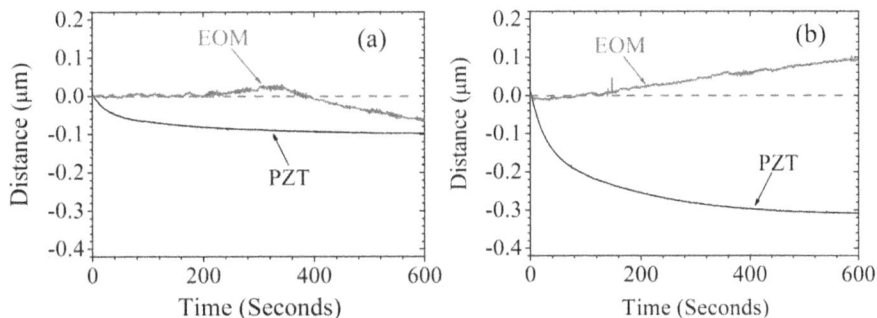

Fig. 10.16. Distance drift using the PZT and EOM sensor at (a) 1.25 kHz and (b) 5 kHz, when the target/mirror was kept fixed.

10.5.4.2. Studying the Stability of the Source Wavelengths

As the system is being operated with four wavelengths hence, any inherent drift of these wavelengths can bring errors in the phase measurement which might result in the observed distance change. Fig. 10.17 depicts the amount of change in λ from its initial value (i.e. $\Delta\lambda$) recorded for 14 hours for the four source wavelengths λ_1, λ_2, λ_3 and λ_4. The observed values of $\Delta\lambda$ are ~0.00018 nm, ~0.0002 nm, ~0.0001 nm and ~0.00018 nm respectively. The values of $\Delta\lambda$ during that 14 hours stayed within 0.0001 - 0.0002 nm which is not much capable of bringing any significant amount of additional distance change during the measurement.

To give an example, the maximum distance at which the target has been placed from the reference for our experimental setup with the EO prototype is ~30 mm. The approximate value of average distance change induced by the $\Delta\lambda$ of the four wavelengths under this condition will be approximately ~3.18 nm after 14 hours. Hence, the previously evaluated instability of 0.0001 - 0.0002 nm of the employed four wavelengths contribute negligibly to the observed distance drift (Fig. 10.16(a) and (b)) in the case for the EO sensor.

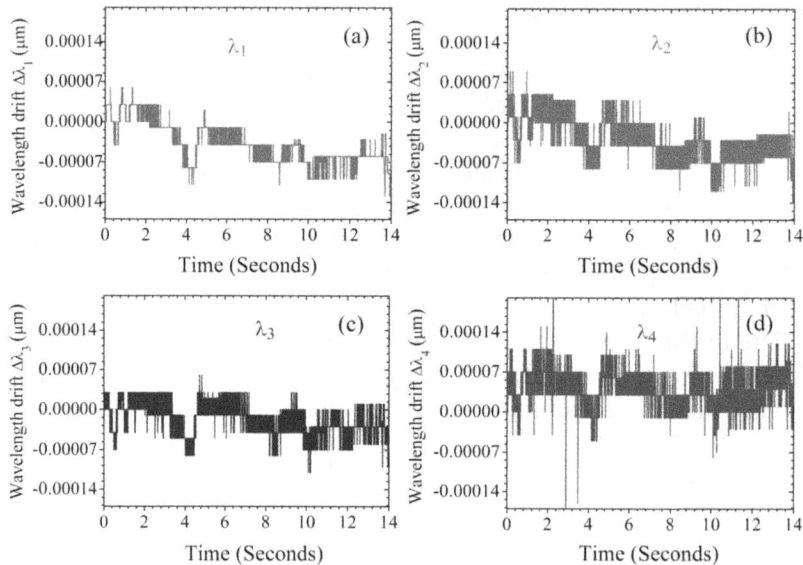

Fig. 10.17. Wavelength drift ($\Delta\lambda$) of the four source wavelengths observed in 14 hours.

10.5.4.3. Synchronicity in Distance Measurement by the Source Wavelengths

The synchronicity refers to the process of checking the similarity among the simultaneously measured distance data by each of the four operating wavelengths. It is a very important method for a system which is being operated with multiple wavelengths in order to find out any measurement abnormality caused due to erroneous phase information by verifying the recorded values for each λ. Hence, the distance drift under EO phase modulation as shown in Fig. 10.16(a) and (b) has been checked for synchronicity (depicted in Fig. 10.18(a) and (b)) at 1.25 kHz and 5 kHz respectively.

The distance data seen by each wavelength are nearly same and consequently the projection of the measurements made by the four wavelengths on the two-dimensional plane of the 'Distance vs. Time' (the ZX plane of Fig. 10.18(a) and (b)) overlapped with each other. The result confirms that the drift is not due to any unusual behavior of the source wavelengths and it is not adding up phase errors in the measurement which might result in the gradual distance change. Hence, the distance drift which has been observed for the EO sensor is not a result of the EOM induced wavelength dispersion in this wavelength region (1530 nm – 1610 nm).

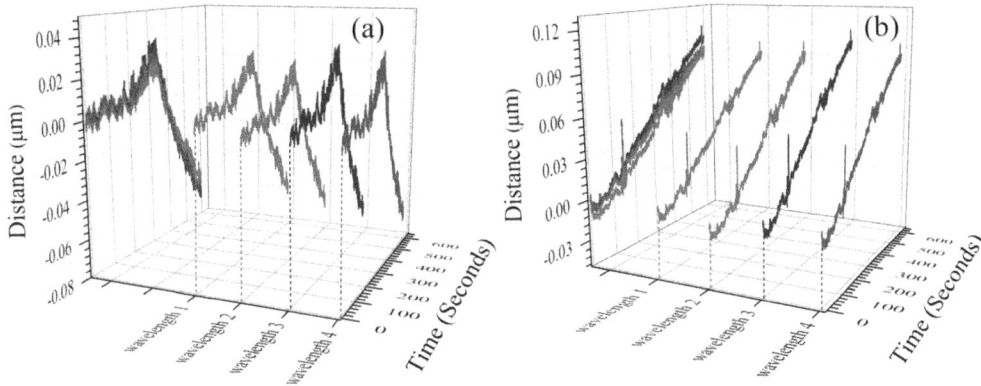

Fig. 10.18. Checking of synchronicity in distance measurement under EO phase modulation process (as shown in Fig. 10.16(a) and (b)) by the four operating wavelengths at (a) 1.25 kHz and (b) 5 kHz.

10.5.4.4. Effects of Environmental Temperature Variation

Out of the remaining factors, the fluctuations in atmospheric temperature can significantly contribute to the optical path length drift. In particular, the current experimental setup with the EO crystal LiNbO$_3$ is very sensitive to environmental temperature variations. However, the PZT sensor is a compact and enclosed system which is consequently less perturbed by such influences. The dependency of interferometric stability on environmental parameters like – variations in room temperature, air turbulence, pressure and humidity are well known, and their influences are capable of bringing low frequency phase drifts based on the type of materials which have been used to manufacture the interferometric system [61-62]. Depending on thermal expansion coefficients different materials bring different amounts of thermal phase disturbances during the measurement process. It is possible to compensate the temperature induced uniform distance change by theoretically evaluating the possible phase drift from the thermal expansion coefficients of the constituting materials within the setup, after minimizing the other perturbing factors.

In order to nullify the influence of temperature, the change of OPD for a material can be theoretically estimated, i.e. $OPD = n.L$. Hence, the rate of change of OPD, i.e. ΔOPD with respect to temperature (T) can be expressed as,

$$\frac{\partial}{\partial T}[OPD] = \left(\frac{\partial}{\partial T}(n) \cdot L\right) + \left(n \cdot \frac{\partial}{\partial T}(L)\right) = (n \cdot \eta_T \cdot L + n \cdot L \cdot \alpha_T), \quad (10.15)$$

where n and L are the initial refractive index of air and length between the reference and target respectively before considering the thermal influence. α_T is the linear expansion coefficient and η_T i.e. $[\frac{1}{n}\frac{\partial n}{\partial T}]$ is the fractional change of refractive index with temperature. As the path L between the reference and the target is in air hence, the n has been taken as unity and the corresponding η_T is almost negligible. For the experimental setup with the

EO crystal the contribution of refractive index in the total OPD drift cannot be neglected. The thermally induced change of refractive index for LiNbO$_3$ crystal causes a significant amount of OPD drift due to the environmental temperature variations. During the experiment, the LiNbO$_3$ crystal has been placed between the reference and the target as shown in Figs. 10.9 and 10.10. Thus, the temperature induced distance drift for this configuration can be approximately predicted by using the following equations

$$\frac{\partial}{\partial T}[OPD] = \frac{\partial}{\partial T}[OPD|_{Cystal}] + \frac{\partial}{\partial T}[OPD|_{Air}]$$

$$= \frac{\partial}{\partial T}[n_e \cdot L_{Crystal}] + \frac{\partial}{\partial T}[(n_{air} \cdot L_{Rail-Crystal})]] - \frac{\partial}{\partial T}[(n_{air} \cdot L_{Crystal})]$$

$$= \left[\left(\frac{\partial(n_e)}{\partial T} \cdot L_{Crystal}\right) + \left(n_e \cdot \frac{\partial(L_{Crystal})}{\partial T}\right)\right]$$

$$+ \left[\left(\frac{\partial(n_{air})}{\partial T} \cdot L_{Rail-Crystal}\right) + \left(n_{air} \cdot \frac{\partial((L_{Rail-Crystal}))}{\partial T}\right)\right]$$

$$- \left[(\frac{\partial(n_{air})}{\partial T} \cdot L_{Crystal}) + (n_{air} \cdot \frac{\partial(L_{Crystal})}{\partial T})\right], \tag{10.16}$$

where, $L_{Crystal}$ is the length of the crystal and $L_{Rail-Crystal}$ is the path in air which can be obtained by subtracting the $L_{Crystal}$ from the total length between the ferule and the mirror. n_e is the extra ordinary refractive index of the crystal and n_{air} is the refractive index of air. However, before including the correction factor for temperature compensation, the design of the LiNbO$_3$ crystal was improved to reduce extra reflection from the front surface in order to avoid additional phase errors. In the previous design of the sensor some amount of light was reflected back from the front surface, but the problem was improved by introducing a 5° cut to it like it has been done for the exit surface as shown in Fig. 10.19.

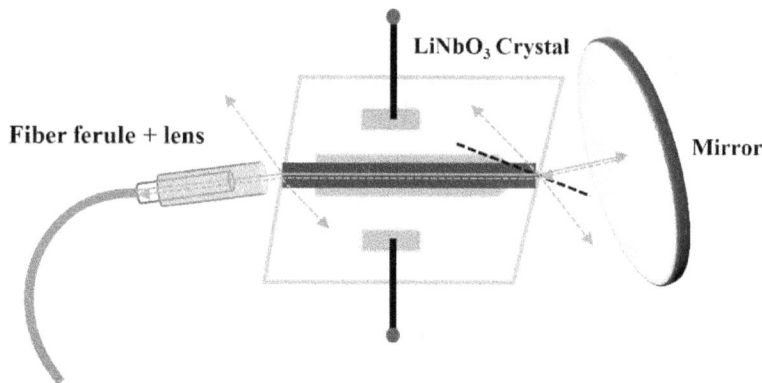

Fig. 10.19. Schematic diagram for the new design of the EO crystal for eliminating additional reflection from front surface.

To check the influence of temperature on the observed distance drift experienced by the experimental setup with the EO sensor a long-term measurement has been done for 14 hours. Fig. 10.20(a) shows the effect of the environmental temperature on the distance measurement. It is clear from the graph that the distance change is correlated with the temperature variation as the data for distance exhibits a linear behavior with the temperature change as shown in Fig. 10.20(b).

However, there is slight non-linearity in the graph which was observed during the first two hours as the rate of distance drift and temperature variation were higher in that region than the rest of the time during the experiment. This might be a result of other external influences as the experimental set up is open and is susceptible of getting affected by the environmental factors like slow response of the thermal sensor to record rapid changes in the room temperature, presence of the experimenter near the set up or, the air turbulence near the EO crystal resulting in erroneous distance drift due to the slight misalignment in the set up.

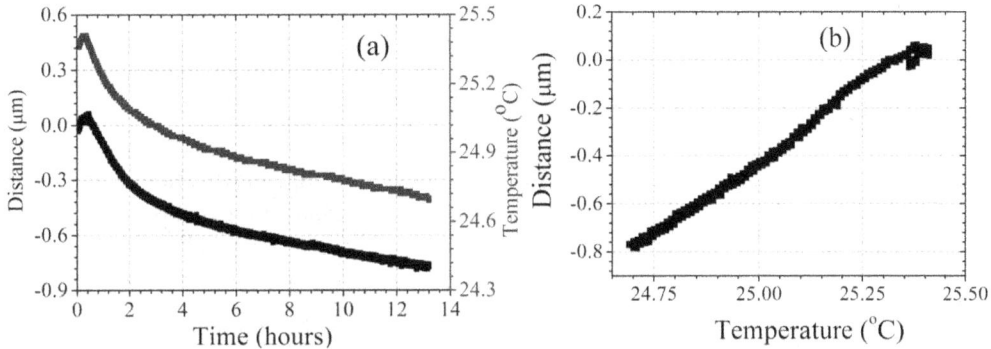

Fig. 10.20. (a) Distance drift (lower black curve) due to external temperature variation (upper blue curve) for the EO sensor; (b) Linear behavior of the observed distance drift with temperature.

To check how well it agrees with the theoretically estimated OPD drift for the experimental setup, $OPD|_{theory}$ was subtracted from the experimentally obtained data ($OPD|_{expt}$). Fig. 10.21 shows the result where during the first two hours the $OPD|_{expt}$ data have not agreed with the $OPD|_{theory}$ because of the previously mentioned non-linearity. However, for the rest 12 hours the $OPD|_{expt}$ almost became equal with the estimated values of $OPD|_{theory}$.

Hence, the next step is to minimize those factors which are responsible for that non-linearity as observed for the temperature induced OPD drift. Apart from that, the EO crystal also needs an encapsulation including other optical components to have a comparatively better linear response with temperature in order to have an error free temperature compensated distance data.

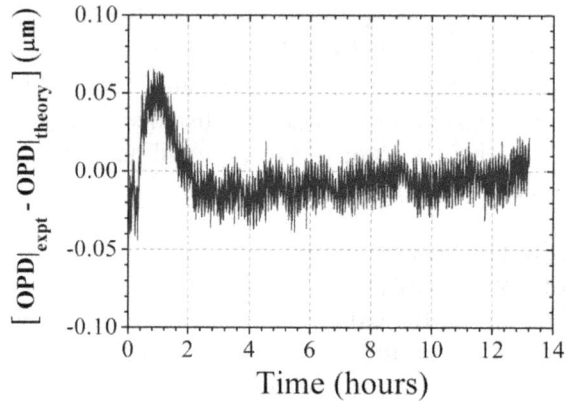

Fig. 10.21. Verifying experimentally obtained values of OPD ($OPD|_{expt}$) with theoretically estimated result $OPD|_{theory}$ using Eq. (10.16).

10.6. Conclusion and Future Prospect

The research field of MWLI serves for various scientific disciplines and specifically has brought major improvements in the area of optical measurement technology. The demand of obtaining high precision in mechanical manufacturing processes e.g. for quality control of fabricated optical elements with only sub-nanometer resolution has increased over the last few decades. Likewise, the techniques of MWLI measurements have improved to meet the needs of technological advancements. As described in this chapter, the MWLI distance measurement process has important applications for changeable measurement requirements where the small and fixed working ranges of conventional interferometry will not be suitable. However, a faster measurement process is equally important for industrial fabrications. Hence, the research for improving the phase modulation process of an MWLI system in order to speed up the measurement process has been presented in this chapter.

Because of the limited working efficiency of the MWLI- mechanical phase shifter probe at higher modulating frequency, a new type of sensor based on the concept of linear EO phase modulation has been proposed. The preliminary prototype of the LiNbO$_3$ based sensor showed comparatively better performance at higher driving frequency which indicates its suitability to replace the current PZT based sensor in order to achieve the goal of faster distance measurement. However, the experimental setup of EO sensor needs further improvements for obtaining better measurement stability particularly from environmental temperature fluctuations. Hence, the prime focus of the future steps will be on the elimination of environmental influences and other perturbing factors along with the process for compensating the temperature induced measurement errors. The MWLI-EO distance sensor is expected to perform almost fifty times faster distance measurement compared to the PZT sensor and it would be a great advantage for the manufacturing processes of optical industry.

Acknowledgements

The authors are thankful to the European Union's Horizon 2020 research and innovation programme (Marie Sklodowska-Curie grant agreement No. 641272) for funding. The authors acknowledge Dr. Ralf Nicolaus, Mr. Thilo May (AMETEK GmbH BU Taylor Hobson/Luphos), Prof. Dr. D. Kip and Dr. C. Haunhorst (Helmut Schmidt-Universität) for suggestions and valuable discussions regarding sensor implementation.

References

[1]. D. Malacara, M. Servín, Z. Malacara, Interferogram Analysis for Optical Testing, 2nd Ed., *Taylor & Francis Group*, 2005.

[2]. P. de Groot, Advances in Optical Metrology, The Optics Encyclopedia: Basic Foundations and Practical Applications, *Wiley-VCH Verlag GmbH,* 2007.

[3]. P. de Groot, Principles of interference microscopy for the measurement of surface topography, *Advances in Optics and Photonics,* Vol. 7, 2015, pp. 1-65.

[4]. J. Yao, A. Anderson, J. P. Rolland, Point-cloud noncontact metrology of freeform optical surfaces, *Optics Express,* Vol. 26, Issue 8, 2018, pp. 10242-10265.

[5]. G. Berkovic, E. Shafir, Optical methods for distance and displacement measurements, *Advances in Optics and Photonics,* Vol. 4, 2012, pp. 441-471.

[6]. P. de Groot, J. Biegen, J. Clark, X. C. de Lega, D. Grigg, Optical interferometry for measurement of the geometric dimensions of industrial parts, *Applied Optics,* Vol. 41, Issue 19, 2002, pp. 3853-3860.

[7]. K. Creath, J. Schmit, J. C Wyant, Optical Metrology of Diffuse Surfaces, Optical Shop Testing, 3rd Ed., *John Wiley & Sons*, 2007.

[8]. P. de Groot, Limitations of Optical 3D Sensors, in Optical Measurement of Surface Topography (R. Leach, Ed.), *Springer-Verlag,* 2011.

[9]. H. Nouira, N. El-Hayek, X. Yuan, N. Anwer, J. Salgado, Metrological characterization of optical confocal sensors measurements (20 and 350 travel ranges), *Journal of Physics: Conference Series,* Vol. 483, 2014, 012015.

[10]. R. K. Leach, C. L. Giusca1, K. Naoi, Development and characterization of a new instrument for the traceable measurement of areal surface texture, *Measurement Science and Technology,* Vol. 20, 2009, 125102.

[11]. J. E. Muelaner, P. G. Maropoulos, Large volume metrology technologies for the light controlled factory, *Procedia CIRP,* Vol. 25, 2014, pp. 169-176.

[12]. A. Karsenty, Y. Lichtenstadt, S. Naeim, Y. Arieli, Improving Interferometry Instrumentation by Mixing Stereoscopy for 2π Ambiguity Solving, *International Journal of Measurement Technologies and Instrumentation Engineering*, Vol. 6, Issue 2, 2017, pp. 43-55.

[13]. C. Yang, A. Wax, R. R. Dasari, M. S. Feld, 2p ambiguity-free optical distance measurement with subnanometer precision with a novel phase-crossing low-coherence interferometer, *Optics Letters,* Vol. 27, Issue 2, 2002, pp. 77-79.

[14]. P. de Groot, Unusual techniques for absolute distance measurement, *Optical Engineering,* Vol. 40, Issue 1, 2001, pp. 28-32.

[15]. K. Meiners-Hagen, R. Schödel, F. Pollinger, A. Abou-Zeid, Multi-wavelength interferometry for length measurements using diode lasers, *Measurement Science Review,* Vol. 9, Issue 1, 2009, pp. 16-26.

[16]. U. Paul Kumar, N. Krishna Mohan, M. P. Kothiyal, Multiple wavelength interferometry for surface profiling, *Proceedings of SPIE,* Vol. 7063, 2008, 70630W.

[17]. M. A Hassan, H. Martin, X. Jiang, Surface profile measurement using spatially dispersed short coherence interferometry, *Surface Topography: Metrology and Properties,* Vol. 2, 2014, 024001.

[18]. Y.-Y. Cheng, J. C. Wyant, Two-wavelength phase shifting interferometry, *Applied Optics,* Vol. 23, Issue 24, 1984, pp. 4539-4543.

[19]. A. Harasaki, J. Schmit, J. C. Wyant, Improved vertical-scanning interferometry, *Applied Optics,* Vol. 39, Issue 13, 2000, pp. 2107-2115.

[20]. T. Suzuki, T. Yazawa, O. Sasaki, Two-wavelength laser diode interferometer with time-sharing sinusoidal phase modulation, *Applied Optics,* Vol. 41, Issue 10, 2002, pp. 1972-1976.

[21]. T. D. Nguyen, J. D .R. Valera, A. J. Moore, Optical thickness measurement with multi-wavelength THz interferometry, *Optics and Lasers in Engineering,* Vol. 61, 2014, pp. 19-22.

[22]. R. Dändliker, K. Hug, J. Politch, E. Zimmermann, High-accuracy distance measurement with multiple-wavelength interferometry, *Optical Engineering,* Vol. 34, Issue 8, 1995, pp. 2407-2412.

[23]. P. de Groot, L. Deck, Surface profiling by analysis of white-light interferograms in the spatial frequency domain, *Journal of Modern Optics,* Vol. 42, Issue 2, 1995, pp. 389-401.

[24]. J. C. Wyant, White light interferometry, *Proceedings of SPIE,* Vol. 4737, 2002, pp. 98-107.

[25]. K. Falaggis, C. E. Towers, Absolute metrology by phase and frequency modulation for multiwavelength interferometry, *Optics Letters,* Vol. 36, Issue 15, 2011, pp. 2928-2930.

[26]. R. Yang, F. Pollinger, K. Meiners-Hagen, J. Tan, H. Bosse, Heterodyne multi-wavelength absolute interferometry based on a cavity-enhanced electro-optic frequency comb pair, *Optics Letters,* Vol. 39, Issue 20, 2014, pp. 5834-5837.

[27]. K. Creath, Step height measurement using two-wavelength phase-shifting interferometry, *Applied Optics,* Vol. 26, Issue 14, 1987, pp. 2810-2816.

[28]. L. Deck, P. de Groot, High-speed non-contact profiler based on scanning white light interferometry, *International Journal of Machine Tools and Manufacture,* Vol. 35, Issue 2, 1995, pp. 147-150.

[29]. J. Petter, Multi wavelength interferometry for high precision distance measurement, in *Proceedings of the Conference SENSOR+TEST,* 2009, pp. 129-132.

[30]. S. Sharma, P. Eiswirt, J. Petter, Electro optic sensor for high precision absolute distance measurement using multi-wavelength interferometry, *Optics Express,* Vol. 26, Issue 3, 2018, pp. 3443-3451.

[31]. S. Sharma, P. Eiswirt, J. Petter, Study on interferometric stability based on modulating frequency, operating wavelengths and temperature using an electro optic multi-wavelength distance sensor, *Sensors & Transducers,* Vol. 225, Issue 9, 2018, pp. 1-7.

[32]. K. J. Gasvik, Optical fibers and fiberoptic communications, in Optical Metrology, 3rd Ed., Chapter 3, *John Wiley & Sons Ltd,* 2002.

[33]. T. G. Brown, Optical fibers and fiberoptic communications, in Handbook of Optics (M. Bass, Ed.), Chapter 10, *McGraw-Hill Inc.,* 1995.

[34]. A. R. Bahrampour, S. Tofighi, M. Bathaee, F. Farman, Optical Fiber Interferometers and Their Applications, Interferometry – Research and Applications in Science and Technology, *Intech,* 2012.

[35]. G. R. Fowles, Introduction to modern optics, 2nd Ed., *Dover Publications Inc.,* New York, 1975.

[36]. E. Hecht, Optics, 5th Ed., *Pearson Education Limited,* 2017.

[37]. Z. Sodnik, E. Fischer, T. Ittner, H. J. Tiziani, Two-wavelength double heterodyne interferometry using a matched grating technique, *Applied Optics,* Vol. 30, Issue 22, 1991, pp. 3139-3144.

[38]. Y. Kuramoto, H. Okuda, High-accuracy absolute distance measurement by two-wavelength double heterodyne interferometry with variable synthetic wavelengths, in *Proceedings of the International Workshop on Future Linear Colliders (LCWS'13)*, Tokyo, Japan, 11-15 November 2013, pp. 1-7.

[39]. R. Dändliker, R. Thalmann, D. Prongue, Two-wavelength laser interferometry using superheterodyne detection, *Optics Letters,* Vol. 13, Issue 5, 1988, pp. 339-341.

[40]. S. Manhar, R. Maurer, H.J. Tiziani, Z. Sodnik, E. Fischer, A. Mariani R. Bonsignori, J. G. Margheri, C. Giunti, S. Zatti, Dual-wavelength interferometer for surface profile measuremts, in Laser/Optoelektronik in der Technik / Laser/Optoelectronics in Engineering(W. Waidelich,Ed.), *Springer,* Berlin, Heidelberg, 1990.

[41]. H. J. Tiziani, Heterodyne Interferometry using two wavelengths for dimensional measurements, *Proceedings of SPIE*, Vol. 1553, 1991, pp. 490-501.

[42]. P. de Groot, Interference Microscopy for Surface Structure Analysis, Handbook of Optical Metrology: Principles and Applications (T. Yoshizawa, Ed.), *CRC Press*, 2017.

[43]. P. de Groot, Design of error-compensating algorithms for sinusoidal phase shifting interferometry, *Applied Optics,* Vol. 48, Issue 35, 2009, pp. 6788-6796.

[44]. O. E. Nyakang'o, G. K. Rurimo, P. M. Karimi, Optical phase shift measurements in interferometry, *International Journal of Optoelectronic Engineering,* Vol. 3, Issue 2, 2013, pp. 13-18.

[45]. J. Novák, New phase shifting algorithms insensitive to linear phase shift errors, *Acta Polytechnica,* Vol. 42, Issue 4, 2002, pp. 51-56.

[46]. A. Dubois, Phase-map measurements by interferometry with sinusoidal phase modulation and four integrating buckets, *J. Opt. Soc. Am. A*, Vol. 18, Issue 8, 2001, pp. 1972-1979.

[47]. J. Millerd, N. Brock, J. Hayes, B. Kimbrough, M. North-Morris, J. C. Wyant, Vibration insensitive interferometry, in *Proceedings of the 6th Internat. Conference on Space Optics (ESTEC'06)*, The Netherlands, 27-30 June 2006, pp. 1-7.

[48]. D. Su, M. Chiu, C. Chen, A heterodyne interferometer using an electro-optic modulator for measuring small displacements, *J. Optics (Paris)*, Vol. 27, Issue 1, 1996, pp. 19-23.

[49]. F. A. Jenkins, H.E. White, Magneto-optics and electro-optics, in Fundamentals of Optics, Chapter 32, *McGraw-Hill Education*, 2001.

[50]. O. Gobert, D. Rovera, G. Mennerat, M. Comte, Linear electro optic effect for high repetition rate carrier envelope phase control of ultra short laser pulses, *Applied Sciences,* Vol. 3, 2013, pp. 168-188.

[51]. C.-C. Shih, A. Yariv, A theoretical model of the linear electro-optic effect, *J. Phys. C: Solid State Phys.*, Vol. 15, 1982, pp. 825-846.

[52]. B. E. A. Saleh, M. C. Teich, Electro-optics, in Fundamentals of Photonics, Chapter 18, *John Wiley & Sons Inc.*, 1991.

[53]. T. A. Maldonado, Electro-optic modulators, in Handbook of Optics Volume II Devices, Measurements and Properties (M. Bass, Ed.), 2nd Ed., Chapter 13, *McGraw-Hill Inc*, 1995.

[54]. B. Kuhlow, Modulators, in Laser Fundamentals. Part 2 Landolt-Börnstein – Group VIII Advanced Materials and Technologies (Numerical Data and Functional Relationships in Science and Technology), Chapter 7.1, *Springer,* Berlin, Heidelberg, 2006, pp. 85-110.

[55]. D. Kip, M. Wesner, Photorefractive Waveguides, in Photorefractive Materials and Their Applications 1, Chapter 10, *Springer-Verlag*, New York, 2006, pp. 289-315.

[56]. M. V. Mantravadi, D. Malacara, Newton, Fizeau, and Haidinger Interferometers, Optical Shop Testing (D. Malacara, Ed.), 3rd Ed., *John Wiley & Sons Inc*, 2007.

[57]. J. E. Greivenkamp, Interference, in Handbook of Optics (M. Bass, Ed.), Vol. 1, 2nd Ed., Chapter 2, *McGraw-Hill Inc*, 1995.

[58]. Y.-T. Chen, M. Ou-Yang, S.-D. Wu, S.-G. Lin, Y.-T. Kuo, C.-C. Lee, Using ensemble empirical mode decomposition to improve the static fringe analysis in optical testing, in

Proceedings of the IEEE International Instrumentation and Measurement Technology Conference (I2MTC'12), Graz, Austria, 13-16 May 2012, pp. 249-253.

[59]. LuphoScan– Interferometric, Scanning Metrology Systems Based on MWLI Technology, https://www.taylor-hobson.com/products/non-contact-3d-optical-profilers/luphos/luphoscan

[60]. S. Sharma, P. Eiswirt, J. Petter, Linear electro optic effect: an alternative method for replacing mechanical phase modulation to improve the speed of absolute distance measurement in a multi-wavelength interferometer, in *Proceedings of 1st International Conference on Optics, Photonics and Lasers (OPAL'18)*, Barcelona, Spain, 9-11 May 2018, pp. 127-130.

[61]. L. Yan, B. Chen, C. Zhang, Y. Liu, W. Dong, C. Li, Measurement of air refractive index fluctuation based on a laser synthetic wavelength interferometer, *Measurement Science and Technology*, Vol. 25, 2014, 095006.

[62]. Y. Nakajima, K. Minoshima, Highly stabilized optical frequency comb interferometer with a long fiber-based reference path towards arbitrary distance measurement, *Optics Express*, Vol. 23, Issue 20, 2015, pp. 25979-25987.

Chapter 11
Microchannel Silicon: A New Insight into Mesoscopic Crystal Structure, Optical and Photonic Phenomena

Gennady Medvedkin

11.1. Introduction

Silicon exists in many solid state modifications but the most in-demand and developed material in the modern electronic industry exploits primarily a single-crystal form. Just to name a few single-crystal applications as memory chips, computer microprocessors, image sensors, radiation detectors, etc. Other solid state forms as polycrystalline, microcrystalline, amorphous, and nanocrystalline modifications of silicon have found their niche and continue expansion in the market, for example, for ecologically clean solar power. These crystalline and amorphous forms of silicon, properly operating in infrared, visible, and ultraviolet optics and optoelectronics, remain opaque to visible light except for very thin films. Practically at a thickness of 10 μm or more, the fundamental bandgap cuts off optical transparency; this happens at $\lambda \approx 1$ μm in mono-/poly-/micro-silicon, and at $\lambda \approx 0.5$ μm in amorphous/nano silicon. Although thin films of submicrometer thickness become partially transparent, all of the just-mentioned silicon materials remain unable to transmit optical radiation from the yellow-green-blue-violet-UV wavelengths. Nanoporous silicon, as one of the exotic forms [1-8], offers another set for optics, but it remains opaque for visible light as well due to disordered nanopores and strong scattering of the light. In addition, the fabrication of a freestanding nanoporous material without a substrate on a large surface of several cm^2 and with thickness of hundreds of micrometers is questionable to date. Therefore, the immediate question arises as to whether it is possible to bring silicon to the table as a transparent material in visible light while preserving the integrity of its monocrystalline diamond structure. We can negatively answer this question within the framework of the classical optics of solids, but mesoscopic insight into the microstructure and optics makes this obstacle easily vincible.

Gennady Medvedkin
General Molded Glass Inc., Torrance, USA

Herein we focus on the mesoscopic crystal structure of silicon the most popular electronic material, which is becoming more efficient in the world of mesoscopic optoelectronics and photonics. With a broader look at semiconductor crystalline materials, it can be noted that the microprocessing of diamond-like crystals in the same way can turn them into mesoscopic materials, which by their physical properties differ significantly and effectively from their classical solid-state analogs.

11.2. Crystal Structure

11.2.1. Why Mesoscopics?

Mesoscopic materials surround us and emerge in new applications [9]. They became an important part of human life. Below are just a few examples:

- Snowflakes are typical mesoscopic structures, they exist in many diverse and amazing shapes and their main symmetry is hexagonal [10, 11].

- CMOS image sensors are another mesoscopic ordered structures which spread everywhere – in smartphone camera, webcam, rear view camera in vehicles, all-round view, surveillance, reconnaissance, medical and dental cameras, etc. [12, 13]. These silicon structures possess a square symmetry in mesoscopic pixel arrays. Up to several millions of pixels can be found in a single color/RGB or monochrome sensor. Silicon wafer of 12-in. typically contains 3, 000 sensors with 1.1 μm megapixel capacity and therefore contains $\sim 10^{10}$ square symmetrical elements. Most of CMOS image sensors have a pixel pitch from 0.9 to 30 μm and a square network for microlens arrays, color filter arrays and Si-pixel arrays. Columns and rows render of a quadratic matrix, and therefore the symmetry of these arrays is quadratic. Recently an inverted pyramid surface has been developed in back-illuminated CMOS image sensor to make the receiving surface enlarged by the factor of $1/\cos(54.7°) = 1.73$ as compared to the flat surface [14].

- Perovskite solar cells emerged for the last few years as devices with the fastest growing PV efficiency; they employ mesoscopic submicron films of TiO_2 in the multilayer structure of the solar cell. This mesoscopic structure is disordered in contrast to the above examples. The mesoporous TiO_2 film plays a key role in the perovskite solar cell performance [15-17].

Below we discuss photonic processes that occur on inclined surfaces and inside a silicon crystal with *p-n* junction starting from simple optics on the surface. Then, considering more sophisticated optical and electronic processes in the device structures with microchannels, we present a few photonic applications, for example, in the ultraviolet spectral range, where silicon photodetectors are not effective enough because their sensitivity is the highest in the visible and near infrared spectral ranges. Other limitations of the known material, such as indirect optical transitions and a rather narrow bandgap can also be overcome by means of mesoscopic transformation in a single crystal structure as shown below in the process section. Thus, microchannel silicon possesses an incredible potential inside the mesoscopic geometry in optical and photonic applications.

11.2.2. Non-traditional Crystal Structure of Silicon

Typical crystal structure of semiconductor crystals, such as silicon, is formed by atoms with covalent bonds between them or partially ion-covalent bonds in compound semiconductor crystals. Some other crystals form in molecular structures with molecules (for example, snowflakes / ice crystals, dry ice / CO_2 crystals) or superlattice materials with larger spatial elements in the form of nanoscale objects of various shapes and dimensions (0D, 1D, 2D, 3D). In the world of nano- and microstructures, such intermediate objects belong to mesoscopic crystalline forms. Microchannel silicon is a well-developed type of mesoscopic crystals and will be described in this chapter as a uniform material demonstrating optical and photonic properties distinct from classic solids.

Passing upstairs from the level of atoms to bulk solids, one can tailor artificial crystal structures by periodical repetition of nano-, meso-, micro-size units and create new artificial solids with multifarious structural geometries. Such crystal structures are utterly scarce or impossible in classical bulk crystalline solids. Fig. 11.1 shows the scaling crystal structures from the atomic level to mesoscopic and macroscopic crystal structures for silicon. The atomic lattice typically measures in Å/nm, mesoscopic lattice made of microchannels and micro/nano slabs – in nm/μm, and macro-dimensional wafer – in mm/inch. Silicon wafer is traditionally cut and polished of the bulk monocrystalline ingot; however the wafer can be alternatively fabricated of microchannel silicon as well.

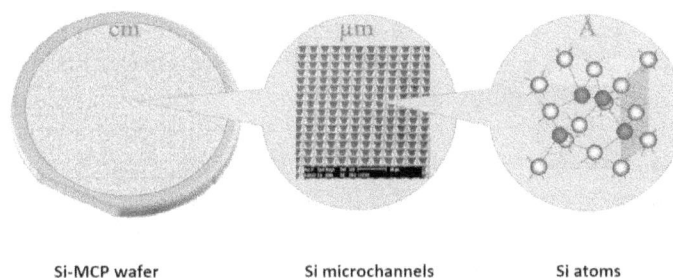

Si-MCP wafer   Si microchannels   Si atoms

Fig. 11.1. Scaling of three-dimensional crystal structures in silicon material.

A number of exotic forms of silicon have been discovered till now, including high-density silicon, black silicon, porous silicon, etc. [1, 18, 19]. Mesoscopic crystals, and especially microchannel silicon, did not get proper attention in literature as uniform continuous material and unfortunately was missed for explicit optical description; therefore this material requires special consideration [20]. Its substantial practical usage and absence of a detailed description as a self-sufficient optical material with a non-traditional crystal structure makes the mesoscopic approach more valuable.

Mesoscopic crystal structure in its entirety is within a definition of crystal lattices with a translation symmetry obeying to a long-range order. Mesoscopic lattice consists of nano or microplanes, of nanowires or microcolumns arranged into this long-range ordered structure with lattice constants in accord with a microchannel pitch of the nm-μm order, cf. with Å-size constants in the diamond lattice of silicon, Fig. 11.1. Any crystalline lattice of classic single crystals in general is anisotropic, even for crystals of the cubic syngony[1], although this anisotropy 10^2 to 10^4 times as low as in uniaxial or double-axial crystals. Amazing is a fact that mesoscopic crystals, which microchannel silicon belongs to, have many and almost all attributes of classic lattices, and comparable by dimensions with molecular crystals whose molecules are of a larger size than individual atoms/ions.

Microchannel silicon is not a porous material in the strict sense because it possesses the long-range ordered structural elements. In the porous silicon, nano-pores or micro-pores are disordered similar texture, because the long-range order is broken. So, nano/microporous silicon can be attributed to disordered or partly ordered solids (by analogy with amorphous or polycrystalline copartners).

Another crystal system of solids with macroscopic coherence is related directly to mesoscopic systems. Examples of the latter are compound superconductors, layered semiconductors, superlattices and other 3D materials. Both mesoscopic and macroscopic systems consist of many atoms but macroscopics obey the laws of classical mechanics, ray optics, and whereas mesoscopics is strongly affected by quantum mechanics and quantum optics rules. In measuring length, the dimension can be defined between angstroms and microns, i.e. all nanoscale systems, especially periodic and coherent, are accommodated inside the area of mesoscopics.

Diversity of silicon mesoscopic systems is suggested to be huge, so this chapter considers just only one silicon material with a microchannel crystal structure. Microchannel material is fabricated from silicon wafers 4 and 6 inches in diameter, representing a uniform continuous material with a microstructure formed by crystalline microplates and surrounded by calibrated vacuum voids. Work in silicon microchannel plates (Si-MCPs) has previously attracted much attention in device development for image amplifiers as alternatives to glass fiber MCPs [21], including many applications in UV, EUV, X-ray, and gamma-ray image photodetectors [22-26] intended for space astronomy, solar-blind operation [27, 28], and lasing [29]. Despite advanced photonic and electro-optic designs, first-hand experimental or modeling data on pure optics for Si-MCPs have been presented in the literature very scarcely [20], just as a solid optical material with a double-crystal structure. The crystalline doubling appears in precisely macromachined single crystals when the crystalline coherency is intact at mesoscopic distances in three dimensions.

---

[1] Syn-gony in Greek: σύν, «accordingly, together», и γωνία, «angle». There are 7 syngonies or crystal systems in crystallography: the highest category (cubic), the intermediate category (tetragonal, hexagonal, rhombohedral), and the lowest category (triclinic, monoclinic, orthorhombic). Silicon crystal belongs to the highest cubic syngony.

11.3. Fabrication of Microchannel Silicon

11.3.1. Formation of Inverted Pyramids

Square based micro pyramids of an inverted orientation are forming on (100) Si-wafer for etching in the KOH agent. At the beginning of process, a pseudo-isotropic etch takes place since the etch speed has not yet a big impact. Anisotropic etch becomes more pronounced with an etching time as shown in Fig. 11.2. Anisotropic crystalline etch simulation (ACES) developed at BUY [30, 31] was used to generate 3-D images of a single inverted pyramid for the 2 min processing run on the 40 μm × 40 μm mask size. The processing for a few minutes makes the directions [010] and [001] prevailed over the [01-1] and [110] ones.

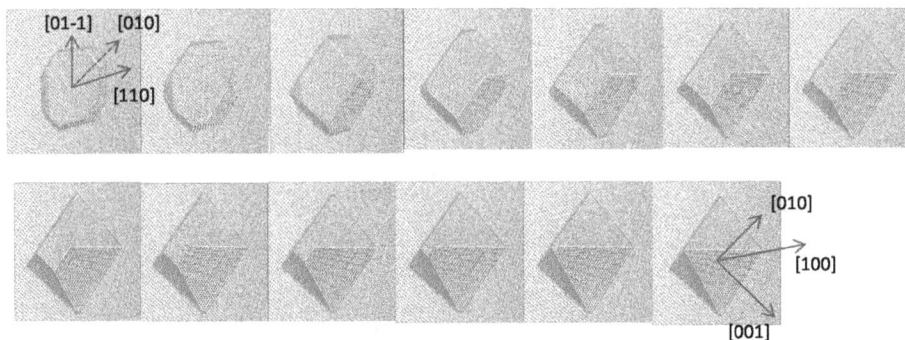

Fig. 11.2. Formation of inverted pyramids for the etching in KOH.

Lateral faces of the inverted micro pyramids are slanted by 54.74° to the (100) plane of Si-wafer and defined by Miller indices {111}, see Fig. 11.3(a). Four crystallographic planes (111), (1-11), (-1-11) and (-111) etch with the same rate and form a regular quadrangle pyramid. Fig. 11.3(b) shows an initial seeding of the inverted pyramidal pit etched in the opening under the lithographic mask. A lithographic pattern defines the micro pyramid pitches over the full wafer. For the mesoscopic crystal lattice in Si-MCP, the square pitch is chosen from 2 μm to 50 μm and the wall thickness from 200 nm to 4 μm. The etch rate in 50 % KOH solution was abt. 25 μm/hr at temperature of 70 °C. Detail about other concentrations and temperature dependencies can be found in [31].

11.3.2. Photoelectrochemical Micromachining

Photoelectric excitation technologies for etching semiconductors and dielectrics have been developed by many groups [32-38], in particular, for low-resistance *n*- and *p*-type silicon wafers with (100), (111), and (110) orientations. In this work, a photoelectrochemical anisotropic etching (PECANE) technology is established to micromachine 2-, 4-, and 6-inch silicon wafers [22-29]. According to PECANE, a silicon wafer (100) is processed in a chemical tank of a photoelectrochemical (PEC) reactor for several hours to produce accurate microchannels with a pixel array evenly distributed over

269

the wafer. Microchannels arranged as strictly parallel light waveguides in the [100] direction are strictly perpendicular or inclined by 2°, 5°, 8°, or 15° to the face plane, as prescribed in specifications. If tilted, the square size and parallelism of the walls are kept constant along the full wafer depth. PECANE uses *n*-type silicon wafers with a low resistance of <200 Ω×cm, a photolithographic pattern for arrays with pitches of 5 to 40 μm, and micropit seeding in the KOH etchant agent at the pre-processing stage. The main process was carried out with the etchant composed of hydrofluoric acid (HF, 49 % concentration), deionized water DI, and ethanol C_2H_6O. A few compositions of the components were employed to vary the wet anisotropic etch. The Pt flat grid electrode was used as a cathode in the PEC reactors. Photos of PEC reactors of generation Gen-2, Gen-3 and Gen-4 are shown in Figs. 11.4 and 11.5. The design of the automated silicon etching station is schematically presented by Solidworks 3D CAD drawing in Fig. 11.6.

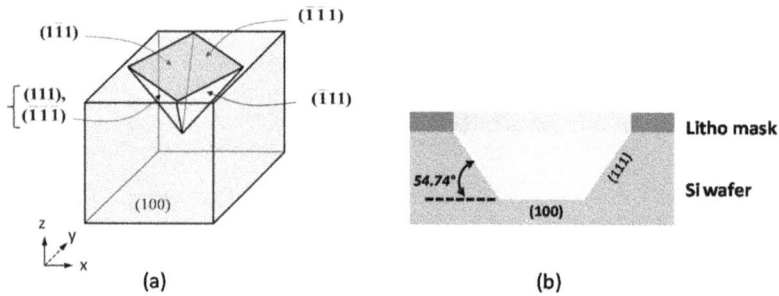

Fig. 11.3. Crystallographic planes of a micropyramidal pit (a), and anisotropic etch geometry with a lithographic mask (b) on Si (100) wafer.

Fig. 11.4. PEC reactors of Gen-2 and Gen-3. Runtime of PECANE process (a); mount of IR LED backlighting unit (b).

The white and infrared backlighting has operated in the reactor to properly excite photoelectrons in bulk Si wafer and work out microchannels in depth, evenly. After completion of the etch process, the wafer was post-processed to open the blind ends and polished on both sides using chemical mechanical planarization (CMP) and finishing procedures. The final thickness was polished down to t = 250-320 μm; the long microchannels of the same length were measured and had an aspect ratio of 8 to 40 and pitches from 5 to 40 μm.

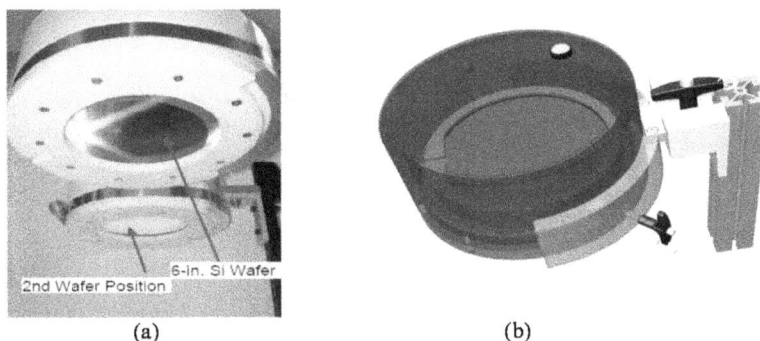

(a)   (b)

Fig. 11.5. Double PEC reactor mount for Gen-4 (a); Inside view of the electrolyte vessel (b).

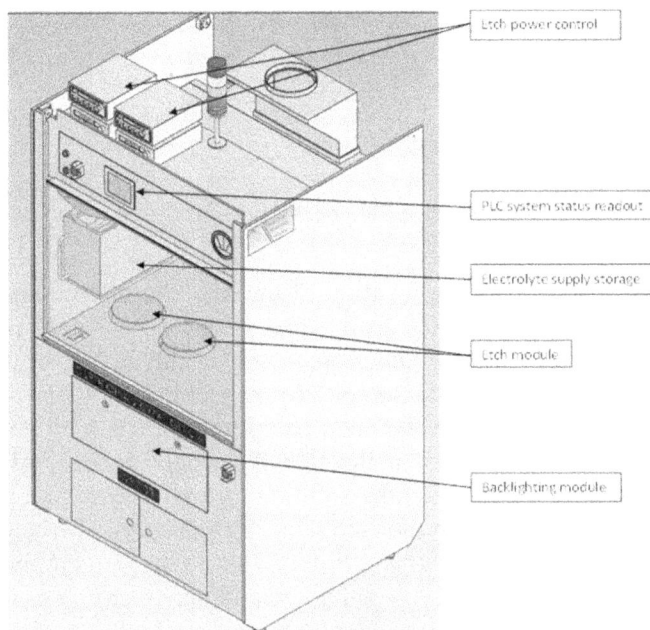

Fig. 11.6. Automated silicon etching station design for Si wafers of 6-in. and 8-in. capacity.

An advanced modification of the double-sided etching reactor has been developed in order to reduce the etching time and better use of the Si-wafer material. Fig. 11.7 demonstrates the creation of microchannel wafers through a double-sided etch process. Chemical extruding the excess material out of the Si-wafer has been developed for various profiles of cross-sections based on the square pixilation defined by the lithographic mask. Some conventional methods for creating channels of various profiles in semiconductor may be limited to certain thicknesses, for example 200-300 μm. Although this may be sufficient for some embodiments, in other embodiments, scintillators may preferably have depths greater than 300 μm, such as 300-1000 μm. The use of the double-sided etching process enable these longer channels to be created. As illustrated in Fig. 11.7, the double-sided etching reactor comprises a vessel, having two transparent windows chemically resistant

271

to the etchant and two ports with inlet and outlet openings. The apparatus further comprises a wafer holder configured to **retain** the preprocessed wafer upright. The reactor further comprises Pt-electrodes, a pair of O-rings disposed to hold Si-wafer in the holder and to electrically isolate it from the holders. In two sink automated station (Fig. 11.6) the holders are placed in the processing vessel so that etchant solution has access to both sides of wafer and so that each side of wafer faces two opposite **windows**.

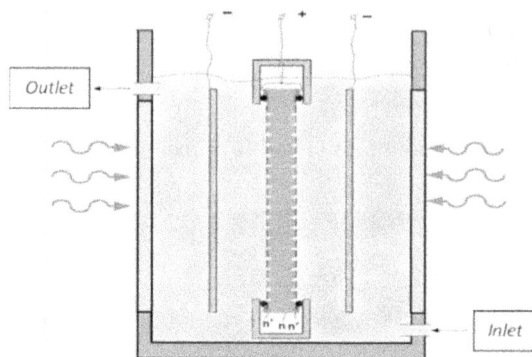

Fig. 11.7. Double-side etching reactor with two backlight illuminations and two Pt-grid cathodes.

The etching process proceeds in the location of the pits, thereby creating an anisotropic etch process where the squared holes grow towards each other until a microchannel is formed. The etching process may produce reaction products that would interfere with further etching if not removed. For example, in an etching process using HF/DDW or HF/Ethanol etchant solutions, some chemical reaction products may be gaseous, such as HF_4, H_2, resulting in bubble formation at the etched surface. Accordingly, inlet and outlet ports are used to provide a flow of etching solution through the vessel, thereby removing the chemical reaction products from the etch sites. Furthermore, the etching process may use relatively large amounts of etchant liquid. For example, etching microchannels through a 500 μm wafer may require several hours and may result in a complete depletion of etchant solution at a rate of about once per hour. Accordingly, a reservoir of the etchant solution in Fig. 11.6 may be provided to allow for circulation of the liquid and replenishment with the fresh fluid.

In PECANE process described with respect to Figs. 11.4 to 11.7, an anisotropic electrochemical etching process is used [23]. In the main wet etching process, an *n*-type silicon wafer positively biased to provide for an anodic dissolution of silicon at the microchannel locations. A basic electrochemical wet etching process may be as follows: (1) holes are generated in the silicon substrate to remove Si-atoms by anodic dissolution and formation of SiF_4; (2) $Si(OH)_2{}^{2+}$ forms on the surface of the wafer; (3) $Si(OH)_2{}^{2+}$ reacts with an agent in the etching solution; and (4) the reaction products are removed into the solution and gaseous atmosphere.

To use Si-wafer in this PECANE process, it may be doped on both sides with phosphorous (P) to provide two n^+-layers on the wafer. These n^+-layers improve the contact resistivity during the etching process, so the doping may occur only where the positive bias contact will be made. In other embodiments, the n^+-layer may further facilitate other aspects of the etching process and may be present on the entireties of the wafer surfaces. The prepared layers may comprise other doped material layers, for example, arsenic, antimony (As, Sb) dopants. The negative electrode may be in direct contact with the wafer during the etching process. In this embodiment, layers may comprise layers of an electrically insulating material, such as Si_3N_4. In the next step, etching pits are prepared using photolithography and wet etching in KOH. In some embodiments, the etching pits are determined according to the eventual use of Si-MCP wafer. For example, to create a Si-microchannel array of the type illustrated in Fig. 11.1, the etched inverted pyramidal pits in Figs. 11.2, 11.3 are spaced in a rectilinear grid according to the desired locations and distributions of the microchannels. In other embodiments, the pits might be spaced in other patterns with 50 % shifted rows according to use, or in other shapes, such as extended micro trenches. As described in more detail with respect to Fig. 11.7, the etched pits provide locations for the anisotropic etching process to begin, so that the etching process comprises microchannels that propagate from both sides of the wafer.

In further embodiments, rather than being disposed in the solution, the negative electrodes may be disposed on both sides of the wafer as Cu-electrodes around the circumference of the wafer. These contacts provide to bias the electrodes coated directly on Si-wafer. In some cases, the negative electrodes may be shaped to cover the portions of wafer that are not to be etched for mechanical hardening purposes. This use of electrodes may further protect the non-etched Si portions from unintentional etching. In these embodiments, Cu-electrodes may be isolated from direct contact with wafer, for example through the use of Si_3N_4 coatings on Si-wafer.

11.3.3. The Full Cycle of Si-MCP Wafer Fabrication

The prime grade n-type (100) silicon wafers are used for the fabrication cycle starting from pre-processing. The next are three main processing steps:

1. The n^+ type thin layer is initially formed by phosphorous diffusion, followed by a low-stress nitride masking layer deposition. Then photolithography is performed to pattern Si_3N_4 and Cr/Au films on the front and back sides, respectively. Finally, inverted pyramid etching is carried out to seed micropits through the Si_3N_4 masking layer on the front-side wafer.

2. To perform the wet etch micromachining of the long slim microchannels, the wafers are installed into PECANE reactors and electric bias is applied between the Pt-cathode and the metallic grid on the back side of the Si-wafer for one-side etching. The PECANE process lasts for 40-70 h in total, for both 4 and 6 inch wafers under the hard-stabilized backlighting and electric current.

3. In the third step, the opening and oxidation procedures are implemented. Grinding and polishing with CMP provided the opening finalizing for all microchannels in the Si wafer. Partial or complete oxidation is performed in thermal oxidation tubular ovens available to process Si wafers up to 8 inches in diameter.

Fig. 11.8 presents a sketch of the microchannel silicon wafer. The second generation of Si-MCP wafers is enabled for the 4 inch and the third generation for the 6 inch diameter wafers. Microchannel pixel arrays counted a multimillion number of tiny vertical channels $N_{ch} = 4.3 \times 10^6$ to 6.2×10^8 per wafer.

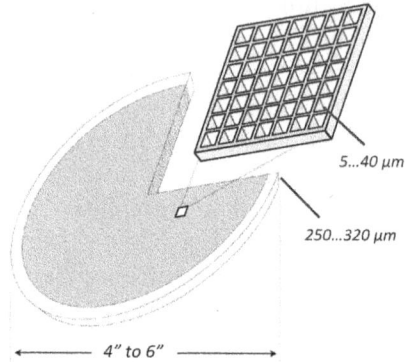

5...40 µm

250...320 µm

4" to 6"

Fig. 11.8. Microchannel silicon wafer with an opened and finished square-cluster microchannel array. Si-MCP wafers 4 inches in diameter have 4 million to 280 million channels.

The wall thickness constancy depends strongly on etching process parameters such as stabilization of photocurrent, uniformity of backlight illumination, and initial wafer homogeneity controlled by the radial distribution of impurities, resistivity, and perfection of the crystal structure. The control of evenness of the walls was a very important and sensitive procedure in PECANE. Upon completion of the micromachining and post-processing, Si-MCP wafers were cut down by IR laser dicing or resizing. Free-standing parts of microchannel silicon can be free from a solid border for further processing in labs or fabs. Small parts of 1 square inch are relatively hard from a mechanical point of view, and can withstand many chemical, thermal, or mechanical procedures; nevertheless, a solid border on large pieces or full Si-MCP wafers is a good support for additional stiffness.

11.4. Methods and Optical Properties

11.4.1. Microscopy and Analysis

A Tescan VEGA II scanning electron microscope (SEM) was used in the transmission, reflection electron, and reflection micro-cathodoluminescence modes. Samples were cleaved in the [010] direction and covered with a thin Au layer in a vacuum to prevent oxidation and create the proper state for draining electric charges for SEM measurements. Another analysis using energy-dispersive X-rays was employed to determine the composition of chemical elements in heterostructures of III-V compounds grown on silicon. Optical microscopy was used for wide-field-view inspection of uniformity and regularity of mesoscopic structures in Si-MCP wafers and heterostructures. A Leitz

Laborlux 12 microscope equipped with a Moticam digital camera and governed by the Motic Images Plus 2.0ML software was used to take images of conoscopic figures in 2D arrays. Olympus FE-190 and FE-230 digital cameras were used to take 2D diffraction images. Spectral and polarization measurements for polarimetric photodiodes and single crystals were conducted with various spectrophotometric instruments, detail are given in references.

11.4.2. Optical Transparency

Single-crystal silicon is a good infrared material with an optical transmission ~52 % in the NIR spectral range $\lambda_T = 1.1\text{-}6.0$ μm. But even the purest, most perfect crystal remains opaque in the visible and ultraviolet ranges of the spectrum and cannot be completely cleared by means of antireflective coatings or by applying a strong electrical bias or mechanical field. Fortunately, the mesoscopic approach allows us to take a fresh look at the problem, using a proper micromachining. With the developed PECANE technological process, we used prime-grade 2 and 4 inch diameter Si wafers to process in an exactly regular way and create industry-acceptable Si-MCP wafers. The process has been successfully upgraded to larger diameters of up to 6 inches with possible further upgrades to 8 inch wafers. In particular, the automated silicon etching station shown in Fig. 11.6 was designed for Si wafers of 6-in. and 8-in. capable to be processed in two sinks simultaneously.

The resulting homogenous microchannel material represents a single-crystal form with a mesoscopic crystal lattice of a typical μm-size periodicity [9]. Strictly speaking, the micro-periodic array exists simultaneously in a macro-regular solid state, so it could bear various names such as mesoscopic, meso-optic, microporous, mid-IR photonic, or metamaterial in accord with inspection conditions or a particular researcher. Note that "meso-optic" is defined by the physical dimension and wavelength and should not be confused with "mesopic", which is defined by light luminance (approximately of 0.001 to 3 cd/m^2). For optical and photonic measurements, the samples were cut from the final Si-MCP wafer to a desired size and orientation. Herein, we consider microchannel silicon as a self-sufficient, bare optical material without any device parts or thin-film structures.

Fig. 11.9 demonstrates a 4 inch Si-MCP wafer processed by PECANE to achieve polychromatic light transmission. The true multicolor transfer through the plate was successfully obtained, particularly for yellow, red, and green colors, as clearly seen in the central area of the Si-MCP. Black and gray colors are also transferred perfectly. The two insets in Fig. 11.9 are the fragments of unopened Si-MCP obtained as top and side views through optical digital microscope (400×) and scanning electron microscope, respectively.

Unlike the direct observation of the backlighted polychromatic image shown in Fig. 11.9, optical spectral measurements by a standard technique on a spectrophotometer (grating, prismatic, or Fourier instruments) may encounter difficulties in retrieving the correct optical data. The microchannel plate constitutes an optical dispersive medium standing in the path of the ray, which is then subjected to a second spectral decomposition on a spectrometric dispersive element with a further complicated transform. Below in paragraph 11.6.1 we consider the Fraunhofer diffraction on Si-MCP plate alone and show a model demonstrating different spatial foci in Fig. 11.20, as well as 2D dispersion and diffraction the plate produces. If Si-MCP is placed in a cuvette compartment of a spectral instrument, a very complicated double dispersion occurs in its dispersive unit because the dispersed and diffracted 2D image is coming from the microchannel plate. To date, there is no detailed description of optical paths or spectral decomposition in a series of such 2D and 3D dispersive elements. We experimentally tested Si-MCP wafers with a thickness t = 250-320 μm to demonstrate the pronounced optical transmission from red to blue wavelengths, as shown in Fig. 11.9. The measurement with a Fourier spectrophotometer is underway for correct setting and interpretation.

Fig. 11.9. Micromachined Si-MCP wafer controlling the visible light integrity (4 inch diameter, thickness 307 μm). Color texts placed behind the Si-MCP are seen distinctly through the transparent achromatic wafer. The round border is the unprocessed silicon and is opaque. Upper inset is the top-view image through an optical microscope in reflection mode; lower inset is an SEM image of the side cleavage of the Si-MCP.

The estimate gives a transmission coefficient as high as 81 % in the visible and wider spectral range λ = 100-1000 nm, which enables MCP to be transparent at $\lambda < d_p/40$, where d_p is the pitch. This extension noticeably exceeds the visible range of λ = 380-760 nm. To the best of our knowledge, such a wide range of transparency with such a high transmission magnitude cannot be offered by any of the bulk semiconductors [39-41] or materials with metasurfaces [42-44]. Furthemore, the microchannel Si-plate can serve an excellent transmissive media for full-color imaging purposes since the every microchannel pixel performs similar to the pixel in CMOS image sensor but functions much simpler than RGB color combinatorics in CIS. The light-shaping with this exotic microchannel form of silicon is even more impressive if one keeps in mind that the electronic and structural properties remain intact inside the Si-MCP walls like in bulk silicon. Despite the well-defined high-efficiency polychromatic transmission, microchannel silicon demonstrates a narrow field of view (FOV), usually a few degrees.

The FOV depends on the aspect ratio and geometric length of the channels, and descends with an increase in the Si-MCP thickness and decrease in pitch.

Although the parallelism of microchannels is determined by the crystal lattice of silicon and is undoubtedly the best in perfection, the evenness of the wall is a variable parameter, as it is subject to changes in the photocurrent and illumination for the etching process. During the long etching process, attention was focused on the current stabilization. Obviously, the quality and smoothness of walls affect the optical transmission, FOV, and polychromatic image transfer because the inclined rays inside the channels are reflected from the wall surfaces. The microchannel plate in Fig. 11.9 has the aspect ratio of 8.5 and FOV = 6.7° without taking into account multiple reflections inside the channels. The observed high optical transmission suggests a good wall quality along these microchannels. One can make an estimate of their smoothness, thickness, and evenness in depth qualitatively by magnifying the cross-cut SEM image in Fig. 11.9. Therefore, the high quality of the fabricated mesoscopic structure is due to the prime-grade silicon wafer, stabilized photocurrent, renewed solution of etchant, and proper backlighting.

11.4.3. Optical Shape Anisotropy

Another feature of Si-MCP is an optical anisotropy for light rays propagated in different crystallographic directions in the mesoscopic lattice. This type of anisotropy can be considered as shape anisotropy since the wavelength is smaller the selected dimensions in the lattice as voids and the length of micro/nano thick walls. In fact, a parallel beam in microchannels propagates without any absorption, refraction, or deflection. Only minor rays deviating from the axis [100] are reflected, refracted, and scattered on the walls. Rays that propagate in a perpendicular direction or directions that are not collinear [100] are absorbed in silicon crystal. The mesoscopic lattice transforms microchannel silicon into an anisotropic substance even within the formalism of classical wave optics. On the one hand, it can be recalled that in the past, Thomas Young (1801) established a relative wave optical effect in his double-slit experiment [45]. On the other hand, modern wet or dry etching techniques such as PECANE and precise photolithography allow the fabrication of multiple mesoscopic apertures at the same time, strictly parallel and evenly distributed throughout the silicon wafer. The microetching technique provides multiplication of double slits in the tens and hundreds of millions over the plate without loss of quality in depth.

Despite the inherent optical isotropy of silicon and other cubic diamond-like crystals ($F\bar{4}3m \equiv T_d^2$ space group), microchannel silicon exhibits a trick with anisotropic effects. One such effect is a well-known Brewster angular anisotropy on flat surfaces [46-48], and another is due to the 3D mesoscopic grid, which causes additional shape optical anisotropy [20]. Si-MCP acquires the optical axis like an ordinary uniaxial anisotropic crystal with tetragonal symmetry. All directions in the 3D coordinate space remain optically opaque except for one longitudinal. To a first approximation, all off-axis directions are equivalent for visible light, with the exception of paraxial or low-angle rays. Mesoscopic optical anisotropy works well in spectral ranges where the wavelength meets a condition of $\lambda \ll d_p$. Unusual 2D diffraction takes effect at longer wavelengths $\lambda \sim d_p$. At $\lambda \gg d_p$, the light ceases to sense microcavities, and the Si-MCP material turns almost isotropic and opaque, with an average refraction index of an effective medium.

Let us numerically estimate the shape optical anisotropy invoked by the mesoscopic crystal lattice in the short wavelength limit $\lambda < d_p$. Consider a cube of material cut from Si-MCP with thicknesses of 300 and 500 μm. For simplicity, one facet is perpendicular to the microchannels, and two others coincide with the wall planes (010) and (001). We use the reflection coefficient $R = 0.30$ for silicon in the visible range, and the transmission coefficient for one wall is

$$T_\perp^{ow} = \frac{(1-R)^2\, e^{-\alpha d}}{1-R^2\, e^{-2\alpha d}}, \qquad (11.1)$$

where α is the optical absorption coefficient and d is the crystal thickness. There are n walls for the light propagating in the perpendicular direction, so $T_\perp = (T_\perp^{ow})^n$. The product of the absorption coefficient and thickness is $\alpha d = 4$ at a wavelength of 500 nm, and $T_\perp = 4.61\times10^{-14}$ ($n = 7.5$) and 5.93×10^{-23} ($n = 12.5$) for two cubic samples, respectively.

The transmission coefficient $T_\parallel$ along the optical axis is 81 % for the dimension 40 μm / 4 μm for pitch / rib. The dichroism coefficient is $D = (T_\parallel - T_\perp) / (T_\parallel + T_\perp) = 99.999$ % in both cubical samples of microchannel silicon. The coefficient D is generally affected by the aspect ratio, and tends to 100 % for all thicknesses or pitches in the actual mesoscopic lattices. It should be noted that pronounced optical features are observed in Si-MCP which strongly contrast with bulk silicon, cf. conventional Si crystals, Si wafers, or device parts [39-41].

11.4.4. Optical Angular Anisotropy

Si-MCP plate fabricated but unprocessed for flatness, polish and finish retains 2D-array of the inverted pyramids on top surface. Every microchannel entrance has four inlet surfaces tilted by 54.74°, see the truncated and full inverted pyramids in Figs. 11.2 and 11.3. Additionally, Fig. 11.9 shows 2D and 3D images of the truncated inverted pyramids in upper and lower insets, respectively. The face morphology of Si-MCP is non-planar, and microscopically relief consists of four tilted facets {111} repeating in the array. So, the incident light is exposed to the Fresnel reflection, refraction and Brewster polarization splitting on every microchannel entrance. Four reflected beams turn into partly polarized beams and take a part in forming the 2D diffraction pattern observed in optical reflection from Si-MCP. The optical angular anisotropy appears on four reciprocally tilted facets (111), (-111), (1-11), (-1-11) simultaneously and cannot be completely averaged because of strong diffraction on 3D periodical structure, which takes place in the optical far-field $d_f \geq 2D^2/\lambda$, which typically is 10 to 10^4 μm. The 2D diffraction pattern will be considered in the next sections below and herein we start from the simplest case of the angular anisotropy on a single planar silicon surface.

Let us consider the case of anisotropy appearing at a flat surface of solids under the oblique incidence of light from air. The unique state of the reflected light when the only *s*-polarization exists was called the Brewster angle after Sir David Brewster (1781-1868), who was the first physicist describing this optical effect in detail [49]. The Brewster law says that the light incident from the air to the crystal surface will be reflected with the only *s*-polarization at the angle equal to $\alpha_B = \arctan(n_2/n_0)$, where n_0 and n_2 are the refraction indices of the air and crystal, respectively. In optical crystals with absorption the *p*-polarized reflection at the Brewster angle α_B never becomes null in contrast to transparent crystals, where *p*-polarization disappears completely at α_B. Therefore for absorbing solids it was called a pseudo-Brewster angle [50].

Applicability of the Brewster effect has surprisingly fast grown up with introduction of lasers. Especially gaseous lasers, like He-Ne laser, have a tube containing a gain gas medium and one or two Brewster clear windows. The windows commit selection of the linear *p*-polarization in the plane of incidence. The efficiency and angular tolerance to mount such polarizing window was analyzed in [51].

11.5. Direct and Reverse Brewster Angular Effect

A regular (direct) angular effect is known to appear when the incident light reflects from a flat surface of solid / liquid. We will not touch the phase change at the reflection from metallic surfaces. The critical incidence angle in optical reflection depends on refraction index of two media, air-solid, and equals to 56° for silica and 74° for silicon. There is also a reverse effect, which occurs when the light penetrates through the boundary of two media, and one of which is an absorbing photosensitive material. The reverse effect is available to observe in silicon but unavailable in silica or conventional glasses. It is not adequately explored and described in literature. Below, the reverse Brewster effect will be considered using an anisotropic photoresponse taken of a semiconductor crystal, which can be implemented in such silicon embodiments as a photoresistor, photodiode, phototransistor, solar cell, CMOS image sensor, CCD or other semiconductor photodetector.

11.5.1. Reverse Brewster Effect in Photosensitive Crystals

In contrast to the regular Brewster effect, which is pure optical, the reverse Brewster effect is opto-electronic or photonic, i.e. associated with both processes: optical absorption and photoelectron generation in semiconductor. Anisotropic photogeneration of charge carriers occurs in photoactive bulk or film materials. Since the anisotropic photogeneration is normally impossible in isotropic solids under the normal incidence of light, in this particular case the anisotropy appears at an inclined interface of two media, air-semiconductor or glass-semiconductor. Interfaces with thin oxide layers like SiO_2-Si can produce a complementary Brewster angle in optics [50] and for the reverse Brewster effect a similar complementary angle in photonics [52]. Two or more minimums in reflection and two or more maximums in photoresponse can be observed and recorded in experiment on such oxidized silicon surfaces. A protective glass on Si and Ge photodiodes produce an additional angular shift as shown below. The reverse Brewster effect also enhances with the incidence angle, which reaches a maximum at angles close to α_B but generally not equal to the Brewster angle. This photonic effect is considered below in detail for silicon surfaces and briefly for other semiconductor materials as Ge, GaAs, ZnS, etc.

Anisotropic photoactive absorption allows handling the linear polarization with isotropic semiconductor crystals for a few applications, e.g. for polarimetric purposes or for space orientation of objects, and eventually various type photodetectors have been developed to date. The simplest example is a silicon single crystal plate tilted with respect to the direction of the light incidence at a proper angle. Whereas the case of split polarizations at the boundary of air-silicon is well-known, herein it will be outlined in terms of a direct coordinate-free tensor method. The method was developed to simplify tangled cases in transparent and absorbing anisotropic crystals with uniaxial or biaxial crystal lattice. The case of isotropic crystal is more straightforward and it will be used to demonstrate usefulness and elegance of the coordinate-free tensor approach. The corresponding calculus math and the method itself was developed by Fedor Fedorov in 1950s for electromagnetic waves and named by him a covariant representation [53, 54]. In spite of its usefulness for various anisotropic phenomena, this approach is slightly known or generally unfamiliar to English-speaking readers since the Fedorov's works were not ever translated from Russian. Exception is only one book published by the author in English translation [55]. It introduces several concepts and parameters previously unfamiliar to the literature of the West. It gives an idea of the powerful mathematical apparatus based on the coordinate-free tensor approach. The covariant methods developed for describing the polarization of radiation, as well as the interaction of radiation with solids, have a number of advantages over the common Jones vectors and matrices, Stokes parameters, Mueller calculus, etc. Herein we apply the covariant representation to introduce the reverse Brewster effect in isotropic semiconductor, which is relevant but different from the classical Brewster angle effect observed in optical reflection only.

Let us consider the air-isotropic semiconductor interface, with linearly polarized radiation (LPR) incident at an angle α_0 (Fig. 11.10). Arbitrary polarization may be presented as the sum of two orthogonal components: perpendicular and parallel with respect to the incident plane. Electric and magnetic vectors of the incident wave in covariant representation [53] are expressed as follows:

$$E_0 = E_0^{\perp} + E_0^{\parallel} = A_0 a' + B_0 [n_0 a'],$$

$$H_0 = H_0^{\perp} + H_0^{\parallel} = A_0 [m_0 \, a'] - B_0 n_0 a', \qquad (11.2)$$

where A_0, B_0 are the scalar amplitudes of the incident wave with corresponding polarization, a', b', q form the orthogonal base as in Fig. 11.10, n_0 is the normal wave vector of the incident ray, n_0, n_2 are refractive indices of media 1 and 2 (air and semiconductor), m_0, m_2 are the refractive vectors of incident and refracted rays. The isotropic semisconductor in Fig. 11.10 is shown with a gradient coloration, which implies the conductivity type gradation, as for example in p-n junction of a photodiode. The refractive index of silicon is relatively insensitive to doping in the visible region of the spectrum, and it is suggested a very small variation <0.01.

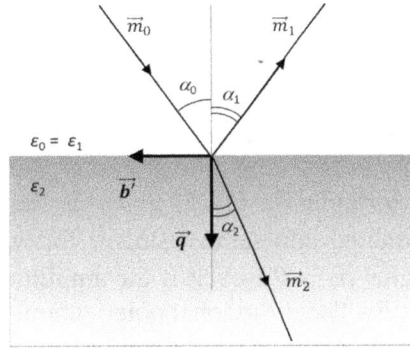

Fig. 11.10. Indicent (m_0), reflected (m_1), and refracted (m_2) rays at the air ($\varepsilon_0 = \varepsilon_1$)– isotropic semiconductor (ε_2) interface.

Let us find the radiation flow intensity for its oblique incidence onto the interface using the radiation power vector W_i and the normal boundary vector q

$$\Phi_i = \langle W_i q \rangle = \frac{c}{4\pi}\langle \mathrm{Re}E_i\rangle^2 m_i q = \frac{c}{4\pi}\langle \mathrm{Re}E_i\rangle^2 \eta_i, \qquad (11.3)$$

where $\eta_i = n_i \cos\alpha_i$; $i = 0, 1, 2$, and $\Phi_i^\perp$, $\Phi_i^\parallel$ polarized components corresponding to linear s- and p-polarizations.

Angular brackets signify the time average value as photosensitive semiconductors usually acquire the time average value of the radiation energy. In CMOS and CCD image sensors the charge accumulation occurs during the integration time.

The incident wave intensities for perpendicular and parallel polarizations of the radiation are expressed by

$$\Phi_0^\perp = \frac{c}{4\pi}A_0^2\eta_0, \quad \Phi_0^\parallel = \frac{c}{4\pi}B_0^2\eta_0 \qquad (11.4)$$

The refracted wave intensities are as follows:

$$\Phi_2^\perp = \frac{c}{4\pi}A_2^2\eta_2, \quad \Phi_2^\parallel = \frac{c}{4\pi}B_2^2\eta_2 \qquad (11.5)$$

Let us use Fresnel formulas in covariant representation for waves refracted at the boundary of isotropic media [53]

$$A_2 = \frac{2\eta_0}{\eta_0+\eta_2}A_0, \quad B_2 = \frac{2n_0n_2\eta_0}{n_2^2\eta_0+n_0^2\eta_2}B_0 \qquad (11.6)$$

These expressions are used to describe the polarizing parameters of an isotropic semiconductor generating anisotropic photoresponse as in the reverse Brewster effect: photodichroism (photopleochroism) and polarization quantum efficiency [47]. Note that the above parameters are similar to common coefficients: optical dichroism or pleochroism in uniaxial or biaxial optical crystals and quantum efficiency in a photosensitive semiconductor, respectively.

11.5.2. Angular Dependencies of the Polarization Quantities P_i and Q_p

Substitution of the Fresnel formulas in (11.5) gives us the obvious result: intensities $\Phi_2^\perp$ and $\Phi_2^\parallel$ of the waves refracted into semiconductor for two orthogonal polarizations differ significantly at all angles $\alpha_0 \neq 0°$, even if the amplitudes for various polarization azimuths are kept equal, $A_0 = B_0$. The orthogonal polarizations ($\perp$ and $\parallel$) in respect to the plane of incidence can be denoted also as *s*-polarization and *p*-polarization, respectively.

The non-equilibrium charge carrier concentration generated in isotropic semiconductor is directly proportional to LPR intensity passing through the interface. Due to the material isotropy, the absorption coefficient α, the quantum photoelectric yield β, and the electron-hole pair division coefficient γ are independent on the polarization of incident radiation and the photoresponse magnitudes will be given in the form

$$I^\perp = \alpha x \beta \gamma \frac{e}{\hbar\omega} \Phi_2^\perp, \quad I^\parallel = \alpha x \beta \gamma \frac{e}{\hbar\omega} \Phi_2^\parallel \tag{11.7}$$

Do not confuse the absorption coefficient α with angles α_0, α_1, α_2, α_B, etc. always bearing sub-indices. There are two main polarization optoelectronic quantities, the photopleochroism coefficient P_i and the polarization quantum efficiency Q_P which are similar to the optical dichroism D coefficient and the quantum efficiency Q, and were introduced especially for polarization sensitive photodevices. The photopleochroism coefficient of the photodetector with regard to (11.7) and (11.5) is

$$P_i = \frac{I^\parallel - I^\perp}{I^\parallel + I^\perp} = \frac{B_2^2 - A_2^2}{B_2^2 + A_2^2}. \tag{11.8}$$

Using (11.6) and assuming that the incident wave intensities with different polarizations are equal ($A_0 = B_0$), and taking into account for the air $\varepsilon_0 = n_1^2 = 1$, we get

$$P_i = \frac{(\varepsilon_2 - 1)(\eta_2^2 - \varepsilon_2 \eta_0^2)}{4\varepsilon_2 \eta_0 \eta_2 - (\varepsilon_2 + 1)(\eta_2^2 - \varepsilon_2 \eta_0^2)} \tag{11.9}$$

The expression for P_i in covariant representation may be written in explicit form as a function of incident angle α_0:

$$P_i = \frac{(\varepsilon_2 - 1)^2 \sin^2 \alpha_0}{(\varepsilon_2 + 1)[2\varepsilon_2 - (\varepsilon_2 + 1) \sin^2 \alpha_0] + 4\varepsilon_2 \cos \alpha_0 \sqrt{\varepsilon_2 - \sin^2 \alpha_0}} \tag{11.10}$$

Here is another important parameter for polarimetric photodetectors, the polarization quantum efficiency equal to the difference of photodetector quantum efficiencies in orthogonal polarizations of radiation

$$Q_p = \frac{I^{\parallel}}{I_0^{\parallel}} - \frac{I^{\perp}}{I_0^{\perp}}, \tag{11.11}$$

where $I_0^{\parallel}$ and $I_0^{\perp}$ are the maximum photoresponses for LPR intensities $\Phi_0{}^{\parallel}$ and $\Phi_0{}^{\perp}$, (11.5). Let us write the equation for the polarization quantum efficiency using expressions (11.7) and (11.5) and Fresnel formulas (11.6), we get the expression for Q_P in covariant representation ($\varepsilon_0 = 1$):

$$Q_p = \frac{\eta_2}{\eta_0}\left[\left(\frac{B_2}{B_0}\right)^2 - \left(\frac{A_2}{A_0}\right)^2\right] = \frac{4\eta_0\eta_2(\varepsilon_2-1)(\eta_2^2-\varepsilon_2\eta_0^2)}{(\varepsilon_2\eta_0+\eta_2)^2(\eta_0+\eta_2)^2} \tag{11.12}$$

The explicit form is as follows:

$$Q_p = \left[\frac{(\varepsilon_2-1)\sin 2\alpha_0}{\varepsilon_2-1+(\varepsilon_2+1)\cos\alpha_0\left(\cos\alpha_0+\sqrt{\varepsilon_2-\sin^2\alpha_0}\right)}\right]^2 \frac{\sqrt{\varepsilon_2-\sin^2\alpha_0}}{\cos\alpha_0} \tag{11.13}$$

It is obvious that at normal incidence the polarization sensitivity is absent, $P_i = 0$, $Q_p = 0$ according to (11.10) and (11.13). The P_i coefficient grows with increasing α_0 and can be described empirically with a simple dependence $P_i = \chi\alpha_0^2$ (within the 2 % to 3 % accuracy). The factor $\chi = 0.010$-0.012 is for materials with $n = 2.5$ to 5.0, which covers the most practically important semiconductors, see Fig. 11.11(a).

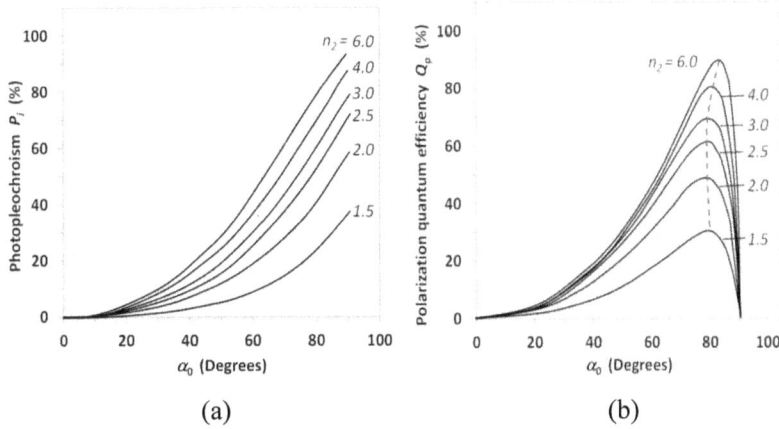

Fig. 11.11. Photopleochroism (a) and polarization quantum efficiency (b) vs. incidence angle for various refractive indices of isotropic semiconductors. Dashed line in (b) panel is the α_0 dependence of Q_p maxima.

In the limiting case $\alpha_0 \rightarrow 90°$ the photopleochroism coefficient tends to a maximal value of

$$P_i = \frac{(\varepsilon_2-1)^2}{(\varepsilon_2+1)^2} \tag{11.14}$$

283

Experiments show that the absolute value of $P_i^{max} = 55$ % to 85 % for a significant part of crystalline semiconductors (IV, III-V, II-VI, I-III-V_2, II-IV-V_2, etc.). The higher values are limited in practice by increasing the reflection at gliding angles of incidence.

Polarization quantum efficiency also monotonously grows with increasing α_0 (Fig. 11.11(b)) and peaks at maximum for all n at angles $\alpha_0 > \alpha_B$ (α_B is the Brewster angle). Though the single s-polarization component is reflected from the surface at $\alpha_0 = \alpha_B$, the refracted ray at any incident angles remains partly polarized. It prohibits the achievement of $Q_p = 100$ % (also $P_i = 100$ %) for the air–isotropic semiconductor interface for all angles including the Brewster angle. The maximum theoretical value Q_p^{max} for semiconductors (Fig. 11.11(b), the curve for $n = 6.0$) is abt. 90 % and falls down fast with $\alpha_0 > \alpha_{rB}$ (α_{rB} is the reverse Brewster angle). The dependence $Q_p^{max} = f(\alpha_0)$ is given by the dashed line. For the main semiconductor group ($n = 2.5$ to 5.0) the maximum of polarization quantum efficiency of 60–85 % peaks at incidence angles of 78° to 81°. This means that in the reverse Brewster effect the Q_p value peaks at incidence angles exceeding α_B in optical reflection, i.e. $\alpha_{rB} > \alpha_B$.

11.5.3. Refraction Index and Spectral Dependencies

Although semiconductors represent a numerous group of solids, their refraction index typically varies in relatively narrow interval from $n_2 = 2.0$ for oxides like ITO to $n_2 = 5.8$ for tellurides like PbTe. There are another materials with meta- or mesostructures like porous, micro/nanochannel ones, their refraction index is usually below $n_2 = 2.0$. We do not consider herein the negative-index materials like meta-crystals, meta-surfaces, etc., whose refractive index has a negative value over the spectral range from UV to NIR. For compositional and non-uniform substances one can use a homogenization theory, for example by approximation of an effective medium developed by Maxwell Garnett [56, 57].

Fig. 11.12 demonstrates the calculation executed for solids in the whole range of refractive indices $n_2 = 1.0$ to 6.0. Two main polarization quantities: polarization quantum efficiency Q_P and photopleochroism P_i grow abruptly at small values of $n_2 < 2.5$. At greater n_2 values the growth reduces, the Q_P and P_i dependences become sublinear and even some decline in the $Q_P(n_2)$ curve is observed. Thus, from the viewpoint of increasing the polarization quantum efficiency the choice of a semiconductor with $n_2 > 4$ is unprofitable at angles less than $\alpha_0 < 70°$. The refractive indices of Si and Ge are marked in Fig. 11.12 by strokes. These materials are very suitable and at the same time well developed to make polarimetric sensors on visible and near IR spectral ranges.

Spectral sensitivity of every individual semiconductor material in the fundamental absorption band follows their complex index of refraction $N_2 = n_2 - ik_2$ (n_2 is the real refractive index, k_2 is the extinction coefficient) and is not smooth over the wavelengths of photosensitivity. Polarization sensitivity of anisotropic semiconductors is usually limited by a narrow spectral range where the optical transitions are strong and obey the polarization selection rule with a large polarization ratio for orthogonal constituents. For example, tetragonal and hexagonal semiconductors demonstrate the photopleochroism band of abt. 0.2-0.3 eV. It is fortunate to isotropic semiconductors that their polarization sensitivity is furnished by another optical effect providing the wider spectral band that is more practical from view for their spectral usage.

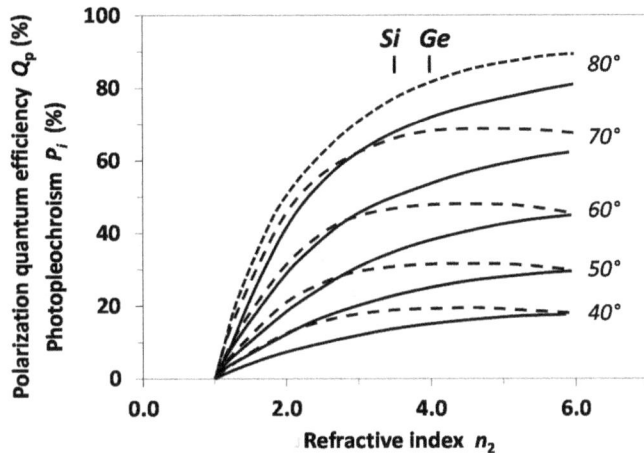

Fig. 11.12. Refractive index dependences of Q_p (dashed lines) and P_i (solid lines) for isotropic semiconductors at various incident angles of LPR.

Experimental data and calculation of Q_P and P_i spectral dependences are given in Fig. 11.13 and adopted from Ref. [47]. The polarization photosensitivity of isotropic semiconductors as a result of anisotropic refraction at the air-semiconductor interface turns out to be weakly spectrally dependent. The spectral band covers the NIR and whole visible range, 0.5 to 3.0 eV. This exceeds the competing anisotropic semiconductors in polarization sensitive photodetectors by the factor of 10-12. Fig. 11.13 presents spectra of Q_P and P_i for well-developed isotropic semiconductors. The complex refractive index $N_2 = n_2 - ik_2$, was used in the calculations. When the incident photon energy increases, the polarization quantum efficiency and the photopleochroism coefficient increase rather slowly and smoothly. This feature is associated with a weak dispersion of n_2 and essential predominance of n_2 over k_2 for semiconductors in the spectral range of their photosensitivity [48, 58]. Evidently for the Q_P and P_i enhancement, the $(\varepsilon_2 - \varepsilon_0)$ difference is a primary in the angular anisotropy. Therefore the air-Si boundary affects the polarization transformation stronger than the SiO_2-Si interface.

Fig. 11.14 shows the dependence of P_i vs. α_0 for both sweeps from negative to positive angles. Experimental points well agree with the theoretical curve (expression 11.10) in the whole range of the incidence angles. Inset shows $P_i (\hbar\omega)$ experimental data for Si p-n junction with antireflective coating (ARC). The curve looks pretty flat over the near-IR and red wavelengths with a low spectral dispersion, so silicon photodiodes are suitable for polarization-sensitive practical applications in the wide spectral range.

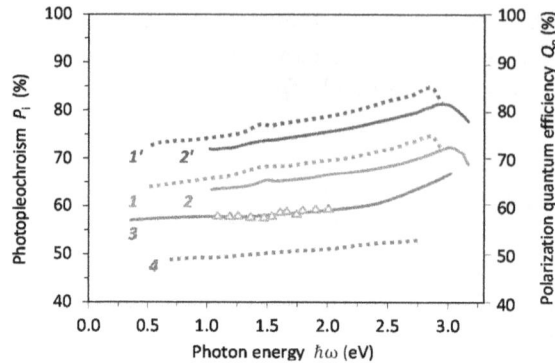

Fig. 11.13. Photon energy dependences of P_i (1 to 4) and of Q_P (1', 2') at T = 300 K. (1, 1') GaAs, α_0 = 80°, (2, 2') InP, α_0 = 80°, (3) Si, α_0 = 75°, (4) ZnS cubic, α_0 = 80°; points – experiment for P_i values measured on a silicon p-n junction at α_0 = 75°.

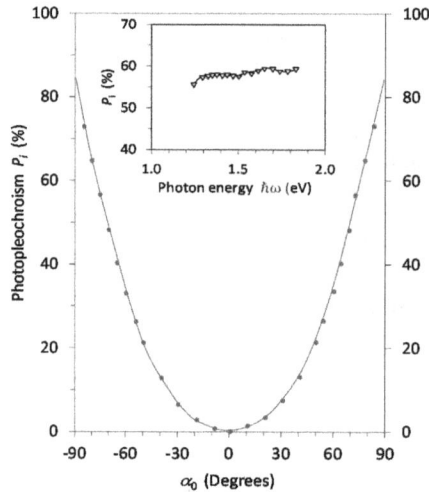

Fig. 11.14. Photopleochroism coefficient vs. incidence angle for Si p-n junction, λ = 950 nm, T = 300 K. Dots – experiment, curve – calculation by (11.10). Insertion – spectrum of photopleochroism coefficient at α_0 = 75°.

11.5.4. Generalized Malus's Law

The law discovered by E. L. Malus in 1808 states that the natural light passing through two polarizing units with their optical transmission axes turned each other by the angle of φ follows the periodical expression of $\cos^2\varphi$ for the light intensity at the exit. In real materials the optical absorption is not zero and therefore the transmission coefficient $T_{\parallel} \neq 100$ %, so the power of the passing radiation Φ expresses through the power of the incident radiation Φ_0 as follows [48]

$$\Phi(\varphi) = \Phi_0 \left(T_{\parallel}\cos^2\varphi + T_{\perp}\sin^2\varphi\right) \tag{11.15}$$

This pure optical relation is working well for transparent or partly transparent materials, however, in the field of the full opaque it cannot be confirmed by optical measurements. A few decades ago we found that expression (11.15) could be applied to fully opaque materials, for example, the heavily absorbing semiconductor crystals [47, 58]. The non-equilibrium concentration of photogenerated charge carriers in semiconductor becomes sensitive to the polarization plane of the incident radiation both in the fundamental and impurity absorption bands. Therefore the expression (11.14) has been extended to various photonic processes as photocurrent, photovoltage and photoluminescence in anisotropic semiconductors II-IV-V_2, I-III-VI_2, II-VI, II-V_2, Mn-III_2-VI_4, etc. and also in isotropic semiconductors Si, Ge, GaAs, CdTe, ZnS, etc. at oblique incidence of the polarized light.

The experiment carried out with a silicon *p-n* junction at $\alpha_0 = 75°$ [47, 52] gives a good agreement between the measured photocurrent i_{sc}, photopleochroism coefficient P_i and the calculated values, see Figs. 11.13 and 11.15. Azimuthal measurements presented in Fig. 11.15 demonstrate experimental data for the short-circuit photocurrent in two cases of oblique and normal incidence and an exact coincidence to the generalized Malus law:

$$i_{sc}(\varphi) = i_{sc}^{\parallel} \cos^2 \varphi + i_{sc}^{\perp} \sin^2 \varphi , \qquad (11.16)$$

where φ is the azimuth angle ($\varphi \equiv \sphericalangle E_0, [n_0 a']$).

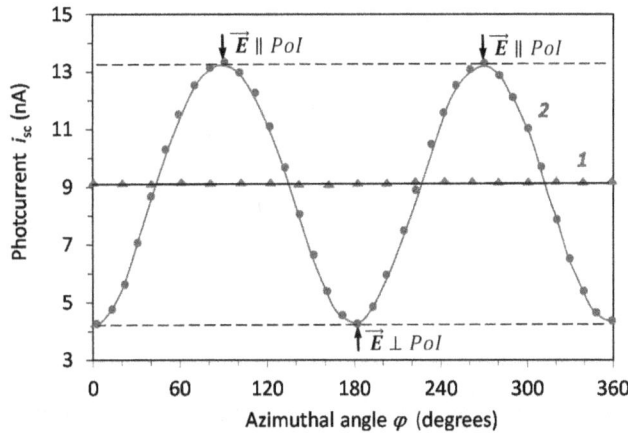

Fig. 11.15. Azimuthal dependencies of photocurrent i_{sc} for silicon *p-n* junction, $\lambda = 950$ nm.
$1 - \alpha_0 = 0°$, $2 - \alpha_0 = 75°$.

Under LPR at $\lambda = 950$ nm and with the flux density of $W_0 = 4 \times 10^{-6}$ W/cm^2, the short circuit photocurrent difference $\Delta i_{ph}^{sc} \cong 10$ nA. Azimuthal sensitivity in the vicinity of $\varphi = 45°$ equals to $di_{ph}^{sc} / d\varphi = 0.2$ nA/degree. The values Δi_{ph}^{sc} and $di_{ph}^{sc} / d\varphi$ have to increase linearly with increasing W_0 and an absolute sensitivity S of the photodetector.

11.5.5. Application of Reverse Brewster Effect

Polarization control, monitoring and imaging of surfaces illuminated with artificial and natural radiation constitutes important task for civil, space and military vehicles and unmovable bodies allowing recognition and identification of targets with enhanced contrast due to LPR. As follows from the calculations and the experimental data, the reverse Brewster effect defines the polarization photosensitivity of isotropic semiconductors at LPR oblique incidence and Q_p^{max} peaks at angles slightly exceeding $\alpha_B = \arctan(n_2/n_0)$. The efficiency of direct analysis of the radiation polarization state at a fixed wavelength is determined by the incidence angle on the receiving semiconductor surface, refractive indices of two media, and the absolute photodetector sensitivity. LPR sensitive photodetectors based on isotropic semiconductors allow avoiding expensive and sometimes complicated polarization optical arrangements to operate with linear polarization for polarimetric analysis, space orientation or transfer information data with the modulated linear polarization.

11.5.6. Photodiodes

Design of polarimetric photodetectors for detection and analysis of the linear polarization has been developing decades ago using both anisotropic and isotropic crystals [48]. Silicon and germanium crystals turned out the first among isotropic semiconductors that suggested to be used in polarimetric photodetectors over the wide spectral range from NIR to UV. Uniaxial semiconductor crystals also demonstrated a high polarization capability, however were inferior to silicon in the spectral band width, material cheapness and design diversity of photosensitive devices. The reverse Brewster effect in isotropic semiconductors allows the immediate practical use of these materials in both modifications – crystalline and amorphous [52, 58]. Semiconductor barrier structures as *p-n* junction, *p-i-n* junction, Schottky barrier, heterostructure, CMOS full-well, etc. additionally provide more efficient separation of electron-hole pairs and low-noise photoresponse. The latter employs such forms as photocurrent, photovoltage or accumulation of electrons (or holes) in CMOS full-well in various polarimetric embodiments.

One of the effective solutions to the problem of analyzing the linearly polarized radiation is the discovered ability of photodiodes made of isotropic silicon and germanium crystals to be used in a photosensitive element set at a certain fixed angle to the axis of the incident beam. With such installation of the photosensitive element, its input surface acts as an analyzer.

The amplitude coefficients of optical transmission on the air-crystal interface are described by Fresnel formulas [46] and for two orthogonal polarizations $E_{||}$ and $E_{\perp}$ are written:

$$\tau_{||}(\alpha) = \frac{2\sin\alpha_0\cos\alpha_2}{\sin(\alpha_0+\alpha_2)\cos(\alpha_0-\alpha_2)}, \quad \tau_{\perp}(\alpha) = \frac{2\sin\acute{a}_2\cos\alpha_0}{\sin(\alpha+\alpha_2)}, \quad (11.17)$$

where α_0, α_2 are the angles of incidence and refraction of the beam at the air-crystal interface; $\alpha_2 = \arcsin(\sin\alpha_0/n_2)$; n_2 is the refractive index of the crystal.

It should be noted that n_2 values in the spectral ranges of photosensitivity of Si and Ge are on average 3.5 and 4, and k_2 values are two to three orders of magnitude as lower, so can be neglected. Let us use the Jones method for the refractive surface, which acts as the optical analyzer. In this case, we make the following assumptions. A beam of rays is approximated by a uniform plane and completely polarized wave. This condition means that the light beam can be described using the Jones vector.

1. The optical system does not contain a depolarizing surface.

2. There are no non-linear optical effects in the optical system.

When LPR is obliquely incident on an isotropic crystal at an angle α_0 and its azimuth of polarization relative to the plane of incidence is φ, the Jones vector of the refracted ray has a form [52]

$$\overrightarrow{E^i} = M_\tau \overrightarrow{E} = \begin{bmatrix} \tau_\parallel(\alpha_0)e^{i\delta_\parallel} & 0 \\ 0 & \tau_\perp(\alpha_0)e^{i\delta_\perp} \end{bmatrix} \begin{bmatrix} E\cos\varphi \\ E\sin\varphi \end{bmatrix} = \begin{bmatrix} E\cos\varphi \ \tau_\parallel(\alpha_0)e^{i\delta_\parallel} \\ E\sin\varphi \ \tau_\perp(\alpha_0)e^{i\delta_\perp} \end{bmatrix}, \qquad (11.18)$$

Here M_τ is the Jones matrix of the refractive surface; $\overrightarrow{E^i}$ and E are the Jones vector of the incident beam and its amplitude, $\delta_\parallel$ and $\delta_\perp$ the phase shift at refraction for corresponding components of the vector $\overrightarrow{E}$.

The radiation power Φ_2 transmitted through the air-crystal interface is proportional to $|E|^2$ and, therefore, by the power Φ_0 of the incident radiation is written as

$$\Phi_2(\alpha_0, \varphi) = \Phi_0\left[\tau_\parallel^2(\alpha_0)\cos^2\varphi + \tau_\perp^2(\alpha_0)\sin^2\varphi\right] \qquad (11.19)$$

The expression (11.19) for the tilted surface is analogous to (11.16) with only difference that transmission τ depends on the incidence angle α_0, whereas in (11.16) no dependence on α_0 is given because the polarizer is always set normally to the incident beam.

If the crystal is a photosensitive element in a diode, then the photocurrent [58] is expressed by the relation with a spectral dependence on wavelength λ of the incident radiation:

$$i = \frac{e\eta\beta}{h\nu}\Phi_2 = \frac{e\eta\beta\lambda}{hc}\Phi_0\left[\tau_\parallel^2(\alpha_0)\cos^2\varphi + \tau_\perp^2(\alpha_0)\sin^2\varphi\right], \qquad (11.20)$$

where e is the electron charge; η is the quantum efficiency of a photodetector; γ is the separation coefficient for pairs of charge carriers; h is the Planck's constant; c is the speed of light in a vacuum.

Therefore, the sensitivity of a photodetector with a receiving surface oriented at the angle α_0 to the incident beam, should be characterized by two values – a sensitivity $S_\parallel$ to the radiation polarized in plane of the beam incident onto the crystal, and a sensitivity $S_\perp$ perpendicular to the plane of incidence.

$$S_\parallel(\lambda, \alpha_0) = \frac{e\eta\beta\lambda}{hc}\tau_\parallel^2(\alpha_0), \quad S_\perp(\lambda, \alpha_0) = \frac{e\eta\beta\lambda}{hc}\tau_\perp^2(\alpha_0) \tag{11.21}$$

The sensitivity dependence on the polarization azimuth also follows the generalized Malus law (let us emphasize that herein it concerns the reverse Brewster effect):

$$S(\alpha_0, \varphi) = S_\parallel(\alpha_0)\cos^2\varphi + S_\perp(\alpha_0)\sin^2\varphi \tag{11.22}$$

Consequently, in optoelectronic measuring systems, where the information parameter is the polarization state of radiation, the signal magnitude is proportional to the effective polarization sensitivity of the photodetector.

$$\Delta S(\lambda, \alpha_0) = S_\parallel(\lambda, \alpha_0) - S_\perp(\lambda, \alpha_0) = \frac{e\eta\beta\lambda}{hc}\left[\tau_\parallel^2(\alpha_0) - \tau_\perp^2(\alpha_0)\right] \tag{11.23}$$

To ensure a high degree of accuracy of polarization measurements, it is required that the photosignal and, therefore, ΔS be sufficiently large. It follows from (11.21) and (11.23) that the dependences $S_\parallel(\alpha_0)$, $S_\perp(\alpha_0)$ and $\Delta S(\alpha_0)$ are completely determined by the dependencies of $\tau_\parallel^2$, $\tau_\perp^2$ and their difference. For isotropic crystals with $n_2 \gg k_2$ the maximum values are $[\tau_\parallel^2(\alpha_0)]_{max} \approx 1$, and $[S_\parallel(\alpha_0)]_{max} \approx e\eta\beta\lambda/hc$.

According to (11.16), (11.21) and (11.23), the sensitivity dependences on the incidence angle α_0 when they are normalized to $[S_\parallel(\alpha_0)]_{max}$ are the followings

$$s_\parallel(\alpha_0) \approx \tau_\parallel^2(\alpha_0) = \left[\frac{2\sin\alpha_2\cos\alpha_0}{\sin(\alpha_0+\alpha_2)\cos(\alpha_0-\alpha_2)}\right]^2, \tag{11.24-1}$$

$$s_\perp(\alpha_0) \approx \tau_\perp^2(\alpha_0) = \left[\frac{2\sin\alpha_2\cos\alpha_0}{\sin(\alpha_0+\alpha_2)}\right]^2, \tag{11.24-2}$$

$$\Delta s(\alpha_0) = \tau_\parallel^2(\alpha_0) - \tau_\perp^2(\alpha_0) \tag{11.24-3}$$

Fig. 11.16 presents the linear polarization sensitivity vs. the incidence angle for Si and Ge crystals. The corresponding Brewster angles are pointed out by blue arrows on top. Two reciprocal polarizations, parallel and perpendicular relatively to the plane of incidence, are shown in red (p-polarization, also denotes $E_\parallel$) and green (s-polarization, also denotes $E_\perp$). The normalized polarization sensitivity, a practically useful parameter, is shown by violet curves. Solid lines according to formulas (11.24-1), (11.24-2), and (11.24-3) depict angular dependencies for naked photodetectors made of Si and Ge, and dashed lines – calculation for the same devices with a protecting glass window ($n = 1.5$). Dots are experiment for Si photodetector with a shallow p-n junction (square 10×20 mm²). The coincidence of the calculation results with measurements indicates that, first, the dependences $s_\parallel$, $s_\perp$ and Δs are almost completely determined by the refraction phenomenon at the air-crystal interface, and second, one can neglect the k_2 values in the approximate calculation of the transmission at this air-crystal interface. Note herein the extinction coefficient has to be taken into accounting at shorter UV wavelengths $\lambda < 380$ nm where n_2 and k_2 become one order of magnitude.

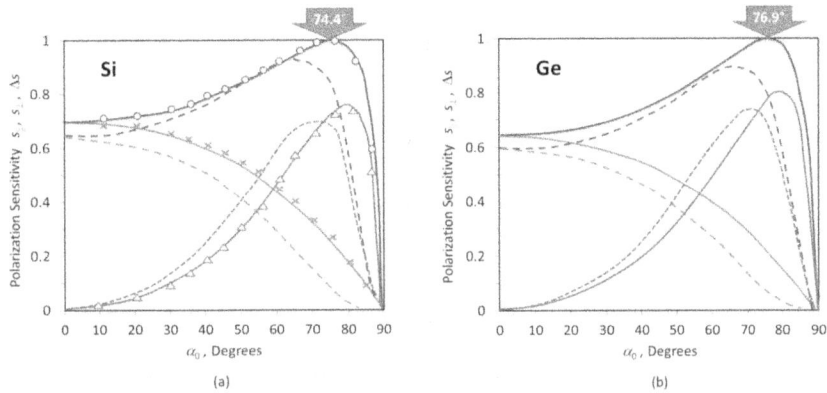

Fig. 11.16. Normalized polarization sensitivity of Si (a) and Ge (b) crystals vs. incidence angle α_0. $s_{||}$ – red, $s_\perp$ – green, and Δs – violet curves: solid and dashed lines are calculation without and with a protection glass ($n = 1.5$), respectively. Curves – calculation by (11.24-1), (11.24-2), and (11.24-3); dots – experiment; arrows on top – the Brewster angles for Si ($\lambda = 950$ nm) and for Ge ($\lambda = 1500$ nm).

The maximum values of the normalized effective polarization sensitivity according to (11.24-3) for silicon photodiodes without ARC equal to $[\Delta s(\alpha_0)]_{max} = 0.77$ at $\alpha_0 = 78°$, and for germanium $[\Delta s(\alpha_0)]_{max} = 0.81$ at $\alpha_0 = 79°$. These digits are greater than that of photodiodes those receiving surfaces oriented normally to the beam, which are 0.69 for Si and 0.64 for Ge. The maximum optical transmission of the polaroid films used in these polarized measurements is of 0.6 to 0.8. So, the efficiency of direct analysis of LPR using Si or Ge based photodetectors those receiving surface is inclined to the axis of the incident beam, at least is not less than that efficiency with use of a classical installation scheme comprising a polarizer and the same photodetector with the normally oriented surface even if their transmission coefficient is enhanced to unity by ARC.

Commercially available photodetectors are of interest for the direct analysis of linear polarization. However, polarizing properties of their surface are affected by ARC coated on the crystal surface and a protective glass (as with or without ARC). For example, modern CMOS image sensors with the 1 to 10 μm pixel pitch employ the double-side ARC coated glass. In Fig. 11.16, dashed lines show the dependences $s_{||}$, $s_\perp$ and Δs for Si and Ge photodetectors which calculated taking into account the effect of clear protective glass made of a plane-parallel plate, $n = 1.5$, and installed in parallel to the receiving crystal surface. With input of the protective glass, maximum of the normalized effective polarization sensitivity $[\Delta s(\alpha_0)]_{max}$ decreases from 0.77 → 0.72 for silicon and from 0.81 → 0.76 for germanium photodetectors. Typical losses in the glass plate oriented by normal to the beam are larger and reach 0.08 on both surfaces. Nevertheless, installing the protective glass parallel to the receiving crystal surface has a number of advantages.

291

First, the maximum of $\Delta s(\alpha_0)$ becomes flatter that reduces the dependence of the signal magnitude on the angle of incidence. Second, the maximum of $\Delta s(\alpha_0)$ is reached at smaller inclination angles of the receiving surface; It makes possible to somewhat reduce its area while keeping the photodetector effective area bigger, i.e. the projection of the receiving surface onto the plane perpendicular to the axis of the incident beam.

Fig. 11.17 shows experimental results for the $s_{||}$, $s_{\perp}$ and Δs dependences vs. incident angles for a series of commercial photodetectors. Variations in antireflective coatings significantly affect the shape of these curves. The polarization dependencies slightly differ from sample to sample, but more for diodes from various batches although their general trend is keeping similar. Sensitivities $s_{||}$ and $s_{\perp}$ are equal each other at $\alpha_0 = 0$ and correspond to datasheets and specs. The maximum values of the normalized effective polarization sensitivity of the studied photodiodes are gathered in Table 11.1. Polarization sensitivities have two maximums more pronounced for FD-11K and FD-7G in Fig. 11.17(a) and almost undistinguished and flat for $s_{||}$ of FD-24K photodiode with a perfect ARC of a deep blue coloration as in Fig. 11.17(b). The second maximum on $s_{||}(\alpha_0)$ and $s_{\perp}(\alpha_0)$ curves are associated with the polarizing effect by the protective glass which have perceptible difference for various models.

When removing the protective glass, the value of $[\Delta s(\alpha_0)]_{max}$ decreases. Measurements taken on various photodiodes equipped with protective glasses show that the effective polarization sensitivity Δs is the greatest for the photodiode FD-11K. However, the receiving surface in this model is considerably deepened into the cage, which is vignetting the light beam at large angles of incidence. Photodiodes of other models (FD-24K, FD-9K, FD-7G) have the smaller value $[\Delta s(\alpha_0)]_{max}$, and a more uneven surface of the receiving crystal, as well as a significant unevenness of sensitivity over the area. This is a significant disadvantage since the receiving surface plays a role of the polarization analyzer. The latter photodiodes are relatively cheap and can be used for polarimetric purposes of a rather moderate accuracy.

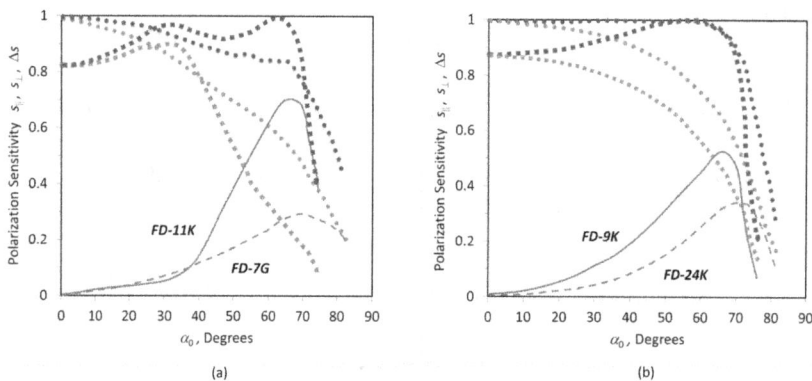

Fig. 11.17. Experimental dependencies $s_{||}$ (red curves), $s_{\perp}$ (green curves), and Δs (violet curves) vs. incidence angle α_0. Photodiode models: FD-11K, $\lambda = 950$ nm and FD-7G, $\lambda = 1500$ nm (a); FD-9K, $\lambda = 950$ nm and FD-24K, $\lambda = 950$ nm (b).

Table 11.1. Parameters of polarization sensitivity of silicon and germanium photodiodes.

Photodiode model	Material	Inclination angle, α_0	$[\Delta s(\alpha_0)]_{max}$
FD-11K	Si	67°	0.77
FD-9K	Si	67°	0.51
FD-24K	Si	72°	0.35
FD-7G	Ge	70°	0.3

Azimuthal dependences $s(\varphi)$ for all studied photodiodes follow the generalized Malus law. The curves for photodiodes with ARC / protective glass or without them are rather well described by expression (11.22).

11.5.7. Comparison of Brewster Angles

A single Brewster angle is observed in the angular optical reflection from a clear surface of solids. However various applications call for surface covered with optical coatings for purposes of antireflection, pass-band, block-band, passivation, protection, etc. Such cases cause the appearance of a few Brewster angles which can be observed at angles always lower than α_B. For the system air-SiO_2-Si multiple Brewster angles are observed in both direct and reverse Brewster effect [50-52].

Fig. 11.18 gives a comparison of two Brewster angles for the direct optical effect recorded by using the reflection R_p (α_0) and for the reverse Brewster effect – the polarization sensitivity $s_{\parallel}(\alpha_0)$. The p-polarizations in both cases produce the second Brewster angle $\alpha_{B2} < \alpha_{B1} \equiv \alpha_B$. The data for the air-$SiO_2$-Si system was obtained under illumination by HeNe laser ($\lambda = 632.8$ nm) of the thermally oxidized Si wafer with SiO_2 film thickness $d = 892.3$ nm. Photoresponse from the photodiode FD-11K with a protection glass was recorded at $\lambda = 950$ nm. The first and second Brewster angles $\alpha_B = 77.42°$, $\alpha_{B2} = 47.58°$ correspond to the first and the second reverse Brewster angles $\alpha_{rB} = 61.2°$, $\alpha_{rB2} = 31.2°$. It can be noted, the gap between these two angles measures by ~30° for both optical reflection and photoresponse. This points out at similar interference impact by SiO_2 film in both cases.

Other higher order Brewster angles are also possible to observe in systems with different ARCs at angles smaller than α_B on both R_p and $s_{\parallel}$ curves. Their appearance is due to the optical interference in SiO_2 thin film on Si crystal, which overlaps with the direct or reverse Brewster effect. Obviously multiple Brewster angles can be observed on the oxidized silicon surfaces with selected thicknesses of thin oxide films and they become more pronounced in the area where the Brewster split is relatively weak, i.e. at $\alpha_0 < \alpha_B$.

However, appearance of multiple Brewster minima and maxima has occasionally a negative effect on applications and the smoothed flat curves are often to be more practical. Fig. 11.19 demonstrates another uncommon feature of R_p (α_0) and $s_{\parallel}(\alpha_0)$ curves as their extended flatness. The flat polarization curve is attractive in practice because the Brewster polarizer and polarimetric photodiode become independent on α_0 within the wider angular sweep as compared to the bare surface of glass or silicon. The most of the used light beams are paraxial or diverging. The p-polarization capability of two devices under consideration remains at the same 0°-level for $s_{\|}(\alpha_0)$ up to $\alpha_0 = 62°$ and for $R_p(\alpha_0)$ up to $\alpha_0 = 72°$. The data for the air-SiO_2-Si system were measured at $\lambda = 632.8$ nm and SiO_2 film had a thickness $d = 892.3$ nm. For the reverse Brewster effect the data were obtained on photodiode model FD-24K with a deep-blue ARC and protection glass, and at the wavelength $\lambda = 950$ nm [52].

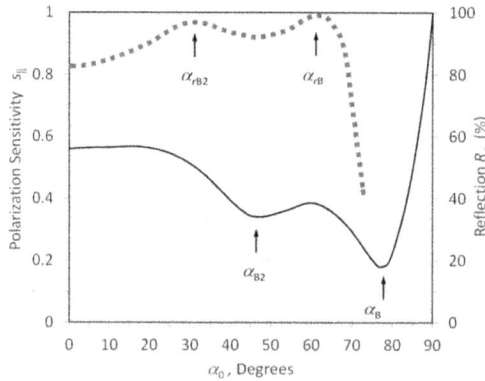

Fig. 11.18. The 1st and 2nd Brewster angles for optical reflection R_p from Si crystal [51] and the 1st and 2nd reverse Brewster angles for polarization sensitivity $s_{\|}$ of Si photodiode [52].

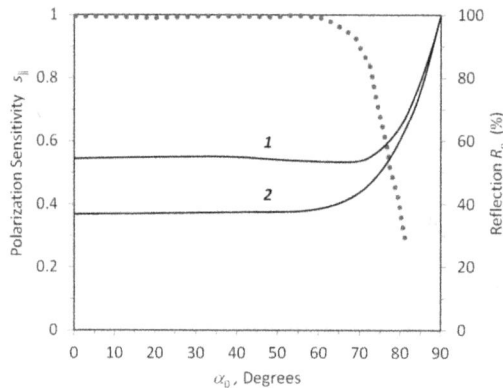

Fig. 11.19. Flat angular dependences for optical reflection R_p from Si crystal [50] and for polarization sensitivity $s_{\|}$ of Si photodiode [52].

The considered optical and photoresponse angular features occur in every pixel in Si-MCP simultaneously on four {111} facets (Fig. 11.3) if the light incident on the {100} principal basal flat. Photodiodes can be created on these facets and be independent each other. Formation of the separated diode structures on top inverted pyramids allows commitment of polarization pixilation in microchannel silicon and development of the advanced polarimetric image sensors and 3D space orientation tools.

11.6. Diffraction and Conoscopy

11.6.1. Fraunhofer Diffraction

Microchannel silicon has 2D gratings on both front and back sides, and so represents a singular plate with a double-diffraction performance. Fraunhofer diffraction can be observed in the Z direction in the far field ($Z \gg d_p^2 / \lambda$) for the reflected and transmitted lights. The transmission mode is more complicated since it is affected by both sides and walls inside, though the output diffraction prevails. For thick enough plates ($W \gg \lambda$), one can neglect the front-side effect. A schematic for a simple 2D model is illustrated in Fig. 11.20, where optical rays pass through Si-MCP and diffract at the outlet. The image appears at the focal plane F of the lens as shifted color patterns, while the image at the rear plane deconvolves into a "white" pattern. The polychromatic image in the far field looks like that obtained if one looks with the naked eye through a Si-MCP plate which is backlit with color texts, as shown in Fig. 11.9. The distance X can be changed to obtain 2D diffraction patterns in any plane at a distance from F to Y. Using the microchannel grating model in Fig. 11.20 and the normal incidence of the input light, we arrive at the fundamental diffraction equation [46]

$$\sin \theta = \frac{m \times \lambda}{d_p}, \tag{11.25}$$

where θ is the diffraction angle, m is the order number of diffraction, λ is the light wavelength, and d_p is the microchannel pixel pitch (with openings typically of 90 % pitch).

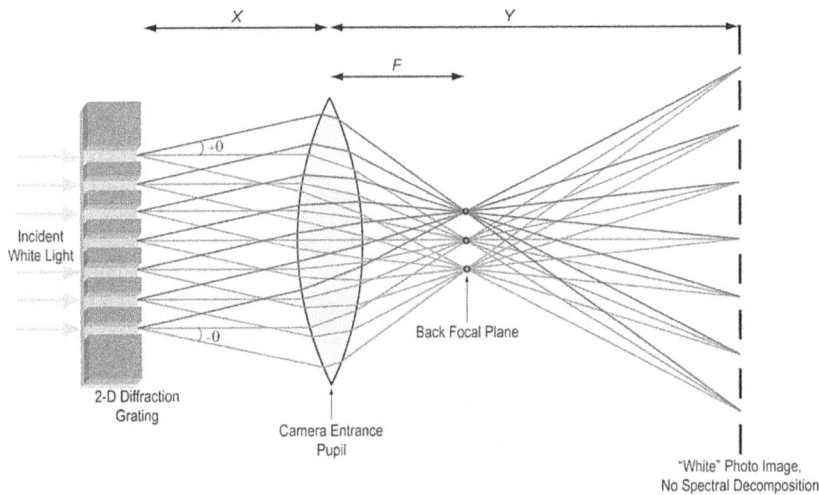

Fig. 11.20. Optical ray-tracing for a silicon microchannel grating model. Diffraction angles $\pm\theta$ vary for different wavelengths and diffraction orders. Red, green, and blue rays converge at the F and Y focus distances for monochromatic and polychromatic images.

The color distribution at the back focal plane F has an amazing in-plane appearance in the form of monochromatic or polychromatic crosses (see Fig. 11.21). Let us briefly estimate some conditions for the Fraunhofer diffraction using Si-MCP with a pitch $d_p = 40$ μm and incident red / green or white rays in a paraxial mode. An exemplar measurement of the ratio X_1 / l, where X_1 is the distance between the diffraction spots on screen and l is the distance from MCP to screen, gives the following values for the red laser ($\lambda = 650$ nm): $X_1 / l = 0.0162 \pm 0.0005$, $\theta = \arctan(X_1 / l) = 0.928° \pm 0.003°$.

The calculation yields $\lambda / d_p = 0.01625$, $\theta = \arcsin(\lambda / d_p) = 0.931°$, which is in good agreement with the experiment. The amplitude of the output light signal at one rectangular aperture is described by the Fraunhofer diffraction with a function $y = (\sin x / x)^2$. The intensity $I(x, y)$ at the in-plane aperture (width $2a$ and height $2b$) alternates with strict maximums and minimums, and obeys the expression [46]

$$I = 16C^2 a^2 b^2 \left[\frac{\sin(\theta_x ka)}{\theta_x ka}\right]^2 \left[\frac{\sin(\theta_y kb)}{\theta_y kb}\right]^2, \tag{11.26}$$

where C is a constant, k is the wave number, and θ_x and θ_y are the angles from the source.

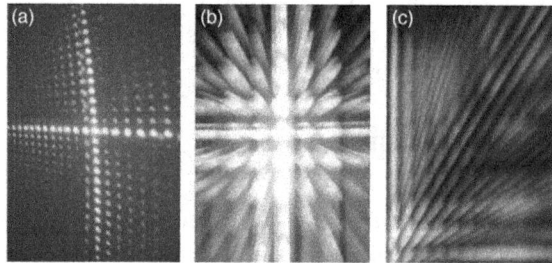

Fig. 11.21. 2D optical diffraction patterns in microchannel silicon. (a) Reflection under illumination with green and red lasers with the convergence angle 3°. (b) Transmission under illumination with white light by sunshine. (c) "Rainbow" 2D dispersion from one quadrant of the Si-MCP plate having the [100] channels, 40 μm pitch, and 4 μm walls in transmission mode.

For a centrally symmetrical source and a square aperture, the intensity is $I = 16C^2 a^4 (\sin \theta ka / \theta ka)^4$. Under simultaneous illumination with green (532 nm) and red (650 nm) lasers, Si-MCP produces two monochromatic cross-shaped patterns in reflection, as shown in Fig. 11.21(a). Paraxial beams of two 5 mW lasers were set with the angle mismatch of ~3°. The phase mismatch due to the wavelength difference $\Delta\lambda = 118$ nm creates a spatial shift for the red and green spots in the screen plane. All the colors in Fig. 11.21(a) look like natural laser ones, except for white spots in the central area of the diffraction crosses. Due to the narrow FOV and high transmission efficiency of Si-MCP, the centered radiation makes the CMOS image sensor overexposed, and the central area bears marks of false colors. In contrast to the previous image, the full polychrome images in Figs. 11.21(b) and (c) were obtained in the transmission mode under natural white light from the California sun on a typical summer day. These iridescent diffraction patterns are shaped as rainbow crosses due to 2D spectral dispersion, with many overlapping spectral orders (more than 10) in the visible domain. Fig. 11.21(b) demonstrates false colors in the center as in the case of monochromatic red / green beams. Fig. 11.21(c) demonstrates the chromatic 2D decomposition with more details from one quadrant of the rainbow cross-shaped pattern. The shown "rainbow" patterns are the unique effect of two-dimensional multi-order spectral dispersion available only in spacial diffraction lattices or 2D gratings.

Similar 2D diffraction effects were observed in photonic crystal fibers with hollow waveguides made of glass with a hexagonal symmetry [59-63]. To date, however, only small sizes were demonstrated for silicon photonic and mesoscopic crystals with a tetragonal symmetry [64-67]. In the present work, large-sized 4 and 6 inch diameter Si-MCP wafers of high optical efficiency for transmission and reflection have been fabricated and demonstrated in monochromatic and polychromatic images for the first time to our knowledge.

Patterns shaped by crosses can be observed in various directions including axial and off-axis, as well as for divergent beams, although in the latter case the cross-shaped symmetry in transmission becomes distorted. This indicates that the 2D diffraction pattern behind the plate is generated mainly by 2D output grating. It is well known that volume phase gratings have to satisfy the Raman-Nath diffraction regimes [68, 69] to produce multiple diffraction orders. Si-MCP with a short interaction length generates multiple diffraction orders, including numerous upshift and downshift ones (as shown in Fig. 11.21), and the cell operates in Raman-Nath mode [69]. Si-MCP with two diffractive surfaces (front and back) represents 3D diffraction plate, where the square-cluster X, Y array serves as a lattice basis and the long microchannels stretched in the Z direction serve as an optical axis. Thus, the diffraction patterns that occur are generally more complicated due to diffraction on both sides and can be regarded as independent in reflection and transmission modes. For both modes, the maximum optical intensity is concentrated at the central output of the plate and is attenuated oscillating outward, which indicates 3D shaping of the light. The calculated ratio of the light intensities in the zeroth order central spot and in a series of the first, second, and third distant spots differ dramatically as 100:4.7:1.7:0.8. This means the intensity of the central peak is about 13 times as high as all peripheral peaks taken together. Such a great ratio can be beneficial to shaping the light in spatial filters, making the image contrast sharper.

As shown previously, microchannel silicon is a uniaxial material with a mesoscopic crystal lattice. Compared with Si-porous material grown on the same starting Si wafer, with equal number of pores, pore size, and aspect ratio, microchannel silicon differs by a long-range order. These two commensurate materials can be attributed to the ordered and disordered mesoscopic solids by analogy with solid-state materials. Optical effects such as diffraction, interference, anisotropy, and conoscopy appear in Si-MCP but are impossible in porous Si.

The observed 2D Fraunhofer diffraction in Si-MCP can be explained in terms of classical optics formalism with 2D peculiarities. The diffraction (and interference) pattern from multiple identical apertures is a product of diffraction (interference) patterns generated by identical single apertures located in regular positions of the array. Diffraction images from a single square aperture and from 2D arrays of identical apertures are described by the same $sinc(x)$ function. The irradiance increases as N_{ch} (the number of repetitions in 2D array) on an equal square. In case of Si-MCP, the irradiance is $(N_{ch})^2$ times the irradiance from a single aperture. Randomly distributed apertures give only a central-symmetric image like a halo in porous Si. This pattern is basically the same as the pattern from a single-round aperture in the Fresnel zone [70]. When the number of repetitions (small apertures) becomes very large, spottiness is not seen. The front surface (100) in microchannel silicon is notably different from a regular silicon surface (100) due to the pronounced 2D diffraction, while the side surfaces (010) and (001) appear to be more similar.

11.6.2. Conoscopy

Light reflected from the face of Si-MCP is to be partially polarized at every microchannel mouth due to multiple reflections of conoscopic rays from the vertical walls. Every microchannel has four long walls reflecting the light in diverse directions. Integrally, the light turns out unpolarized in the far field. In reflection geometry, microscopic images can be watched through an optical microscope in the form of conoscopic figures. Typical 2D conoscopic figures are presented in Fig. 11.22. As the focus plane moves deep into the microchannels, the optical figures gradually change from a rectangular grid to a cross-wheel throughout the whole 2D array. The arrow shows embedding of the focus planes into microchannels. The reflection conoscopic figures look the same as in uniaxial optically transparent crystals inspected by conoscopic microscopy [71-73], but no polarizers are used for Si-MCP in the reflection mode.

The main difference is in multiple repetitions of regular conoscopic figures, as shown in Fig. 11.22 for 2D images with various foci #270 to #275. The presented conoscopic arrays reveal in depth the optical equivalence of microchannel pixels over the wafer surface. The observed transformation in depth also manifests a transition from the Fresnel to Fraunhofer (far-field) diffraction at square slits [46, 70]. These amazing conoscopic figures are seen in the isotropic direction [100] due to acquired polarization and interference during multiple reflections inside the microchannels. Fig. 11.23 shows the ray-tracing of reflected rays in deep microchannels to explain the appearance of conoscopic figures in reflection. Their shapes vary when changing the foci from F_1 to F_4, as shown by the red lines.

The micropits were etched at the preprocessing stage as inverted micro-pyramids; they are not deep and therefore cannot form conoscopic figures. Images of the inverted micro-pyramids are seen in the upper inset of Fig. 11.9. For deep microchannels, one can observe in depth a smooth transfiguration of one conoscopic figure to another with changing foci. Conoscopic shaping of the light in Si-MCP is strongly contrasted with a regular uniaxial optical crystal because the equal conoscopic figures appear in all microchannel pixels simultaneously. As a single "Maltese cross" figure is the characteristic of uniaxial crystals, numerous "Maltese cross" figures also deliver evidence of the uniaxiality of microchannel silicon. The mesoscopic optical axis is oriented along the [100] direction, as clearly determined by centering the "Maltese cross" in every pixel conoscopic figure. Further study of the similarity of 2D conoscopic figures in various cross-sections of Si-MCP can give valuable information on uniformity in depth and in the plane of the full wafer.

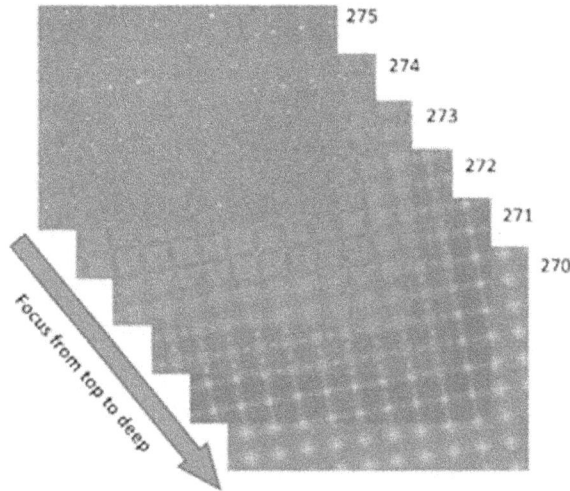

Fig. 11.22. 2D arrays of conoscopic images of Si-MCP. Six snapshots overlap and are offset relative to each other to show the optical MCP face transformation when the optical focus is embedded into microchannels 300 μm in length. The arrow denotes a transition from the near field to the far field inside the microchannels.

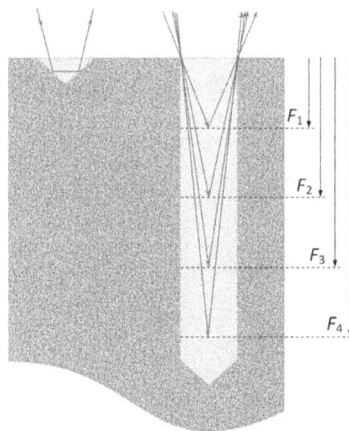

Fig. 11.23. Reflection ray-tracing inside micropits and deep microchannels; $F_1...F_4$ are foci for different conoscopic figures.

11.6.3. Surface and Weight

Mesoscopic systems are halfway from the classic to the quantum world due to dimensions, and the extended boundaries between the crystal and vacuum / air matters. Microchannel silicon has a total surface of flat walls with the Miller indices (010), (0-10), (001), and (00-1), significantly exceeding the surface of conventional (100) Si wafers widely used in industry for the mass production of semiconductor chips. All these planes are equal to each other by crystallography and electronic properties, and are able to accommodate semiconductor device structures up to the ultra large scale integrated circuit (ULSIC). Comparing the microchannel and conventional materials, one can see that the former is superior to the latter in terms of the total area available for the devices' allocation. Si-MCP offers a significantly increased area as the pitch becomes smaller (see details in Table 11.1). The wall typically has a thickness ~10 % of the pitch value and, for practical pitches, the single-crystal wall of 0.5 to 4 μm is accurately reproduced inward on the Si wafer to the full microchannel depth. To estimate, we use the ratio MCP / mono, which comes to 54 for 4 inch wafers (cf. Si-MCP full area and Si wafer one-side area); see Table 11.2. The smaller the pitch, the higher the aspect ratio, and this makes the MCP / mono ratio greater. For a pitch of 8 μm and wall of 1 μm, the ratio exceeds 2 orders of magnitude. This, along with excellent payload characteristics and area-to-weight ratio, makes microchannel silicon an outstanding candidate for space and flight applications as well as lightweight devices for new-generation silicon photonics.

Microchannel silicon proffers advancement to 3D architecture in the areas of 2D area and weight. For decades, conventional silicon has been used to allocate micro- and nanoscale devices on the front and back sides of (100) Si wafers. The modern concept of 3D die stacking employs a multilayer sandwich structure, with the vias between thin silicon layers or silicon-on-insulator. This technology produces monolithic 3D integrated circuits of a very high packing density, but still remains within the uniplanar architecture on the (100) and (-100) planes. Alternative vertical wet / dry micromachining allows the creation of nano- and micro-devices on vertical walls parallel to each other with remarkably increased total areas. This 3D concept can be attractive in the near future. Nowadays, it is important to note that a highly precise control of the wall flatness remains an important task for microchannel silicon wafers to compete with traditional Si wafers.

Table 11.2. Dimensional parameters and number of microchannels in 4 inch Si-MCP wafer.

Thickness, μm	Pitch, μm	N_{ch}	Ratio*	Total Area, mm²
300	40	4.9×10^6	27	2.1×10^5
300	20	2.0×10^7	54	4.2×10^5
300	8	1.2×10^8	131	1.0×10^6

*Ratio is MCP / mono, i.e., a total area of microchannel walls in the Si-MCP wafer relative to the one-side top area of a conventional Si wafer.

Another aspect, a very light weight, is already available today to fabricate the advanced material. The mesoscopic structure, ~4/5 air and ~1/5 thin-walled crystal, leads to weight reduction by a factor of 5.0-5.2 as compared to bulk silicon. The lattice rib thickness can be adjusted to be further reduced to tens of nanometers while retaining mechanical strength and rigidity at a level acceptable for industrial processing – in particular, to meet fab design rules, photolithographic accuracy, and precise photo electrochemical etching.

The lightweight, single-crystal-rib framework provides a mechanical rigidity and hardness that produces a high optical transparency, unique to silicon materials, that is attainable in Si-MCP but unattainable in other well-known forms of silicon. The fabricated Si-MCP wafers of up to the 320 μm thickness possess a moderate rigidity sufficient to reliably manipulate the wafers in a cleanroom environment for a variety of semiconductor processes and tests. Microchannel silicon is regarded as an exotic form of silicon due to these outstanding opto-mechanical properties.

11.7. Photonic Properties

11.7.1. Photoelectron Generation

Shaping the light between different spectral ranges, e.g., the light conversion from UV into visible (VIS), is enabled using Si-MCP in two steps. This technique consists of a UV photon-to-electron conversion followed by a reverse electron-to-photon conversion to the visible portion of light. This paragraph briefly describes a photocathode layer grown on Si-MCP as a means of achieving the optoelectronic light conversion. The method was employed in MCP intensifiers [21, 74] for imaging in various spectral ranges invisible to silicon photodetectors [26-28, 75-77] and optical visualization of secondary electron clouds [22, 25].

In the long term, it was tempting to amplify faint light fields in images using an MCP intensifier. The photocathode at the MCP entrance generates photoelectrons under incident UV light, and then the electrons are accelerated by a strong electric field ($V = 1$ to 1.5 kV) in the vacuumed microchannels. Dozens of electron collisions at walls result in a secondary electron avalanche generation, i.e., the electron amplification. Silicon MCP was developed to a proper generation of dense electron flows in microchannels with aspect ratio of 10 to 30 and pitch of 8 to 40 μm [26-28]. The total electron avalanche at the outlet of Si-MCP can reach 10^5-10^6 e- per one photo-electron, comparable with the best glass-fiber MCPs [74, 75]. The multiplication factor follows the expression

$$M = \frac{1}{1-|V/V_{BR}|^n},\qquad (11.27)$$

where V and V_{BR} are the voltage applied to the MCP structure and the break voltage, respectively.

To generate UV photoelectrons, the photocathode layer was directly grown on the front surface of the Si-MCP [24-28]. The photocathode architecture is comprised of multiple layers, with a top layer made of GaN, AlGaN, or GaInN binary semiconductors [78, 79]. A matching sublayer is also necessary to meet the chemical and crystallographic requirements for epitaxy on profiled silicon surfaces. AlN/AlGaN/GaN/AlGaN heterolayers were grown on {111} facets of Si-MCP using thin AlN/AlGaN buffers with a gradient composition and thickness of $d_{buf} = 10$-20 nm. The buffer layer inserted between the silicon and the Mg-doped AlGaN photocathode layer serves as a crystallographic matcher for crystal lattices of two semiconductors. Additionally, it serves as a wide bandgap barrier to prevent electronic back-diffusion into the MCP substrate interfacial region, where the defect densities are expected to be higher and the nonradiative recombination of photoexcited electrons larger. Since the microchannel front mouth possesses several facets with general Miller indices {111} and {100}, the growth conditions are different for the same growing compound material. This task has been recently solved and superior layers were grown by a hydride vapor phase epitaxy (HVPE) process [79]. In case of polycrystalline growth, the conditions are smoother and good results are achieved.

Along with square-shaped pixel arrays, the micromachining technology enables us to modify the profile to a star-shaped one, with the microchannel orientation perpendicular or inclined up to 15° to the (100) plane [24-27]. Fig. 11.24 demonstrates high-resolution images of Si-MCP with star-shaped microchannels coated with GaN/AlN poly-layers on top. The images were taken in SEM micro-cathodoluminescence (CL) mode. Uniform growth of the photocathode layers was achieved on differently oriented facets, as seen in Fig. 11.24(a). The right-hand panels in Fig. 11.24(b) demonstrate images with a magnification from bottom-right to upper-left. The CL read-out mode promotes the discernment of nanocrystalline growth on the MCP facets and in deep microchannels [23]. The nano-dimensional growth is properly seen in the upper-left quadrant in Fig. 11.24(b). The pulled-out nanowires measure 30-80 nm wide and 1.5-2.0 μm long. There also exist the quantum confined effects in needle-shaped AlGaN nanocrystals grown inside Si-MCP hollows in a precisely regular mesoscopic lattice. Proper doping of AlGaN poly-layers makes it possible to achieve the negative electron affinity for efficient UV photogeneration using both the continuous photocathode and nanowire layers. Then the electron micro-avalanches at the output are converted into secondary light fields from every MCP pixel, thereby making the light signal conversion from the UV / EUV to VIS spectral ranges via this two-step optoelectronic process.

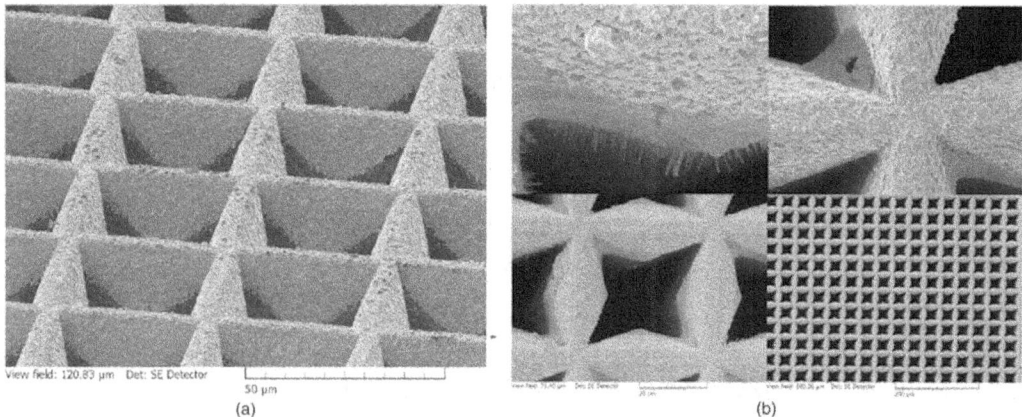

(a)   (b)

Fig. 11.24. Micro cathodoluminescent images of Si-MCP photocathode structures. (a) Top {111} facets are uniformly grown over with GaN/AlN layers; (b) Star-shaped microchannel apertures with various space resolutions are depicted in four quadrants. Magnification increases from right to left and down to up. Nano-crystalline columns are in the upper left quadrant.

11.8. 3D Model Simulation

2D and 3D modeling of the light-field distribution was conducted using Matlab software. The model simplifies the mesoscopic lattice down to a flat grid, with the pitch exactly corresponding to that of microchannel silicon. Herein we point out the experimental result whereby the cross-like shape in intensity distribution can be successfully modeled with a simplified 2D diffraction grid. The model image is recorded at the exit of the mesoscopic lattice along the direction [100]. The following Matlab algorithm was used for 3D modeling:

% Create a grid of x and y points;

>>[X, Y] = meshgrid(−10:.1: 10);

% Define the function $R = f(x, y)$;

>> R = sqrt($X.^2 + Y.^2$)+ eps;

% Create the surface plot using the mesh command;

>> mesh(abs($\sin(R)./R)^2$).

Since the intensity of light is essentially positive, we used its absolute value. The mesh(R) draws a wireframe mesh using $X = 1{:}n$ and $Y = 1{:}m$, where $[m, n] = $ size(R). The height, R, is a single-valued function defined over a rectangular grid. Color is proportional to surface height.

Fig. 11.25 demonstrates the light-intensity fields simulated for the model Si-MCP ($11{\times}11$) grid illuminated with a monochromatic beam. Staining in false colors is given specifically to highlight the spatial features and 3D capability of the microchannel plate in shaping the light by square aperture arrays. Fig. 11.25(a) shows the top view for the cross-shaped image superposed with the grid on a square piece of Si-MCP wafer. As the angle of view in Figs. 11.25(b)-(d) increases, the grid disappears and the 3D distribution of light fields is clearly visible in various orthogonal projections.

2D charts of intensity $I(x)$ due to the Fraunhofer diffraction are plotted in Figs. 11.25(e) and (f). The curves are described by the function $sinc(x)$ and Eq. (11.26). The central lobe prevails over the outer ones. The semi-logarithmic scale in Fig. 11.25(f) is given to show that side lobes are not vanishingly small, although their intensity rolls off by orders of magnitude. The central lobe or zero-order maximum stands by the narrow-band powerful beam over the full optical aperture of Si-MCP, and demonstrates 3D redistribution of the light intensity in the plane projection. The central-lobe bandwidth determined by the spacing between the first two minima is equal to the opening width, and full width at half maximum (FWHM) = 0.875.

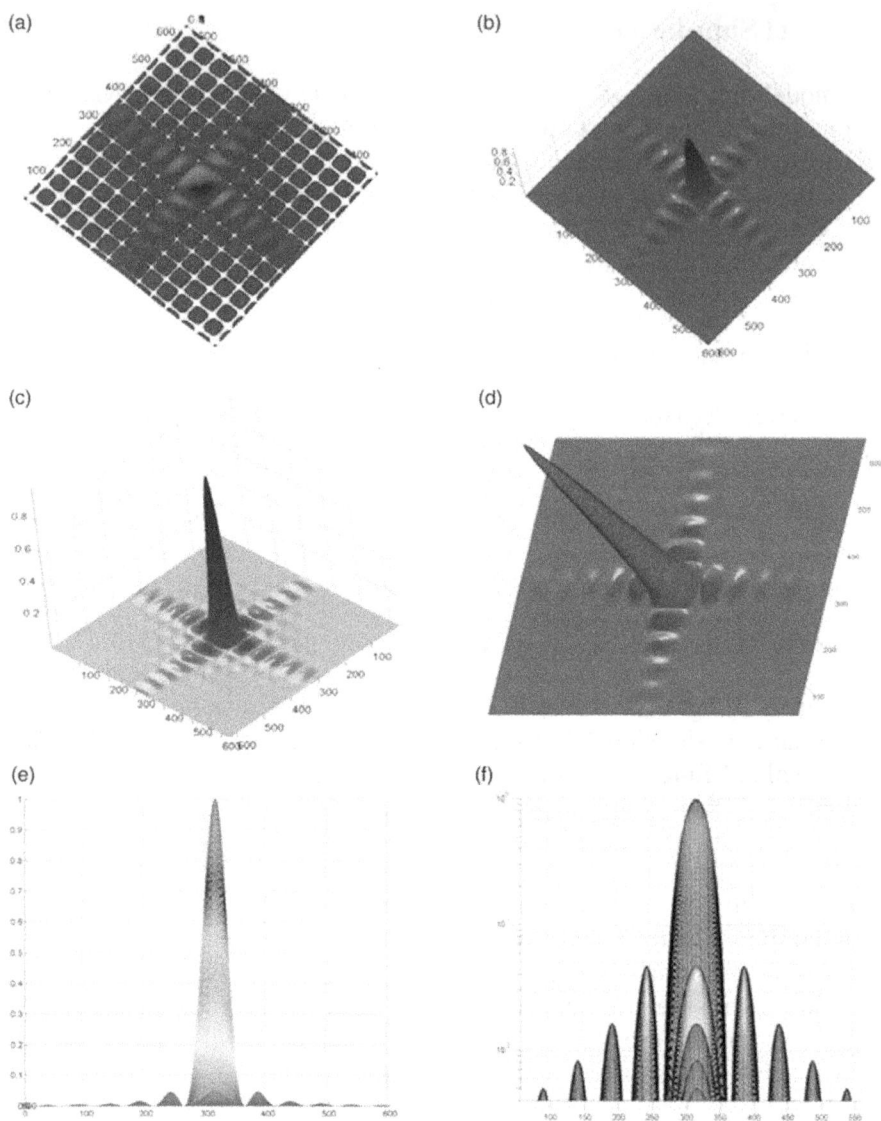

Fig. 11.25. 3D light intensity distribution at Si-MCP outlet. (a) Top view in (100) plane at 45° azimuth; (b), (c) Increasing angles of view at 45° azimuth; (d) Orthographic view at azimuth 90°; (e), (f) Linear and semi-logarithmic distributions of intensity in (010) plane in accord with Eq. (11.26).

11.9. Conclusion

Microchannel silicon offers a new approach to creating optically transparent media while maintaining semiconductor properties characteristic of crystalline silicon.

Although there have been some previous achievements with meta-interfaces and interfaces, similar to the meta-boundary between silicon and air (29.6 % transmission

304

efficiency in narrow FOV [43], 45 % for vortex converter, and 36 % for beam control device [45]), all of them remain well below the level achieved in this work. Optical anisotropy was observed in nanostructured porous silicon as the in-plane birefringence [80, 81], which was found to be 10^4 times stronger than in bulk silicon and appeared as a result of the transformation of crystallographic symmetry. In comparison with silicon metasurfaces and nanostructured surfaces, microchannel silicon enables achievement of higher optical transmission and anisotropy in a wide spectral range, covering the entire spectrum from UV to NIR.

Based on the accurate micromachining of prime-grade silicon wafers, we have developed and fabricated a double-crystalline form of silicon, and one of those is the mesoscopic crystal lattice. Microchannel silicon can be regarded as a new optical material with unique optical, opto-mechanical and photonic properties. Square plates ranging in size from 1 to 2 inches and round 4 to 6 inch Si wafers made of microchannel silicon exhibit the extraordinary high optical transmission T > 80 %, strong uniaxial optical anisotropy with dichroism $D \approx 100$ %, as well as polychromatic cross-shaped Fraunhofer diffraction and 2D array of conoscopic patterns.

The capability for 3D shaping of the light flux with diffraction and spectral dispersion makes microchannel silicon the material of choice for optical and photonic applications. Si-MCP remains a semiconductor, with all the inherent bulk electronic properties of silicon inside the mesoscopic lattice slabs, surrounded by the air / vacuum microcavities. Its payload characteristics with fivefold lighter weight make microchannel silicon a very promising material for use in mobile vehicles and electronic components in space. Future Si-MCP wafer designs, due to the significantly expanded usable square, can promote the creation of a 3D architecture for ULSIC and Telecom devices. Therefore, microchannel silicon can be regarded as a potential base material for 3D chips used in artificial intelligence.

Acknowledgements

The author acknowledges Drs. Michael Gertsenshteyn (BAE Systems) and Tomasz Jannson (Physical Optics Corporation) for critical ideas and fruitful discussions, Keith Shoemaker (Physical Optics Corporation) for multifaceted engineering support, Dr. Alex Usikov (TDI, Oxford Instruments) for technological processing of the photocathode layers, and Dr. Anton Tremsin (University of California, Berkeley) for useful discussions regarding Si-MCP properties and applications. The author would also like to note fruitful and collaborative work with the Photonic Systems team at POC (Torrance, California) and TDI at Oxford Instruments (Silver Spring, Maryland) in the joint development of epitaxial growth for the wide-gap III-V nitrides. The author is also thankful to the diagnostics centers at SEAL Labs (El Segando, California) and Tescan USA Inc. (Warrendale, Pennsylvania) for the SEM, EDAX, and CL topograms. The early-stage work was supported by NASA, DARPA, and AFRL with SBIR/STTR grants from 2007-2010.

References

[1]. P. C. Taylor, Exotic forms of silicon, *Phys. Today*, Vol. 69, 2016, pp. 34-39.

[2]. X. Li, P. W. Bohn, Metal-assisted chemical etching in HF/H_2O_2 produces porous silicon, *Appl. Phys. Lett.*, Vol. 77, Issue 16, 2000, pp. 2572-2574.

[3]. S. P. Scheeler, S. Ullrich, S. Kudera, C. Pacholsk, Fabrication of porous silicon by metal-assisted etching using highly ordered gold nanoparticle arrays, *Nanoscale Res. Lett.*, Vol. 7, 2012, 450.

[4]. Y. Qu, L. Liao, Y. Li, H. Zhang, et al., Electrically conductive and optically active porous silicon nanowires, *Nano Lett.*, Vol. 9, Issue 12, 2009, pp. 4539-4543.

[5]. S. Chattopadhyay, X. L. Li, P. W. Bohn, In-plane control of morphology and tunable photoluminescence in porous silicon produced by metal-assisted electroless chemical etching, *J. Appl. Phys.*, Vol. 91, Issue 9, 2002, pp. 6134-6140.

[6]. S. Matthias, F. Müller, J. Schilling, U. Gösele, Pushing the limits of macroporous silicon etching, *Appl. Phys. A*, Vol. 80, 2005, pp. 1391-1396.

[7]. A. Santos, T. Kumeria, Electrochemical etching methods for producing porous silicon, Chapter 1, in Electrochemically Engineered Nanoporous Materials, Methods, Properties and Applications (D. Losic, A. Santos, Eds.), *Springer*, 2015.

[8]. L. Pascual, R. J. Martín-Palma, A. R. Landa-Cánovas, P. Herrero, J. M. Martínez-Duart, Lattice distortion in nanostructured porous silicon, *Appl. Phys. Lett.*, Vol. 87, 2005, 251921.

[9]. Y. Murayama, Mesoscopic Systems: Fundamentals and Applications, *Wiley-VCH*, 2001.

[10]. A. Reiter, A local cellular model for snow crystal growth, *Chaos, Solitons and Fractals,* Vol. 23, 2005, pp. 1111-1119.

[11]. J. Gravner, D. Griffeath, Modeling snow-crystal growth: A three-dimensional mesoscopic approach, *Phys. Rev. E*, Vol. 79, 2009, 011601.

[12]. Image Sensors and Signal Processing for Digital Still Cameras (J. Nakamura, Ed.), *Taylor & Francis*, 2006.

[13]. High Performance Silicon Imaging. Fundamentals and Applications of CMOS and CCD Sensors (D. Durini, Ed.), *Woodhead Publishing,* 2014.

[14]. S. Yokogawa, I. Oshiyama, H. Ikeda, Y. Ebiko, T. Hirano, S. Saito, T. Oinoue, Y. Hagimoto, H. Iwamoto, IR sensitivity enhancement of CMOS Image Sensor with diffractive light trapping pixels, *Sci. Reports,* Vol. 7, 2017, 3832.

[15]. T. Feleki, G. Bex, R. Andriessen, Y. Galagan, F. Di Giacomo, Rapid and low temperature processing of mesoporous TiO_2 for perovskite solar cells on flexible and rigid substrates, *Materials Today Commun.*, Vol. 13, 2017, pp. 232-240.

[16]. M. I. H. Ansari, A. Qurashi, M. K. Nazeeruddin, Frontiers, opportunities, and challenges in perovskite solar cells: A critical review, *J. Photochem. Photobiol. C: Photochem. Reviews,* Vol. 35, 2018, pp. 1-24.

[17]. P. Vivo, J. K. Salunke, A. Priimagi, Hole-transporting materials for printable Perovskite solar cells, *Materials*, Vol. 10, 2017, E1087.

[18]. J. S. Kasper, S. M. Richards, The crystal structures of new forms of silicon and germanium, *Acta Crystal.,* Vol. 17, 1964, pp. 752-755.

[19]. H. Jansen, M. de Boer, R. Legtenberg, M. Elwenspoek, The black silicon method: a universal method for determining the parameter setting of a fluorine-based reactive ion etcher in deep silicon trench etching with profile control, *J. Micromech. Microeng.*, Vol. 5, Issue 2, 1995, pp. 115-120.

[20]. G. Medvedkin, Shaping light with microchannel silicon, *J. Opt. Soc. Amer. B*, Vol. 35, Issue 5, 2018, pp. 993-1003.

[21]. M. Lampton, The microchannel image intensifier, *Sci. Amer.*, Vol. 245, 1981, pp. 62-71.

[22]. T. Jannson, R. Pradhan, M. Gertsenshteyn, V. Grubsky, I. Marienko, G. Medvedkin, W. Mengesha, V. Romanov, Y. Yang, Quantum Imaging System and Mode of Operation and Method of Fabrication, U.S. Patent 7781739, March 13, *USA*, 2010.

[23]. V. Grubsky, T. Jannson, E. M. Patton, V. Romanov, G. Medvedkin, P. Shnitser, K. Shoemaker, X-ray Imaging System and Method, U.S. Patent 8705694, April 22, *USA*, 2014.

[24]. P. Shnitser, G. Medvedkin, Photodetector with Nanowire Photocathode, U.S. Patent 9818894, March 2, *USA*, 2017.

[25]. R. D. Pradhan, M. Gertsenshteyn, Y. Yang, V. Grubsky, W. Mengesha, V. Romanov, I. Mariyenko, G. Medvedkin, I. Berezhnyy, T. P. Jannson, Gamma-ray detection by optical visualization of secondary electron clouds, in *Defense, Security and Sensing, Session 11, Radiological and Nuclear Sensing (SPIE'09)*, Vol. 7304-63, 2009, p. 56.

[26]. M. Gertsenshteyn, P. Shnitser, G. Medvedkin, Silicon Microchannel Plate Large Area UV Detector, SBIR Project NNX08CB84P, *NASA GSFC,* 2009.

[27]. M. Gertsenshteyn, P. Shnitser, G. Medvedkin, V. Grubsky, Highly Efficient FUV Photodetector with AlGaN Nanowire Photocathode, SBIR Project NNX09CD49P, *NASA GSFC*, 2010.

[28]. G. Medvedkin, Ultraviolet Focal Plane Array Detector, SBIR Project N10PC20064, *DARPA,* 2010.

[29]. G. Medvedkin, Silicon Microchannel Nanocrystal Multiresonator Laser, SBIR Project AF083-190, *AFRL,* 2009.

[30]. H. Seidel, L. Csepregi, A. Heuberger, H. Baumgartel, Anisotropic etching of crystalline silicon in alkaline solutions, *J. Electrochem Soc.,* Vol. 137, Issue 11, 1990, pp. 3612-3632.

[31]. KOH Etching, BYU Cleanroom, https://cleanroom.byu.edu/KOH

[32]. V. Lehmann, Electrochemistry of Silicon, *Wiley-VCH Verlag*, 2002.

[33]. X. G. Zhang, Electrochemistry of Silicon and Its Oxide, *Kluwer Academic*, 2001.

[34]. G. T. A. Kovacs, N. I. Maluf, K. E. Petersen, Bulk micromachining of silicon, *Proceedings of IEEE,* Vol. 86, 1998, pp. 1536-1551.

[35]. Photonic Crystals (K. Busch, S. Lolkes, R. B. Wehrspohn, H. Foll, Eds.), *Wiley-VCH Verlag*, 2004.

[36]. J.-H. Lourtioz, H. Benisty, V. Berger, J.-M. Gerard, D. Maystre, A. Tchelnokov, Photonic Crystals. Towards Nanoscale Photonic Devices, *Springer*, 2005.

[37]. E. V. Astrova, A. A. Nechitailov, Boundary effect under electro-chemical etching of silicon, *Semiconductors*, Vol. 42, 2007, pp. 480-484.

[38]. M. Hoffmann, E. Voges, Bulk silicon micromachining for MEMS in optical communication systems, *J. Micromech. Microeng.*, Vol. 12, 2002, pp. 349-360.

[39]. Landolt-Börnstein Group III Condensed Matter, Semiconductors Group IV Elements, IV-IV and III-V Compounds. Part B – Electronic, Transport, Optical and Other Properties, Vol. 41A, Issue 1b, *Springer*, 2002, pp. 1615-1925.

[40]. Optical properties (D. E. Aspnes, Ed.), in Properties of Crystalline Silicon, EMIS Data Reviews Series No. 20 (R. Hull, Ed.), *INSPEC,* 1999, p. 677.

[41]. S. M. Sze, K. K. Ng, Physics of Semiconductor Devices, 3rd Ed., *Wiley*, 2007.

[42]. C. M. Roberts, T. A. Cook, V. A. Podolskiy, Metasurface-enhanced transparency, *J. Opt. Soc. Amer. B*, Vol. 34, Issue 7, 2017, pp. D42-D45.

[43]. M. I. Shalaev, J. Sun, A. Tsukernik, A. Pandey, K. Nikolskiy, N. M. Litchinitser, High-efficiency all-dielectric metasurfaces for ultracompact beam manipulation in transmission mode, *Nano Lett.*, Vol. 15, 2015, pp. 6261-6266.

[44]. A. Arbabi, Y. Horie, M. Bagheri, A. Faraon, Dielectric metasurfaces for complete control of phase and polarization with subwavelength spatial resolution and high transmission, *Nature Nanotech.,* Vol. 10, 2015, pp. 937-943.

[45]. T. Young, On the theory of light and colours (The 1801 Bakerian Lecture), *Philos. Trans. Royal Soc. of London,* Vol. 92, 1802, pp. 12-48.

[46]. M. Born, E. Wolf, Chapter VIII, in Principles of Optics, 7th Ed., *Cambridge Univ. Press,* 2002.

[47]. G. A. Medvedkin, Yu. V. Rud, The parameters of polarization photosensitivity of isotropic semiconductors, *Phys. Stat. Solidi (a),* Vol. 67, 1981, pp. 333-337.

[48]. G. A. Medvedkin, Optical sensors sensitive to the light polarization: Phenomenological consideration through electronic parameters, *Jap. J. Appl. Phys.,* Vol. 39, 2000, pp. 355-356.

[49]. David Brewster, On the laws which regulate the polarisation of light by reflection from transparent bodies, *Phil. Trans. Royal Soc. of London,* Vol. 105, 1815, pp. 125-159.

[50]. R. M. A. Azzam, T. F. Thonn, Pseudo-Brewster and second-Brewster angles of an absorbing substrate coated by a transparent thin film, *Appl. Opt.,* Vol. 22, 1983, pp. 4155-4165.

[51]. R. M. A. Azzam, Angular sensitivity of Brewster-angle reflection polarizers: an analytical treatment, *Appl. Opt.,* Vol. 26, 1987, pp. 2847-2850.

[52]. V. V. Korotaev, G. A. Medvedkin, E. D. Pankov, Yu. V. Rud, Direct analysis of linearly polarized radiation by Ge and Si photoreceivers, *J. Optical Technology,* Vol. 11, 1981, pp. 14-16.

[53]. F. I. Fedorov, Theory of Gyrotropy, *Nauka i Tekhnika,* Minsk, 1976.

[54]. F. I. Fedorov, V. V. Filippov, Light Reflection and Refraction by Transparent Crystals, *Nauka i Tekhnika, Minsk,* 1976.

[55]. Fedor I. Fedorov, Theory of Elastic Waves in Crystals, *Springer,* 1968.

[56]. J. C. M. Garnett, Colours in metal glasses and in metallic films, *Philos. Trans. R. Soc. London,* Vol. 203, 1904, pp. 385-420.

[57]. J. C. M. Garnett, Colours in metal glasses, in metallic films, and in metallic solutions II, *Philos. Trans. R. Soc. London,* Vol. 205, 1906, pp. 237-288.

[58]. G. A. Medvedkin, Yu. V. Rud, M. Tairov, Semiconductor Crystals for Photodetectors of Linearly Polarized Radiation, *FAN,* Tashkent, 1992 (in Russian).

[59]. P. S. J. Russell, Photonic-crystal fibers, *J. Lightwave Technol.,* Vol. 24, 2006, pp. 4729-4749.

[60]. P. Russell, Photonic crystal fiber: finding the Holey Grail, *Opt. & Photon. News,* Vol. 18, 2007, pp. 26-31.

[61]. H. Imam, Metrology: Broad as a lamp, bright as a laser, *Nature Photonics,* Vol. 2, 2008, pp. 26-28.

[62]. M. Qi, E. Lidorikis, P. T. Rakich, S. G. Johnson, J. D. Joannopoulos, E. P. Ippen, H. I. Smith, A three-dimensional optical photonic crystal with designed point defects, *Nature,* Vol. 429, 2004, pp. 538-542.

[63]. P. E. Barclay, K. Srinivasan, O. Painter, Nonlinear response of silicon photonic crystal microresonators excited via an integrated waveguide and fiber taper, *Opt. Express,* Vol. 13, 2005, pp. 801-820.

[64]. S. Matthias, F. Müller, Asymmetric pores in a silicon membrane acting as massively parallel Brownian ratchets, *Nature,* Vol. 424, 2003, pp. 53-57.

[65]. S. Matthias, F. Müller, U. Gösele, Simple cubic three-dimensional photonic crystals based on macroporous silicon and anisotropic posttreatment. *J. Appl. Phys.,* Vol. 98, 2005, 023524.

[66]. S. Matthias, F. Müller, J. Schilling, U. Gösele, Pushing the limits of macroporous silicon etching. *Appl. Phys. A,* Vol. 80, 2005, pp. 1391-1396.

[67]. D. T. Pierce, R. L. Byer, Experiments on the introduction of light and sound for the advanced laboratory, *Amer. J. Physics,* Vol. 41, 1973, pp. 314-325.

[68]. M. G. Moharam, T. K. Gaylord, R. Magnusson, Criteria for Raman-Nath regime diffraction by phase gratings, *Optics Commun.,* Vol. 32, 1980, pp. 19-23.

[69]. S.-T. Chen, M. R. Chatterjee, A numerical analysis and expository interpretation of the diffraction of light by ultrasonic waves in Bragg and Ramam-Nath regimes using multiple scattering theory, *IEEE Trans. on Educ.,* Vol. 39, Issue 1, 1999, pp. 56-68.

[70]. E. Hecht, Optics, 4th Ed., *Addison Wesley*, San Francisco, 2002.

[71]. W. J. Patzelt, Polarized Light Microscopy, *Ernst Leitz, Wetzlar GmbH,* 1985.

[72]. Yu. E. Sirotin, M. P. Shaskolskaya, Fundamentals of Crystal Physics, *Nauka*, Moscow, 1979.

[73]. N. H. Hartshorne, A. Stuart, Crystals and the Polarizing Microscope, *Arnold*, London, 1970.

[74]. L. Wiza, Microchannel plate detectors, *Nuclear Instr. Methods*, Vol. 162, 1979, pp. 587-601.

[75]. G. J. Timothy, Microchannel plates for photon detection and imaging in space, *Observing Photons in Space*, Vol. 22, 2013, pp. 365-390.

[76]. T. Gys, Micro-channel plates and vacuum detectors, *Nucl. Instr. Meth. Res. Phys. A*, Vol. 787, 2015, pp. 245-260.

[77]. O. H. W. Siegmund, J. V. Vallerga, J. McPhate, A. S. Tremsin, Next generation microchannel plate detector technologies for UV Astronomy, *Proceedings of SPIE,* Vol. 5488, 2002, pp. 789-800.

[78]. A. M. Dabiran, A. M. Wowchak, P. P. Chow, O. H. W. Siegmund, J. S. Hull, J. Malloy, A. S. Tremsin, Direct deposition of GaN-based photocathodes on microchannel plates, *Proceedings of SPIE*, Vol. 7212, 2009, pp. 131-137.

[79]. A. Usikov, V. Soukhoveev, L. Shapovalova, A. Syrkin, O. Kovalenkov, A. Volkova, V. Sizov, V. Ivantsov, V. Dmitriev, New results on HVPE growth of AlN, GaN, InN and their alloys, *Phys. Stat. Solidi (c)*, Vol. 5, Issue 6, 2008, pp. 1825-1828.

[80]. L. Pascual, R. J. Martín-Palma, A. R. Landa-Cánovas, P. Herrero, J. M. Martínez-Duart, Lattice distortion in nanostructured porous silicon, *Appl. Phys. Lett.*, Vol. 87, 2005, 251921.

[81]. D. Kovalev, G. Polisski, J. Diener, H. Heckler, N. Kunzner, V. Yu. Timoshenko, F. Koch, Strong in-plane birefringence of spatially nanostructured silicon, *Appl. Phys. Lett.*, Vol. 78, 2001, pp. 916-918.

Chapter 12
Pattern Recognition with Log-polar Joint Transform Correlation

M. Nazrul Islam

12.1. Introduction

Pattern recognition is the process of automatically identifying a pattern of interest in an unknown input scene. Image processing techniques usually utilize the spatial features of target object, including shape, size, and color, for comparing with the input scene objects. Requirements for a pattern recognition technique include simple architecture, very fast operation for real-time applications, efficient detection of all targets in variable conditions, like scale and rotation distortion, noisy input scene, and generation of no false alarms.

Digital systems process the input scene and reference images pixel by pixel to perform correlation and detect the target, which makes operation slow for real-time applications. On the otherhand, optical pattern recognition systems employ optical optics to process the images in parallel resulting in an instantaneous decision on target detection [1]. Optical techniques are based on two major algorithms, VanderLugt filter [2] and joint transform correlation (JTC) [3-4]. The JTC algorithm outperforms the former one by not requiring any complex optical filter and having capability of operating at video frame rates. In addition, it allows real-time updating of the reference image and permits parallel Fourier transformation of the reference image and input scene. However, the classical JTC technique suffers from poor correlation discrimination, wide sidelobes, conjugate correlation peak, strong zero-order correlation, and failure to perform in noisy environment. Several modifications of the classical JTC technique have been proposed in the literature to alleviate these problems, which include binary JTC [5], phase-only JTC [6], fringe-adjusted JTC (FJTC) [7], shifted phase-encoded JTC [8], and complementary-reference and complementary-scene JTC [9]. Though these techniques show efficiency in some respects of pattern recognition, however, majority of these techniques are not as efficient in detecting targets with sufficiently high discrimination from non-target objects and also not having a simple system architecture and not being able to perform at high

M. Nazrul Islam
State University of New York at Farmingdale, USA

speed operation for real-time pattern recognition applications. Additionally, the target detection performance is not invariant to distortions due to scale and rotation variations of the targets in the input scene.

The objective of this chapter is to present an efficient pattern recognition technique invariant to scale and rotation variations having high discrimination between target and non-target objects. It develops a multiple phase-shifted-reference JTC (MRJTC) technique that processes four phase-shifted reference images with the same input scene and hence removes all the unwanted correlation terms and produces a single and highly discriminant correlation peak for each potential target [10]. It also incorporates a log-polar transformation to images that are fed to the MRJTC algorithm in order to make the correlation output invariant to scale and rotation variations. Performance of the proposed technique is investigate via computer simulation involving real-life patterns which shows that it is successful in recognizing the potential targets present in a real-life input scene with complex background conditions.

12.2. Log-polar Transformation

The log-polar transformation is a conformal mapping of a mathematical function from Cartesian coordinate plane to a logarithmic polar coordinate plane [12]. The mapping is described as follows

$$f_t(\rho, \theta) = LPT\{f(x, y)\}, \tag{12.1}$$

where LPT denotes a log-polar transformation, and ρ and θ are the logarithm of the radius and the angular position in the log-polar coordinate system, respectively, calculated using the following formulas:

$$\rho = \log \sqrt{x^2 + y^2}, \tag{12.2}$$

and

$$\theta = \tan^{-1}\left(\frac{y}{x}\right), \tag{12.3}$$

The opposite transformation, log-polar to Cartesian, can be obtained as:

$$x + iy = e^{\rho + i\theta} = e^{\rho}(\cos\theta + i\sin\theta) \tag{12.4}$$

Fig. 12.1 illustrates the basic principle of log-polar transformation.

12.3. Optical Pattern Recognition Technique

Fig. 12.2 shows the schematic diagram of the target detection system. The reference image of the target, $r(x, y)$, in the Cartesian coordinates is first transformed to the log-polar coordinate system

$$r_t(\rho, \theta) = LPT\{r(x, y)\}, \tag{12.5}$$

where the coordinates, ρ and θ, calculated using the following formulas:

$$\rho = log\sqrt{(x - x_0)^2 + (y - y_0)^2},$$ (12.6)

and

$$\theta = tan^{-1}\left(\frac{y - y_0}{x - x_0}\right),$$ (12.7)

where (x_0, y_0) is the center pixel of the image in the Cartesian coordinate. The transformed reference image of Eq. (12.5) is then phase-shifted by 0, 90, 180 and 270 degrees, respectively and fed to four parallel processing channels.

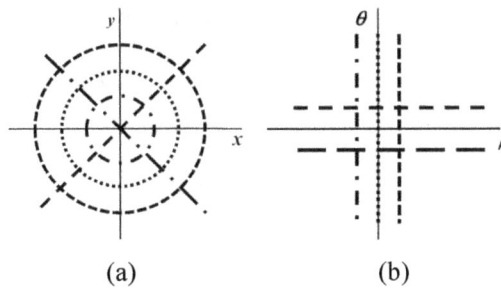

Fig. 12.1. Log-polar transformation: (a) function in Cartesian coordinates, (b) function in log-polar coordinates.

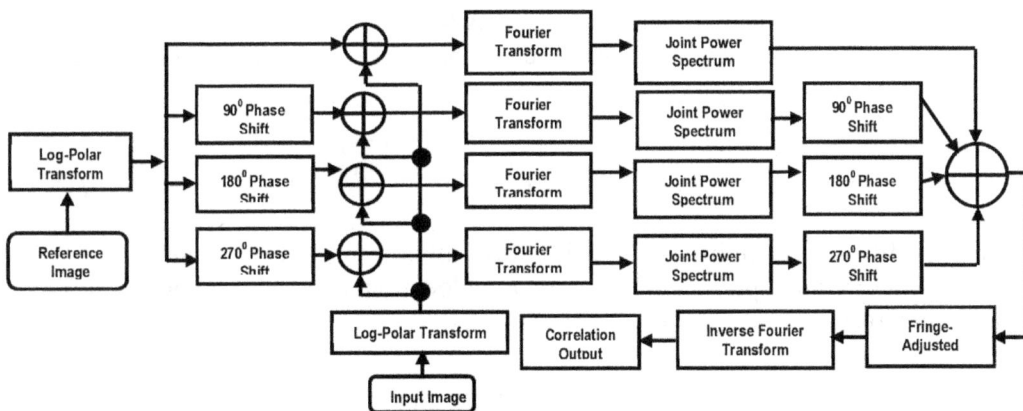

Fig. 12.2. Block diagrams of the proposed pattern recognition system.

The given input scene, $t(x, y)$, containing unknown objects is also transformed to the log-polar domain

$$t_t(\rho, \theta) = LPT\{t(x, y)\}$$ (12.8)

The transformed input image of Eq. (8) is introduced to transformed reference image in four separate channels and four joint images are formed as follows:

$$f_1(x, y) = r_t(x, y) + t_t(x, y), \tag{12.9}$$

$$f_1(x, y) = jr_t(x, y) + t_t(x, y), \tag{12.10}$$

$$f_1(x, y) = -r_t(x, y) + t_t(x, y), \tag{12.11}$$

$$f_1(x, y) = -jr_t(x, y) + t_t(x, y) \tag{12.12}$$

These joint images are now Fourier transformed, where from four different joint power spectra (JPS) are obtained:

$$P_1(u, v) = |F_1(u, v)|^2 = |R_t(u, v)|^2 + |T_t(u, v)|^2 +$$
$$+R_t(u, v)T_t^*(u, v) + R_t^*(u, v)T_t(u, v), \tag{12.13}$$

$$P_2(u, v) = |F_2(u, v)|^2 = |R_t(u, v)|^2 + |T_t(u, v)|^2 +$$
$$+jR_t(u, v)T_t^*(u, v) - jR_t^*(u, v)T_t(u, v), \tag{12.14}$$

$$P_3(u, v) = |F_3(u, v)|^2 = |R_t(u, v)|^2 + |T_t(u, v)|^2 -$$
$$-R_t(u, v)T_t^*(u, v) - R_t^*(u, v)T_t(u, v), \tag{12.15}$$

$$P_4(u, v) = |F_4(u, v)|^2 = |R_t(u, v)|^2 + |T_t(u, v)|^2 -$$
$$-jR_t(u, v)T_t^*(u, v) + jR_t^*(u, v)T_t(u, v) \tag{12.16}$$

Each of the JPS terms in equations (12.13)-(12.16) contains a number of terms to produce auto-correlation and cross-correlation terms as is the case for a classical JTC technique. However, the JPS signals can be combined by incorporating an algebraic operation as given by

$$P(u, v) = P_1(u, v) + jP_2(u, v) - P_3(u, v) - jP_4(u, v) =$$
$$= 4R_t^*(u, v)T_t(u, v) \tag{12.17}$$

The modified JPS signal in Eq. (12.17) eliminates all unwanted correlation terms and also produces the correlation peak exactly at the target location. Next a fringe-adjusted filter is incorporated to enhance the correlation performance.

$$H(u, v) = \frac{C(u,v)}{D(u,v) + |R(u,v)|^2}, \tag{12.18}$$

where $C(u,v)$ and $D(u,v)$ are either constants or functions of u and v. The parameter $C(u,v)$ is adjusted to avoid having an optical gain greater than unity, while $D(u,v)$ is used to overcome the pole problem otherwise associated with a normal filter. Since the power spectra of the reference image can be pre-calculated and stored, implementation of the FAF also will not adversely impact the processing speed of the pattern recognition system.

The JPS in Eq. (12.17) is multiplied by the FAF transfer function in Eq. (12.18) to yield an enhanced JPS as given by

$$P_f(u, v) = \frac{4C(u,v)4R_t^*(u,v)T_t(u,v)}{D(u,v)+|R(u,v)|^2} \qquad (12.19)$$

Inverse Fourier transformation of this signal results in a single and very sharp correlation peak for each potential target of the reference class present in the input scene.

Finally, the target detection decision is made by measuring the peak correlation value, $c_p(x, y)$, and correlation values around the peak, $c_i(x, y)$. Then the peak-to-side lobe ratio (PSR) is determined as

$$PSR = \frac{c_p(x,y)}{\sum_{i = 1, i \neq p} c_i(x,y)} \qquad (12.20)$$

The PSR value is compared to a threshold value to determine if the correlation peak belongs to a target.

12.4. Simulation Results

The proposed target detection system is simulated using MATLAB program. Fig. 12.3(a) shows a binary image with a size of 32×32 pixels. Then the image is rotated by 30 and 60 degree, respectively, as depicted in Fig. 12.3(b)-(c). These images are then log-polar transformed as shown in Fig. 12.3(d)-(f). It can be observed that the in-plane rotation of the image in the spatial domain does not affect the transformed image significantly.

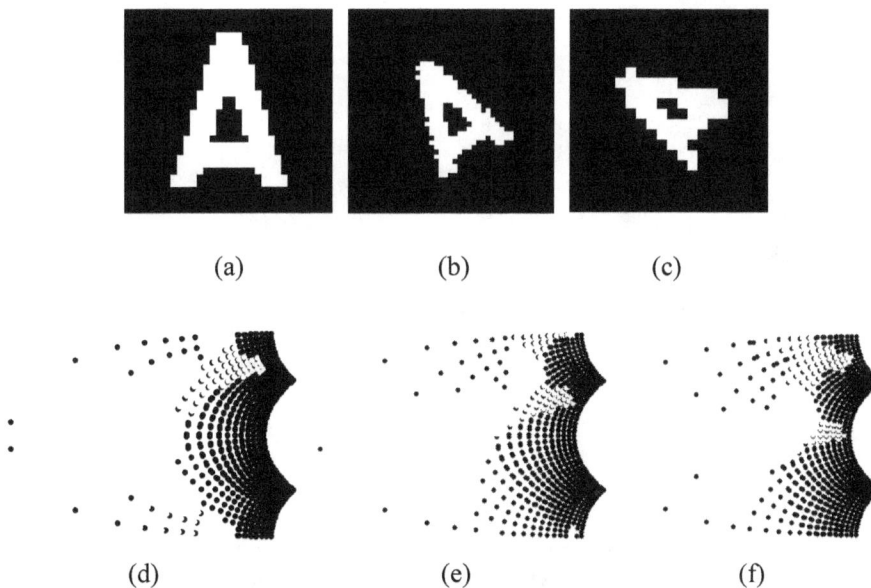

(a)   (b)   (c)

(d)   (e)   (f)

Fig. 12.3. Log-polar transformation of images: (a)-(c) spatial domain images, (d)-(f) transformed images.

The correlation performance of the proposed technique is evaluated using some test images. Fig. 12.4(a) shows the joint image where the top half of the image contains the reference object from Fig. 12.3(a) and the bottom half contains the unknown input scene. It is obvious from the correlation output shown in Fig. 12.4(b) that the proposed system is highly efficient in producing a very sharp correlation peak corresponding to the target object present in the input scene. Next, we used a rotated image in the input scene while keeping the same reference object as shown in Fig. 12.3(c). The correlation performance shown in Fig. 12.3(d) indicates a failure in detecting the target with distinction.

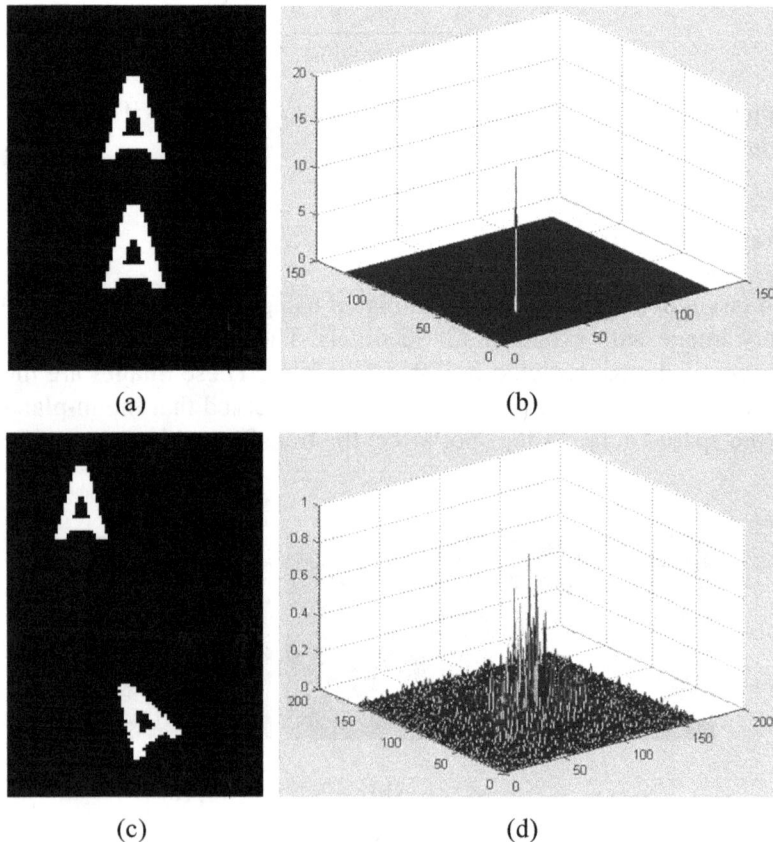

(a)

(b)

(c)

(d)

Fig. 12.4. Correlation performance for images in the spatial domain: (a), (c) joint input scene; (b), (d) correlation output.

Then the simulation experiment was carried out employing the log-polar transformation. As before, a joint image was formed using the transformed reference image and unknown input scene in the top half and lower half of the plane, respectively. Figs. 12.5(a) and 12.5(c) show two test joint images. The corresponding correlation outputs are depicted in Figs. 12.5(b) and 12.5(d), respectively. It can be obvious that the proposed target detection technique has been successful in detecting a target even with a significant amount of rotational distortion.

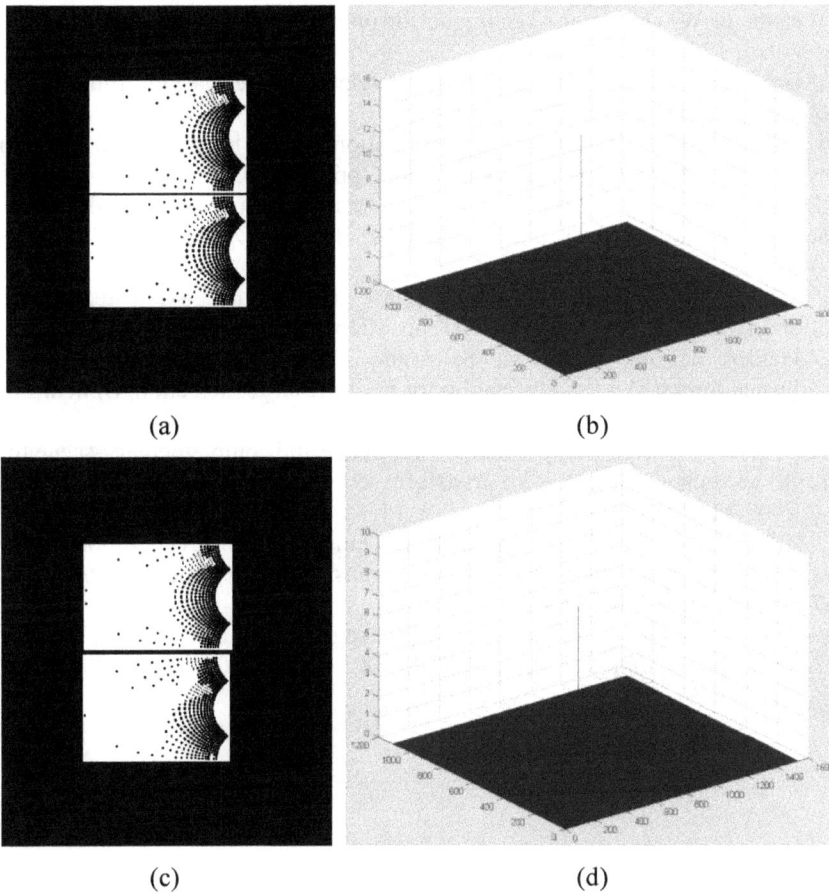

Fig. 12.5. Correlation performance in the transformed domain: (a), (c) joint input scene; (b), (d) correlation output.

12.5. Conclusions

This chapter presents a robust and effective pattern recognition system, which involves a rather simple architecture and hence can be easily implemented using optics for fast and precision performance. Computer simulation results demonstrate that the proposed technique can detect a target is various practical conditions, including scale and rotation variations. The algorithm can be built with optics, which yields a promising candidate for real-life applications in the field of target detection and tracking, biometric authentication and industry automation.

References

[1]. J. W. Goodman, Introduction to Fourier Optics, 2nd Ed., *McGraw-Hill Co*, 1996.
[2]. A. VanderLugt, Signal detection by complex spatial filtering, *IEEE Transactions on Information Theory*, Vol. 10, 1964, pp. 139-146.

[3]. C. S. Weaver, J. W. Goodman, Technique for optically convolving two functions, *Applied Optics*, Vol. 5, 1966, pp. 1248-1249.

[4]. M. S. Alam, M. A. Karim, Joint-transform correlation under varying illumination, *Applied Optics*, Vol. 32, Issue 23, 1993, pp. 4351-4356.

[5]. B. Javidi, C. Kuo, Joint transform image correlation using a binary spatial light modulator at the Fourier plane, *Applied Optics*, Vol. 27, 1988, pp. 663-665.

[6]. G. Lu, Z. Zhang, S. Wu, F. T. S. Yu, Implementation of a non-zero-order joint transform correlator by use of phase-shifting techniques, *Applied Optics*, Vol. 36, Issue 2, 1997, pp. 470-483.

[7]. A. K. Cherri, M. S. Alam, Reference phase-encoded fringe-adjusted joint transform correlation, *Applied Optics*, Vol. 40, 2001, pp. 1216-1225.

[8]. M. R. Haider, M. N. Islam, M. S. Alam, Enhanced class associative generalized fringe-adjusted joint transform correlation for multiple target detection, *Optical Engineering*, Vol. 45, Issue 4, 2006, 048201.

[9]. H. A. Kamal, A. K. Cherri, Complementary-reference and complementary-scene for real-time fingerprint verification using joint transform correlator, *Optics and Laser Technology*, Vol. 41, 2009, pp. 643-650.

[10]. M. N. Islam, K. V. Asari, M. A. Karim, Optical pattern recognition using multiple phase-shifted-reference fringe-adjusted joint transform correlation, *Optical Engineering*, Vol. 47, Issue 9, 2008, 097204.

[11]. R. Matungka, Y. F. Zheng, R. L. Ewing, Image registration using adaptive polar transform, *IEEE Transactions on Image Processing*, Vol. 18, Issue 10, 2009, pp. 2340-2354.

Index

www.ingramcontent.com/pod-product-compliance
Lightning Source LLC
Chambersburg PA
CBHW080928220326
41598CB00034B/5716

* 9 7 8 8 4 0 9 0 9 0 1 4 3 *